VOLUME ONE HUNDRED AND SIXTY THREE

INTERNATIONAL REVIEW OF NEUROBIOLOGY

Essential Tremor: Current Concepts and Controversies

INTERNATIONAL REVIEW OF NEUROBIOLOGY

VOLUME 163

SERIES EDITOR

PATRICIA JANAK
Janak Lab, Department of Neuroscience
Dunning Hall, John Hopkins University
Baltimore, MD, USA

PETER JENNER
Division of Pharmacology and Therapeutics
GKT School of Biomedical Sciences
King's College, London, UK

VOLUME ONE HUNDRED AND SIXTY THREE

INTERNATIONAL REVIEW OF NEUROBIOLOGY

Essential Tremor: Current Concepts and Controversies

Edited by

SHENG-HAN KUO

Department of Neurology,
College of Physicians and Surgeons;
Initiative for Columbia Ataxia and Tremor,
Columbia University, New York, NY, United States

ELAN D. LOUIS

Department of Neurology,
University of Texas Southwestern Medical Center,
Dallas, TX, United States

ELSEVIER

ACADEMIC PRESS
An imprint of Elsevier

Academic Press is an imprint of Elsevier
50 Hampshire Street, 5th Floor, Cambridge, MA 02139, United States
525 B Street, Suite 1650, San Diego, CA 92101, United States
The Boulevard, Langford Lane, Kidlington, Oxford OX5 1GB, United Kingdom
125 London Wall, London, EC2Y 5AS, United Kingdom

First edition 2022

ISBN: 978-0-323-89974-1
ISSN: 0074-7742

For information on all Academic Press publications
visit our website at https://www.elsevier.com/books-and-journals

Publisher: Zoe Kruze
Acquisitions Editor: Sam Mahfoudh
Developmental Editor: Federico Paulo Mendoza
Production Project Manager: Abdulla Sait
Cover Designer: Matthew Limbert

Typeset by STRAIVE, India

Contents

Contributors

Steven Bellows
Baylor College of Medicine, Houston, TX, United States

Hyun Joo Cho
Human Motor Control Section, National Institute of Neurological Disorders and Stroke, National Institutes of Health, Bethesda, MD, United States

Stephanie Cosentino
Cognitive Neuroscience Division of the Taub Institute for Research on Alzheimer's Disease and the Aging Brain; Department of Neurology, Columbia University Irving Medical Center, New York, United States

Phyllis L. Faust
Department of Pathology and Cell Biology, Columbia University Vagelos College of Physicians and Surgeons and the New York Presbyterian Hospital, New York, NY, United States

Steven J. Frucht
The Marlene and Paolo Fresco Institute for Parkinson's and Movement Disorders, New York University Langone Health, New York, NY, United States

Alexandre Gironell
Movement Disorders Unit, Department of Neurology, Sant Pau Hospital, Autonomous University of Barcelona, Catalonia, Spain

Adrian Handforth
Neurology Service (W127), Veterans Affairs Greater Los Angeles Healthcare System, Los Angeles, CA, United States

Joohi Jimenez-Shahed
Icahn School of Medicine at Mount Sinai, New York City, NY, United States

Sheng-Han Kuo
Department of Neurology, College of Physicians and Surgeons; Initiative for Columbia Ataxia and Tremor, Columbia University, New York, NY, United States

Eric J. Lang
Department of Neuroscience and Physiology, Neuroscience Institute, New York University, New York, NY, United States

Elan D. Louis
Department of Neurology, University of Texas Southwestern Medical Center, Dallas, TX, United States

Mario Manto
Unité des Ataxies Cérébelleuses, Service de Neurologie, CHU-Charleroi; Service des Neurosciences, Université de Mons, Belgium

William Ondo
Houston Methodist Hospital, Weill Cornel Medical School, Houston, TX, United States

Sanjay Pandey
Department of Neurology, Govind Ballabh Pant Postgraduate Institute of Medical Education and Research, New Delhi, India

Alex Rajput
Division of Neurology, Saskatchewan Movement Disorders Program, University of Saskatchewan/Saskatchewan Health Authority, Saskatoon, SK, Canada

Giulietta M. Riboldi
The Marlene and Paolo Fresco Institute for Parkinson's and Movement Disorders, New York University Langone Health, New York, NY, United States

Ludy C. Shih
Department of Neurology, Parkinson's Disease and Movement Disorders Center, Boston Medical Center; Department of Neurology, Boston University School of Medicine, Boston, Massachusetts, United States

Aparna Wagle Shukla
Department of Neurology, Fixel Institute for Neurological Diseases, University of Florida, Gainesville, FL, United States

SECTION I

What's in a Name? Nomenclature of essential tremor: Syndrome or family of diseases?

Elan D. Louis[a] and Sheng-Han Kuo[b,c]

[a]Department of Neurology, University of Texas Southwestern Medical Center, Dallas, TX, United States
[b]Department of Neurology, College of Physicians and Surgeons, Columbia University, New York, NY, United States
[c]Initiative for Columbia Ataxia and Tremor, Columbia University, New York, NY, United States

As part of the current series, which highlights current controversies in the essential tremor (ET) field, we commissioned authors to submit papers with opposing viewpoints, one using available data to put forth the argument that ET is a syndrome and the other, using available data to put forth the argument that ET is a family of diseases (Pandey, 2022; Riboldi & Frucht, 2022). We are grateful to both sets of authors for thoughtfully making their cases and here we provide additional commentary.

It should not go unnoticed that both views start from the same place; namely, that not all ET patients express the same pattern of clinical features and mode or rate of clinical progression. That is, that there is clinical heterogeneity across patients. Rather than the monosymptomatic entity that ET was largely viewed as 20 years ago (Elble, 2002), the field has evolved over the past two decades, with a recognition that there is a myriad of clinical features, both motor and non-motor, differentially expressed across ET patients (Louis, 2021). Indeed, it is the universal recognition of this heterogeneity among scholars and clinicians that was a motivator behind the need to revise the Consensus Statement on classification of tremor several years ago (2018) (Bhatia, Bain, Bajaj, et al., 2018). The prior Consensus statement had been written 20 years previously (1998) prior to the publication of an ever-burgeoning data on the heterogeneity in ET (Deuschl, Bain, & Brin, 1998).

It is also important to note that on some level both views recognize that heterogeneity in ET extends beyond the purely clinical arena, and encompasses the realms of etiology (Aridon et al., 2008; Kovach, Ruiz, Kimonis, et al., 2001), pathophysiology (Holtbernd & Shah, 2021; Louis, Mazzoni, Ma, et al., 2012; Wong, Wang, Lin, et al., 2021) and pharmacological response phenotype (Chung, Kwon, Lee, et al., 2013).

The two views begin to differ when they move beyond their observations of clinical heterogeneity. One view and school of thought uses this heterogeneity to make the argument that ET is not one entity but rather a syndrome (Elble, 2020). The other uses the same heterogeneity to make the argument that ET is not one entity but is a family of diseases—"the essential tremors" (Lenka & Louis, 2020; Louis, 2021, 2014). Common to both views is both the recognition of heterogeneity as well as the acceptance that ET is not one entity. Hence, there is broad agreement on several fronts, and this deserves some acknowledgement. Where the views differ is both conceptual and nomenclatural—whether ET should be regarded as a "syndrome" or rather, a "family of diseases".

As discussed in detail elsewhere (Lenka & Louis, 2020), a "syndrome" is a recognizable complex of symptoms and physical findings that indicate a specific condition for which a direct cause is not necessarily understood. That is, syndromes describe a specific collection of symptoms and signs which recurrently co-occur (Lenka & Louis, 2020). Once medical science identifies the causative agents (i.e., etiologies, including genetic or environmental factors) and pathogenesis (i.e., tissue-based physiological or pathological derangements) of a particular condition, the term "syndrome" tends to be replaced by "disease" (Lenka & Louis, 2020). Thus, while it is clear that "syndrome" and "disease" overlap in terms of the requirement of recurring symptoms and signs, they diverge because the term "syndrome" refers to recurrent co-occurrence of a set of signs and symptoms in the absence of a robust understanding of the underlying pathogenesis and etiology, whereas the term "disease" has additional characteristics in terms of information on etiology and pathogenesis (Lenka & Louis, 2020). Thus, diseases are entities that have identifiable causes and underlying pathophysiologies and are more than an arrangement of symptoms and signs. There are three additional features of disease that should be highlighted (Lenka & Louis, 2020). The first of these is that a disease is a state that places individuals at increased risk of adverse consequences (i.e., additional morbidity and mortality). In the case of ET, this would include significantly increased risks of dementia and Parkinson's disease (Louis, Joyce, & Cosentino, 2019; Tarakad & Jankovic, 2018). Second, "disease" generally connotes a progressive disorder or one in which deterioration and decline occur; this is clearly what one observes in ET (Putzke, Whaley, Baba, Wszolek, & Uitti, 2006). Third, based on the fact that the pathogenesis of a disease is better elucidated compared to that of a syndrome, for diseases, therapies are more often biologically based and specifically targeted toward pathogenic mechanisms, as we have seen in recent years in ET (Ondo, 2020; Papapetropoulos, Lee, Boyer, & Newbold, 2019).

With knowledge now of specific genetic susceptibility factors for ET (Muller, Girard, Hopfner, et al., 2016) as well as the repeated observation of both pathological (Louis & Faust, 2020) and physiological changes in the ET brain (Wu et al., 2021), knowledge of the disease etiology and pathogenesis is not a blank slate; rather, an understanding is emerging. Granted, clear separations between specific diseases within the hypothetical ET family of diseases has not been achieved, as there is no validated separation of entities that are each defined along an etiological-pathophysiological-clinical axis. However, the role of future research is to achieve this.

One should not lose sight of the commonalities of thought. Both views agree that they are seeing heterogeneity and that it is difficult to fully encapsulate all of ET under one name. Furthermore, what's in a name? Whether one refers to ET as a syndrome or as a family of diseases might be a regarded as a mere academic debate. However, the issue of applied names does have some import. First, such nomenclatural issues do have significant repercussions—a condition with the label of "syndrome" may not be recognized as a serious problem, may remain in the shadows, may be plagued by diminished public awareness, and may not garner funds for research that a condition with the label "disease" would (Engel Jr., 2010; Lenka & Louis, 2020). Second, at the heart of this debate is whether one uses knowledge of clinical features and underlying etiology and pathophysiology to define groupings as "syndromes" (i.e., using as a starting point the clinical constellations and then as a separate and secondary step considering underlying cause), as was the approach taken in the recent consensus paper (Bhatia et al., 2018), or whether one uses the more traditional approach in medicine, dating back to the mid-eighteenth century, of defining "disease" as an integrated clinical-pathological entity (i.e., from the start, recognizing holistically that pathophysiological disarrangements are the seats of clinical constellations, and that "disease" is the integrated entity that results from this compounded and inseparable clinical-pathological composite) (Louis, 2016). For the past several centuries, this traditional approach, handed down by Morgagni, has been at the heart of our attitude toward and nomenclature around human diseases (Louis, 2016). This should not be taken lightly.

References

Aridon, P., Ragonese, P., De Fusco, M., Salemi, G., Casari, G., & Savettieri, G. (2008). Further evidence of genetic heterogeneity in familial essential tremor. *Parkinsonism & Related Disorders, 14*, 15–18.

Bhatia, K. P., Bain, P., Bajaj, N., et al. (2018). Consensus statement on the classification of tremors. From the task force on tremor of the International Parkinson and Movement Disorder Society. *Movement Disorders, 33*, 75–87.

Chung, S. J., Kwon, H., Lee, D. K., et al. (2013). Neuroanatomical heterogeneity of essential tremor according to propranolol response. *PLoS One, 8*, e84054.

Deuschl, G., Bain, P., & Brin, M. (1998). Consensus statement of the Movement Disorder Society on tremor. Ad hoc scientific committee. *Movement Disorders, 13*(Suppl 3), 2–23.

Elble, R. J. (2002). Essential tremor is a monosymptomatic disorder. *Movement Disorders, 17*, 633–637.

Elble, R. J. (2020). Do we belittle essential tremor by calling it a syndrome rather than a disease? No. *Frontiers in Neurology, 11*, 586606.

Engel, J., Jr. (2010). Do we belittle epilepsy by calling it a disorder rather than a disease? *Epilepsia, 51*, 2363–2364.

Holtbernd, F., & Shah, N. J. (2021). Imaging the pathophysiology of essential tremor-a systematic review. *Frontiers in Neurology, 12*, 680254.

Kovach, M. J., Ruiz, J., Kimonis, K., et al. (2001). Genetic heterogeneity in autosomal dominant essential tremor. *Genetics in Medicine, 3*, 197–199.

Lenka, A., & Louis, E. D. (2020). Do we belittle essential tremor by calling it a syndrome rather than a disease? Yes. *Frontiers in Neurology, 11*, 522687.

Louis, E. D. (2014). 'Essential tremor' or 'the essential tremors': Is this one disease or a family of diseases? *Neuroepidemiology, 42*, 81–89.

Louis, E. D. (2016). De Sedibus et Causis Morborum: Is essential tremor a primary disease of the cerebellum? *Cerebellum, 15*, 233–234.

Louis, E. D. (2021). The essential tremors: Evolving concepts of a family of diseases. *Frontiers in Neurology, 12*, 650601.

Louis, E. D., & Faust, P. L. (2020). Essential tremor within the broader context of other forms of cerebellar degeneration. *Cerebellum*.

Louis, E. D., Joyce, J. L., & Cosentino, S. (2019). Mind the gaps: What we don't know about cognitive impairment in essential tremor. *Parkinsonism & Related Disorders*.

Louis, E. D., Mazzoni, P., Ma, K. J., et al. (2012). Essential tremor with ubiquitinated intranuclear inclusions and cerebellar degeneration. *Clinical Neuropathology, 31*, 119–126.

Muller, S. H., Girard, S. L., Hopfner, F., et al. (2016). Genome-wide association study in essential tremor identifies three new loci. *Brain, 139*, 3163–3169.

Ondo, W. G. (2020). Current and emerging treatments of essential tremor. *Neurologic Clinics, 38*, 309–323.

Pandey, S. (2022). Is essential tremor a family of diseases or a syndrome? A syndrome. *International Review of Neurobiology, 163*, 31–59.

Papapetropoulos, S., Lee, M. S., Boyer, S., & Newbold, E. J. (2019). A phase 2, randomized, double-blind, placebo-controlled trial of CX-8998, a selective modulator of the T-type Calcium Channel in inadequately treated moderate to severe essential tremor: T-CALM study design and methodology for efficacy endpoint and digital biomarker selection. *Frontiers in Neurology, 10*, 597.

Putzke, J. D., Whaley, N. R., Baba, Y., Wszolek, Z. K., & Uitti, R. J. (2006). Essential tremor: Predictors of disease progression in a clinical cohort. *Journal of Neurology, Neurosurgery, and Psychiatry, 77*, 1235–1237.

Riboldi, G. M., & Frucht, S. J. (2022). Is essential tremor a family of diseases or a syndrome? A family of diseases. *International Review of Neurobiology, 163*, 7–29.

Tarakad, A., & Jankovic, J. (2018). Essential tremor and Parkinson's disease: Exploring the relationship. *Tremor and Other Hyperkinetic Movements, 8*, 589.

Wong, S. B., Wang, Y. M., Lin, C. C., et al. (2021). Cerebellar oscillations in familial and sporadic essential tremor. *Cerebellum*. (Epub ahead of print).

Wu, Y. C., Louis, E. D., Gionco, J., Pan, M. K., Faust, P. L., & Kuo, S. H. (2021). Increased climbing Fiber lateral crossings on Purkinje cell dendrites in the cerebellar hemisphere in essential tremor. *Movement Disorders, 36*, 1440–1445.

CHAPTER ONE

Is essential tremor a family of diseases or a syndrome? A family of diseases

Giulietta M. Riboldi and Steven J. Frucht*
The Marlene and Paolo Fresco Institute for Parkinson's and Movement Disorders, New York University Langone Health, New York, NY, United States
*Corresponding author: e-mail address: steven.frucht@nyulangone.org

Contents

Abstract

It is now well-established that essential tremor (ET) can manifest with different clinical presentations and progressions (i.e., upper limb tremor, head tremor, voice tremor, lower limb tremor, task- or position-specific tremor, or a combination of those). Common traits and overlaps are identifiable across these different subtypes of ET, including a slow rate of progression, a response to alcohol and a positive family history. At the same time, each of these manifestations are associated with specific demographic, clinical and treatment-response characteristics suggesting a family of diseases rather than a spectrum of a syndrome. Here we summarize the most important clinical, demographic, neuropathological and imagingfeatures of ET and of its subtypes to support ET as a family of identifiable conditions. This classification has relevance for counseling of patients with regard to disease progression and treatment response, as well as for the design of therapeutic clinical trials.

International Review of Neurobiology, Volume 163
ISSN 0074-7742
https://doi.org/10.1016/bs.irn.2022.02.001

1. Introduction

Essential tremor (ET) represents one of the most prevalent movement disorders worldwide, found in about 5.79% of the population after the age of 65 and further increases with age (Deuschl, Bain, & Brin, 1998; Louis & McCreary, 2021). The prevalence of essential tremor also changes according to geographic latitudes (Welton et al., 2021). The impact of this condition on patients' daily life is not always disruptive, and many affected subjects never seek medical attention. Large cohort studies have defined the prevalence, the incidence as well as the clinical, demographic, neuropathological and imaging features of this disorder. Through a broader understanding of this condition, the definition of ET has evolved as well (Bhatia et al., 2018; Deuschl et al., 1998). The last consensus statement classified tremor according to two axes: Axis 1—based on clinical characteristics, and Axis 2—based on etiology (Bhatia et al., 2018). According to this classification the different presentations of ET (such as upper limb tremor, head tremor, voice tremor, task-specific tremor) are considered separate entities under Axis 1, while additional neurological features (referred to as "soft symptoms" and including, among the others, balance impairment, subtle dystonia and mild memory deterioration) have been included in a category termed "ET plus."

The debate of ET as a syndrome vs a family of diseases has been ongoing for many years. We argue that the different forms of tremor have a common origin, and that clinical features, brain networks, imaging findings, and response to therapy are most consistent with a family of diseases.

1.1 Syndrome vs disease: General concepts

A clinical syndrome refers to clinical signs and symptoms present in different combinations, where each single trait has no (or not-yet-defined) prognostic or etiologic value by itself (Calvo, Karras, Phillips, Kimball, & Wolf, 2003). For example, in corticobasal syndrome (CBS) different neurological manifestations have been described, including parkinsonism, eye movement abnormalities, alien limb phenomenon, clonus and cognitive impairment (Constantinides, Paraskevas, Paraskevas, Stefanis, & Kapaki, 2019). The presence or absence of any of these symptoms and signs is not known to influence the clinical course and demographic features. At the same time, a well-defined correlation between pathological and clinical features is lacking (Constantinides et al., 2019). Other disorders (e.g., breast cancer) clearly represent a family of diseases. Each subtype represents a distinct entity, with unique demographic, epidemiological, genetic, clinical, prognostic and neuropathological features, although overlap between subtypes occurs. Based on our current knowledge, we argue that ET is best accounted for as a family of diseases. ET can manifest with many different presentations (upper limb tremor, head tremor, vocal tremor, limb tremor, primary writing tremor or a combination of these), each with distinct demographic, clinical, neuroimaging, and, in certain cases, pathologic traits. Research in different field of ET (including clinical, pathological, genetics, and therapy-outcome studies) have nowadays largely contributed in better defining these features in ET and thus qualifying this condition as a disease more than a syndrome (Lenka & Louis, 2020; Welton et al., 2021).

The definition of ET as a family of diseases is also helpful in describing underlying pathological mechanisms, prognosis and response to treatments, while this would not be the case if considering ET as a syndrome (Lenka & Louis, 2020). In this chapter we present evidences supporting the concept of ET as disease, and, in particular, a family of diseases, based on the description of unique epidemiological and clinical features (Table 1).

2. Demographic and clinical features of the ET-family—A common origin with different branches

Tremor is defined by three cardinal features: a (1) regular (2) oscillation of a body part (3) around an axis. A positive family history and response to alcohol are typically shared by many patients with ET, independent of the topography of the disorder (Fig. 1). In ET, tremor is mostly postural and kinetic, although tremor at rest may be present as well (Cohen, Pullman, Jurewicz, Watner, & Louis, 2003). While upper limb tremor represents the most common presentation, different topographic localizations of ET are possible, presenting either in isolation or in combination in the same patient. In this section we will review the clinical profiles of the different subtypes of ET, highlighting shared traits and differences.

2.1 Overview of ET subtypes

The last consensus statement on ET routes the classification of ET subtypes on two main Axes: Axis 1 includes data relative to the clinical features of tremor (such as demographic characteristic, disease progression and concomitant other clinical or family history), characteristics of tremor (distribution, frequency, and triggers), additional clinical and instrumental signs; Axis

Table 1 ET-subtypes.

		Demographic features			Neurophysiological features		Response to treatments					
	Affected body part	**Gender**	**Age onset (decade)**	**Family history**	**Frequency**	**Progression**	**Propranolol**	**Primidone**	**Botulinum toxin**	**Alcohol**	**DBS**	**HiFU**
Upper limb tremor	Upper limbs (pronation-supination, flexion-extension)	M=F	2nd / 6th	Yes	7 Hz (accelerometer), 8–12 Hz early and 4–8 late (EMG)	Yes (with spreading to head and or voice)	Yes	Yes	Yes	Yes	Yes (VIM)	Yes
Head tremor	Head and neck	F>M	6–7th	Yes	3–8 Hz	Mild	Poor	Poor	Yes	Yes	Yes (VIM)	NA
Vocal tremor	Laryngeal > pharyngeal > respiratory muscles	F>M (75–93% female)	6th (19–74 y)	Yes (38–45%)	4–5, 4–8, 4–10 Hz	Yes	No	Poor	Yes (56 %)	Yes	NA	NA
Primary writing tremor	Distal upper limb (usually unilateral)	NA	6th	Yes (not always)	5–7 Hz	Possible	Poor	Yes	Yes	Yes	Yes (VIM)	NA
Embouchure tremor	Mimic, tongue, laryngeaband respiratory muscles	NA	6th	Yes	5, 3–8 Hz	No progression	Yes	NA	NA	Yes	NA	NA
Musician tremor	Proximal and distal upper limbs muscles (i.e., bowing tremor)	M>F (?)	5th	Yes (not always)	3–8, 4.7–6.7 Hz (mean 5.7–5.8 Hz)	Possible	Yes	Yes	Yes	Yes	NA	NA

Characterization of demographic, neurophysiological, and clinical traits, as well as treatment outcomes of the most frequent subtypes of ET. Lower limb tremor was not included, as reports from the literature are limited. Recurrent features across the different subtypes are highlighted in red. DBS: deep brain stimulation; EMG: electromyography; F: females; HiFU: High-Intensity Focused Ultrasounds; M: males; NA: not available; VIM: ventralis intermediate nucleus (of the thalamus).

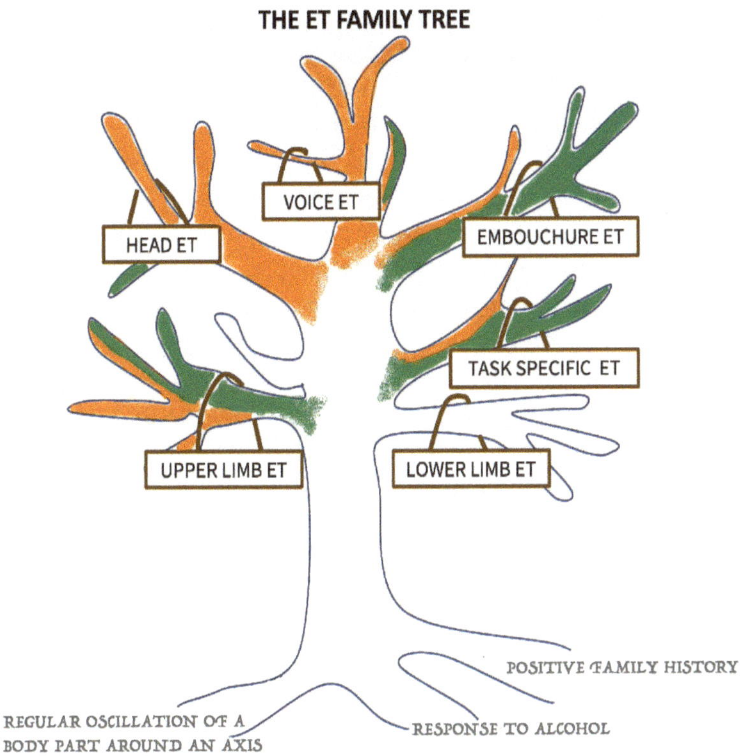

Fig. 1 Essential tremor: a family of diseases. This schematic representation summarizes the shared common features of subtypes of tremor (included in the "roots" of the three), and the branching of different subtypes of tremor based on topographical clinical presentation. The colors in the branches represent the proportion of males (green) vs females (orange) subjects in the different tremor subtypes. Not enough data are available for the lower limb ET.

2 based on supposed etiology (acquired, genetics, and idiopathic) (Bhatia et al., 2018). As further detailed in the below paragraph, topographic presentation of the tremor best helps distinguishing across the different ET subtypes, although overlaps between the different forms are frequent either within the same subject or across different subjects in the same family.

The most common and well-recognized form of ET is upper limb tremor. One of the first documented reports of a subject with essential tremor affecting the upper limbs is the founding father Samuel Adams, who also had a history of vocal and head tremor (Louis, 2001). Upper limb ET is usually symmetric, although asymmetric presentations also occur (Louis, Wendt, Pullman, & Ford, 1998). Tremor is exacerbated by posture and action, mildly worse on reaching the target (such as during finger-nose--finger test, or when bringing a spoon to the mouth) (Louis, 2011). A rest tremor can be present in ET, with no additional signs or symptoms typical of a diagnosis of Parkinson's disease (Cohen et al., 2003).

Other forms of ET are represented by head tremor and voice tremor. These two forms of tremor can present in isolation or be combined. Certain subjects present with tremor of the upper limb only when performing specific tasks, such as writing or approaching the paper with the pen ("primary writing tremor") (Bain et al., 1995), when playing musical instruments such as wind instruments like trombone or windpipe ("embouchure tremor") (Ragothaman, Swaminath, Pal, & Muthane, 2007), string instruments (Lee, Furuya, & Altenmüller, 2014) ("primary bowing tremor") (Lederman, 2012; Lee, Chadde, Altenmüller, & Schoonderwaldt, 2014; Lee, Tominaga, Furuya, Miyazaki, & Altenmüller, 2013), as well as in the context of other professions (such as hair dressers or dentists) (Bain et al., 1995; Rosenbaum & Jankovic, 1988; Soland, Bhatia, Volonte, & Marsden, 1996). Task-specific tremor is different from focal or task-specific dystonia (FTSD) in terms of phenomenology and certain clinical features, which make it a distinct entity. Indeed, compared to FTSD, task specific tremor present with a regular oscillation of the affected body part when attempting to perform the trigger task. In adjunct, progression is absent or very slow, and it can be associated with essential tremor of other body part or in other members of the family. All these features are usually not associated with FTSD.

Additional signs and symptoms can be detected in subjects with ET as the disease progresses. These include mild ataxia, nonmotor symptoms, dystonic posturing, tremor at rest, eye movement abnormalities and cognitive complaints (Bermejo-Pareja et al., 2007; Bologna et al., 2019; Chatterjee, Jurewicz, Applegate, & Louis, 2004; Helmchen et al., 2003; Louis, 2011; Miller et al., 2007; Schneier, Barnes, Albert, & Louis, 2001; Tan et al., 2005). The term "ET-plus" was proposed to classify subjects with these additional symptoms, although with some controversy (Bhatia et al., 2018; Louis et al., 2020). Indeed, some studies suggest that these additional features may be indicative of the progression of the disease, especially within certain ET subtypes, rather than a separate entity with distinct neuropathological, prognostic or treatment response features (Louis et al., 2020). For example, cognitive and affective disorders are more prevalent in patients with ET and head tremor compared to subjects with ET without head tremor (Bologna et al., 2019; Peng et al., 2020). Progressive involvement of the cerebellum may be responsible for these additional manifestations (Bologna et al., 2019; Peng et al., 2020). The development of additional cerebellar features (e.g., ataxia) is also consistent with the primary involvement of the cerebellum in the pathogenesis of ET (Fasano et al., 2010).

2.2 Epidemiology

2.2.1 *Age*

The age of onset of ET has a bimodal distribution, with a peak in the 2nd decade and a second peak in the 6th decade (Lou & Jankovic, 1991) (Fig. 2). Upper limb ET can present at any age, starting in childhood (Louis, Ford, & Barnes, 2000; Prasad et al., 2019). Age of onset is delayed in subjects with voice and head tremor, also when these occur in combination with other subtypes of ET (Barkmeier-Kraemer, 2020; Prasad et al., 2019). Indeed, head tremor usually starts after the age of 40, more frequently in the 60s independently from ET duration when they present in combination with other forms of tremor (Lenka, Bhalsing, Jhunjhunwala, Chandran, & Pal, 2015; Louis, 2013). Age of onset is also later for task- or position-specific tremor, such as primary writing tremor and embouchure tremor, where the mean age of onset is between 50 and 60 years (Datta, Batra, & Pandey, 2021; Lee, Voget, Furuya, Morise, & Altenmüller, 2016) or between

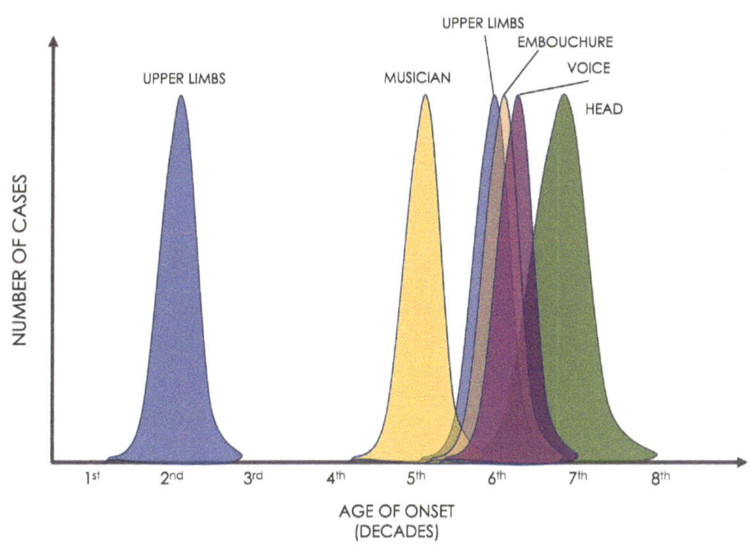

Fig. 2 Age of onset of different subtypes of ET. The different subtypes of tremor are coded with a different color. Upper limb tremor presents two peaks (around the 2nd and 6th decades).

40 and 50 years for bowing tremor (Lederman, 2012; Lee, Chadde, et al., 2014). On the contrary, a younger age of onset has been reported in subjects with a family history of tremor (Tallón-Barranco et al., 1997).

2.2.2 Gender

While ET of the upper limbs shows a balanced distribution between males and females (Louis, Ferrer, Eliasen, Gaini, & Petersen, 2019) with only a possible preference in males (Hubble, Busenbark, Pahwa, Lyons, & Koller, 1997), head and voice tremor are significantly more frequent in females, with up to 93% female cases in certain cohorts (Barkmeier-Kraemer, 2020; Chen et al., 2017; Hubble et al., 1997; Louis, Ford, & Frucht, 2003; Patel & Frucht, 2015; Tallón-Barranco et al., 1997) (Fig. 1). Given the rarity of this condition and the limited number of reported cohorts, it is not possible to establish a clear gender preference for task or position-specific ET, such as embouchure tremor, primary writing tremor or bowing tremor. Considering the few cases reported of musician tremor, male gender may be more common (Lederman, 2012; Lee, Chadde, et al., 2014).

2.2.3 Family history

A positive family history is reported in about half of cases in subjects with vocal tremor and task-specific or postural tremor (i.e., musician tremor) (Barkmeier-Kraemer, 2020; Lee, Chadde, et al., 2014; Patel & Frucht, 2015) although it is less consistent in subjects with primary writing tremor and primary bowing tremor (Klawans, Glantz, Tanner, & Goetz, 1982; Lederman, 2012).

2.2.4 Progression

The rate of progression of ET has been assessed by only a few studies. Normally, ET progression is slow, although older age of onset is associated with faster progression (Louis et al., 2009, 2000). The progression of ET can be associated with a spread of tremor to other regions, mostly according to an appendicular to axial gradient. For example, subjects with a long history of hand tremor may develop mild head and voice tremor with aging (Chen et al., 2017). On the contrary, task- or position-specific ET, such as embouchure tremor and primary writing tremor, usually do not progress to involve other activities in the same topographic areas (such as eating or speaking for embouchure tremor, or typing or sewing for primary writing tremor), while these forms can be associated with ET in other parts of the body

(Bain et al., 1995; Ragothaman et al., 2007). Spread to other activities has been instead reported in subjects with primary bowing tremor (Lederman, 2012; Lee, Chadde, et al., 2014). Similarly, additional features described under the umbrella term "ET plus" (such as cerebellar ataxia and dysmetria, dystonia, rest tremor, and cognitive changes) appear to be age- and ET-stage related, therefore suggesting that "ET plus" may be part of ET progression in certain subjects and may define a stage of ET rather than a subtype (Louis, Huey, & Cosentino, 2021).

2.2.5 Response to treatment

The therapeutic approach commonly used in ET includes oral medications (beta-blockers, such as propranolol, GABAergic medications such as primidone and gabapentin, topiramate, and anticholinergic medications), as well as botulinum toxin injection, and surgical interventions such as Deep Brain Stimulation (DBS) and High-Intensity Focused Ultrasounds (HiFU). A positive response to botulinum toxin injection has been reported for all different subtypes of ET except for embouchure tremor, while response to pharmacological treatments is more variable (Bain et al., 1995; Byrnes, Mastaglia, Walters, Archer, & Thickbroom, 2005; Ondo & Satija, 2012; Papapetropoulos & Singer, 2006). Primidone is more effective than propranolol on vocal tremor but less beneficial than botulinum toxin injection (Justicz et al., 2016; Sulica & Louis, 2010). In primary writing tremor, improvements have been reported with primidone, while less satisfactory results are seen with propranolol and anticholinergics such as trihexyphenidyl (Bain et al., 1995; Elble, Moody, & Higgins, 1990; Kachi, Rothwell, Cowan, & Marsden, 1985; Ondo & Satija, 2012; Papapetropoulos & Singer, 2006). DBS of the ventralis intermediate nucleus (Vim) of the thalamus has a very positive and sustained benefit in upper limb ET, primary writing tremor and head tremor (Blomstedt, Fytagoridis, & Tisch, 2009; Cury et al., 2017; Flora, Perera, Cameron, & Maddern, 2010; Lu et al., 2020; Lyons, Shneyder, & Evidente, 2013; Minguez-Castellanos et al., 1999; Ondo & Satija, 2012; Putzke et al., 2004; Racette, Dowling, Randle, & Mink, 2001; Wang et al., 2020). DBS has been shown to improve vocal tremor when the procedure was performed for other types of ET (Erickson-DiRenzo, Sung, Ho, & Halpern, 2020; Kundu, Schrock, Davis, & House, 2018). In both head and voice tremor, better outcomes are observed after bilateral VIM-DBS (Deuschl, Raethjen, Hellriegel, & Elble, 2011). HiFU of the thalamus has been performed mostly for upper limb tremor with very positive outcomes

(Elias et al., 2016; Krüger et al., 2021). A recent systematic evidence-based review of treatment options for ET classified primidone, propranolol and topiramate (>200 mg/day) as clinically useful, while alprazolam, botulinum toxin A, VIM-DBS, and MRI-guided focused ultrasound thalamotomy were classified as possibly useful (Ferreira et al., 2019). This study only considered the therapeutic response in upper limb tremor, as large studies were not available for head and voice tremor. It has been hypothesized that the different degrees of response to available treatments is a reflection of distinct neuropathological changes in the brains across ET subtypes (Gironell et al., 1999; Louis et al., 2007; Louis & Vonsattel, 2008; Lyons & Pahwa, 2008).

At the same time, the shared treatment response in different subtypes of ET suggest that Backspace certain underlying pathogenic processes may overlap. For example, most subtypes of ET improve upon consumption of moderate amounts of alcohol (Bain et al., 1995; Jimenez-Jimenez et al., 1998; Koller & Martyn, 1986; Massey & Paulson, 1985; Patel & Frucht, 2015; Sulica & Louis, 2010). This feature strongly supports the pathogenic role of brain networks regulated by alcohol, most likely GABA-mediated, across the different subtypes of ET, as previously demonstrated in animal models (Kralic et al., 2005; Nietz et al., 2020). The strong response of ET to sodium oxybate (SBX) also supports a central role of the GABAergic system (Frucht & Riboldi, 2020).

3. ET-family neuropathology

Neuropathological and neurophysiological studies have provided important insights in understanding brain regions and circuits possibly involved in ET. Initial studies suggested that ET was caused by an increased burst activity of the inferior olivary nucleus (ION) reflecting an increased inhibitory, synchronous output firing from the cerebellum (Llinás & Volkind, 1973). However, these observations were mostly based on animal models of acute action tremor and were not confirmed. More recent observations based on neuropathological and neurophysiological findings have reoriented the focus to different networks and distinguished between primary (i.e., Purkinje cells swelling, degeneration and death) and compensatory (i.e., increased arborization and hypertrophy of GABAergic basket cell and Purkinje cells) neuropathological changes (Louis, Babij, Cortés, Vonsattel, & Faust, 2013; Louis & Faust, 2020).

3.1 Neuropathology

Neuropathology studies from cohorts of subjects with ET (encompassing a total of about 69 subjects with ET and 55 controls, across different studies) highlight the prominent involvement of the cerebellum and Purkinje cells (PC). In particular, degeneration of both the PC (dendritic swelling, dendritic spine loss and pruning, and torpedoes), afferents to the PC (such as basket cell and climbing fibers), and of the locus coeruleus correlate with disease severity (Babij et al., 2013; Erickson-Davis et al., 2010; Lin et al., 2014; Louis et al., 2007, 2014; Shill et al., 2008). Cerebellar volume loss has been confirmed by multiple imaging studies, although some authors consider PC loss in the cerebellum a normal effect of aging in these patients (Rajput, Robinson, & Rajput, 2004; Symanski et al., 2014). A reduction of the GABA(A) and GABA(B) receptors particularly in the dentate nucleus was reported in a study analyzing the brains of 16 patients with ET and 10 controls (Paris-Robidas et al., 2012).

To date, there are few studies assessing neuropathological differences across ET subtypes (i.e., voice tremor, primary writing tremor, musician tremor). Head tremor, with or without upper limb tremor, is the only subtype of ET, other than upper limb tremor, with specific postmortem findings. One study showed that subjects with head tremor had an increased PC loss, but no changes in the other parameters of ET or cellular morphological features (Kuo et al., 2017). No clear differences have been identified between familial and sporadic cases of ET (Louis et al., 2017). Similar results were reported from the analysis of brains of subjects with early ($n = 30$ subjects) vs late-onset ($n = 30$ subjects) ET (Kuo et al., 2017). The authors suggested that age of onset is not mirrored in cerebellar pathology, but it can affect ET distribution, amplitude and frequency of tremor oscillation (see below). Interestingly, some reports also showed the presence of Lewy bodies in patients with ET (Louis et al., 2005). However, the neuropathological variability found in cases of ET further supports the hypothesis that ET is a collection of distinct entities or diseases.

4. ET-family neurophysiology

The frequency of oscillation of ET is usually around 8 Hz. Pioneer work recorded the frequency of upper limb ET at 7 Hz when measured with accelerometers, while EMG recordings showed a frequency of 8–12 Hz in early stage ET that dropped to 4–8 Hz in later stages (Elble, 1986; Wade,

Gresty, & Findley, 1982). In vocal tremor, the frequency is around 4–5 Hz (Hachinski, Thomsen, & Buch, 1975; Koda & Ludlow, 1992; Paige et al., 2018) or 4–8 Hz according to other cases series (Patel & Frucht, 2015). Interestingly, different degrees of involvement of the intrinsic and extrinsic laryngeal muscles (Koda & Ludlow, 1992), pharyngeal muscles, and in few cases respiratory muscles have been appreciated (Koda & Ludlow, 1992; Patel & Frucht, 2015). The few reported cases of embouchure tremor showed a frequency of 5 Hz (Ragothaman et al., 2007). In one report, subjects presented with 5 Hz embouchure tremor and slightly faster tremor (5–5.5 Hz) in the upper limbs (Ragothaman et al., 2007).

Primary writing tremor was reported with a frequency of 5–7 Hz, mostly involving pronator-supinator or flexor-extensor muscles (Rothwell, Traub, & Marsden, 1979). A wider range of frequencies, spanning from 3 to 8 Hz or between 4.7 and 6.7 Hz (mean 5.7–5.8 Hz), has been reported for primary bowing tremor (Lee, Chadde, et al., 2014; Lee et al., 2013). There are few studies reporting the neurophysiological characteristics of essential head tremor. The study assessing the largest cohort of these patients reported a range of frequency between 3 and 8 Hz, with EMG recording of the sternomastoid and splenius capitis muscles (Valls-Solé et al., 1997). Overall, the frequency of ET decreases with age (Lou & Jankovic, 1991), possibly explaining why subtypes of ET with later age of onset (i.e., voice tremor, head tremor) have lower frequency compared to hand tremor in subjects with earlier age of onset, and with progression of the disease (Elble, 2000). On the contrary, tremor amplitude increases with age (Louis et al., 2011).

4.1 Neuronal networks

The neurophysiological basis of ET has been extensively explored in animal models and in clinical studies. As previously discussed, the cerebellum and its connections with the thalamus, the sensory-motor cortex (primary and supplementary motor cortex), and the brainstem (Neely et al., 2015; Schnitzler, Münks, Butz, Timmermann, & Gross, 2009; Tsuboi et al., 2021) play a key role in the tremor network. The well-established improvement of tremor after DBS of the VIM nucleus of the thalamus further supports the role of the cortico-ponto-cerebellar-thalamic-cortical circuit in ET. Previous studies showed an oscillatory network with coupling between the primary motor cortex and muscle activity (Schnitzler et al., 2009). These circuitries differ from those of tremor associated with aging and tremor in patients with

Parkinson's disease or other types of tremor (Helmich, Toni, Deuschl, & Bloem, 2013; Muthuraman et al., 2018). There are limited data about networks in other forms of ET. In primary writing tremor, TMS recordings identified differences in intracortical inhibition compared to controls and to subjects with classical ET (Modugno et al., 2002).

5. Evidence from genetics: Interfamilial variability and overlap

A genetic contribution to ET is well established. Very frequently patients report the presence of the condition in multiple family members, and relatives of patients with ET have increased risk of tremor (Louis, Ford, Frucht, & Ottman, 2001). Nevertheless, despite the clear familial component of this disease, strong genetic determinants have not been identified yet, even after numerous linkage analysis studies of large pedigrees. The latest works in this field highlighted genes such as *LINGO1, CACNA1A, SHF, NOTCH2NLC, NOS3, FUS* (Alakbarzade et al., 2019; Deng, Wu, & Jankovic, 2019; Liao et al., 2020; Ng et al., 2020; Sun et al., 2020; Tan, 2010). ET is transmitted in an autosomal dominant fashion with reduced penetrance (Bain et al., 1994). Concordance in monozygotic twins was reported in up to 63%, or even up to 99% when more stringent criteria for the diagnosis of ET were applied (Lorenz et al., 2004; Tanner et al., 2001). By leveraging a method called genome-based restricted maximum likelihood—LD- and MAF-stratified genomic-relatedness (GREML-LDMS), which consider the contribution of rare and common genetic variants to a specific trait, the hereditability of ET has been estimated to be of 75.5% (Diez-Fairen et al., 2019). The same study also highlighted that, based on the results of this large cohort ($n = 1751$ cases of ET vs $n = 5311$ controls), genetic hits were found on chromosome 6 and 21 (Diez-Fairen et al., 2019).

Members of the same family frequently report different topographic distribution of ET (Louis, Hernandez, Clark, & Ottman, 2013). Unfortunately, the lack of definitive known causative genetic variants prevents formulating genotype-phenotype observations between different families. A large number of genes associated with other neurodegenerative conditions, such as Parkinson's disease (i.e., *LRRK2*, *GBA*, *SNCA* genes), other neurodegenerative diseases (i.e., *MAPT*, *TREM2*, *FUS*), as well as new genetic hits (such as *LINGO1, LINGO2, STK32B, SLC1A2, PPARGC1A, CTNNA3*), have been tested in ET cases, but few have been associated to the a susceptibility of the disease across different studies (Siokas et al., 2020). GGC repeat expansions, such as in the *NOTCH2NLC* gene, were identified only in sporadic cases of ET but not in familial cases (Ng et al., 2020). Long-term follow-up of these cases was limited only to two subjects (Ng et al., 2020). Interestingly, while the age of onset of ET can normally span between the second and seventh decade, in familial case of ET family members all share a similar age of onset (Louis, Hernandez, Rabinowitz, Ottman, & Clark, 2013).

6. Evidence from imaging studies

The central role of the cerebellum in ET is supported by imaging studies, demonstrating volume loss as well as changes in connectivity or function of specific brain areas. Plain brain MRI does not reveal structural alterations in patients with ET. Instead, alterations of the cerebellar peduncles and dentate nucleus are documented by diffusion tension imaging studies (DTI), although volumetric imaging studies failed to consistently show loss of gray matter in the cerebellum (summarized in Holtbernd & Shah, 2021) (Nicoletti et al., 2010). An interesting study comparing different subtypes of tremor affecting various cranial regions have showed gray matter loss in discrete cerebellar lobules, suggesting the importance of more in depth structural analysis in order to detect specific changes (Dyke, Cameron, Hernandez, Dydak, & Louis, 2017). Resting state functional MRI reported impairment of the cerebellar and cerebello-thalamo-cortical activation/connectivity (summarized in Holtbernd & Shah, 2021). The studies mainly assessed subjects with upper limb ET, although whether the tremor was present elsewhere in the same patient was inconsistently reported.

In subjects with primary writing tremor, the cerebellum (especially the connections with the medial cerebellum), and the frontal cortex, particularly the cingulum, are preferentially involved (Berg, Preibisch, Hofmann, & Naumann, 2000; Lenka et al., 2017). Loss of tracts through the cerebellum and the superior cerebellar peduncle were seen in DTI studies.

In patients with voice tremor, a recent study compared brain functional MRI (fMRI) showing involvement of the primary sensorimotor cortex, superior/inferior parietal lobules, the inferior temporal gyrus across different types of tremor, while the activation of the cerebellum was specific for ET of the voice compared to other forms of voice tremor (de Lima Xavier & Simonyan, 2020). Finally, the posterior lobe of the cerebellum and of the cerebello-cortical areas are mostly affected in subjects with head tremor, while in subjects with upper limb tremor an involvement of the thalamo-cortical path has been showed (Wang et al., 2018).

7. Conclusions

Patient A walks into the office with a tremor of her head, described as a regular oscillation of 5–6 Hz. Patient B comes to the office complaining of a 10-year history of tremor of the hands, mostly noticed when eating or holding cups, with a frequency of 5–7 Hz.

Using currently available rubrics and understanding, both Patient A and Patient B have ET (a regular oscillation around an axis with a frequency of 5–8 Hz). As presented in this chapter, understanding of the different subtypes of tremor allows us to predict that Patient A is most likely be a woman in her 60s or 70s with a subtle vocal tremor with head tremor. Patient B is likely to be a subject in their 50s, male or female, with ET of the hands since his/her 20s with possible (but not necessarily) axial manifestations (vocal and head tremor). Patient A does not respond to beta-blockers, but is successfully treated with primidone. Patient B responds well even to a low dose of propranolol. Both Patient A and Patient B report a family history of tremor, and they both report improvement of symptoms with moderate alcohol consumption.

This brief example supports the idea that ET is a disease, a well-defined clinical condition, characterized by a mean core of central features consisting of a regular oscillation of a body part around an axis with a frequency of 5–8 Hz, usually familial, alleviated by alcohol. Clinical subtypes of ET represent separate entities of a family of diseases, each well characterized in the literature, allowing prediction of progression, response to treatments, and patient counseling.

References

Alakbarzade, V., Iype, T., Chioza, B. A., Singh, R., Harlalka, G. V., Hardy, H., et al. (2019). Copy number variation of LINGO1 in familial dystonic tremor. *Neurology. Genetics*, *5*(1), e307. https://doi.org/10.1212/NXG.0000000000000307.

Babij, R., Lee, M., Cortés, E., Vonsattel, J.-P. G., Faust, P. L., & Louis, E. D. (2013). Purkinje cell axonal anatomy: Quantifying morphometric changes in essential tremor versus control brains. *Brain: A Journal of Neurology*, *136*(Pt. 10), 3051–3061. https://doi.org/10.1093/brain/awt238.

Bain, P. G., Findley, L. J., Britton, T. C., Rothwell, J. C., Gresty, M. A., Thompson, P. D., et al. (1995). Primary writing tremor. *Brain: A Journal of Neurology*, *118*(Pt. 6), 1461–1472. https://doi.org/10.1093/brain/118.6.1461.

Bain, P. G., Findley, L. J., Thompson, P. D., Gresty, M. A., Rothwell, J. C., Harding, A. E., et al. (1994). A study of hereditary essential tremor. *Brain: A Journal of Neurology, 117*(Pt. 4), 805–824. https://doi.org/10.1093/brain/117.4.805.

Barkmeier-Kraemer, J. M. (2020). Isolated voice tremor: A clinical variant of essential tremor or a distinct clinical phenotype? *Tremor and Other Hyperkinetic Movements (New York, N.Y.), 10*. https://doi.org/10.7916/tohm.v0.738.

Berg, D., Preibisch, C., Hofmann, E., & Naumann, M. (2000). Cerebral activation pattern in primary writing tremor. *Journal of Neurology, Neurosurgery, and Psychiatry, 69*(6), 780–786. https://doi.org/10.1136/jnnp.69.6.780.

Bermejo-Pareja, F., Louis, E. D., Benito-León, J., & Neurological Disorders in Central Spain (NEDICES) Study Group. (2007). Risk of incident dementia in essential tremor: A population-based study. *Movement Disorders: Official Journal of the Movement Disorder Society, 22*(11), 1573–1580. https://doi.org/10.1002/mds.21553.

Bhatia, K. P., Bain, P., Bajaj, N., Elble, R. J., Hallett, M., Louis, E. D., et al. (2018). Consensus statement on the classification of tremors. From the task force on tremor of the International Parkinson and Movement Disorder Society. *Movement Disorders: Official Journal of the Movement Disorder Society, 33*(1), 75–87. https://doi.org/10.1002/mds.27121.

Blomstedt, P., Fytagoridis, A., & Tisch, S. (2009). Deep brain stimulation of the posterior subthalamic area in the treatment of tremor. *Acta Neurochirurgica, 151*(1), 31–36. https://doi.org/10.1007/s00701-008-0163-7.

Bologna, M., Berardelli, I., Paparella, G., Ferrazzano, G., Angelini, L., Giustini, P., et al. (2019). Tremor distribution and the variable clinical presentation of essential tremor. *Cerebellum (London, England), 18*(5), 866–872. https://doi.org/10.1007/s12311-019-01070-0.

Byrnes, M. L., Mastaglia, F. L., Walters, S. E., Archer, S.-A. R., & Thickbroom, G. W. (2005). Primary writing tremor: Motor cortex reorganisation and disinhibition. *Journal of Clinical Neuroscience: Official Journal of the Neurosurgical Society of Australasia, 12*(1), 102–104. https://doi.org/10.1016/j.jocn.2004.08.004.

Calvo, F., Karras, B. T., Phillips, R., Kimball, A. M., & Wolf, F. (2003). Diagnoses, syndromes, and diseases: A knowledge representation problem. *AMIA .. Annual Symposium Proceedings. AMIA Symposium, 2003*, 802.

Chatterjee, A., Jurewicz, E. C., Applegate, L. M., & Louis, E. D. (2004). Personality in essential tremor: Further evidence of non-motor manifestations of the disease. *Journal of Neurology, Neurosurgery, and Psychiatry, 75*(7), 958–961. https://doi.org/10.1136/jnnp.2004.037176.

Chen, W., Hopfner, F., Szymczak, S., Granert, O., Müller, S. H., Kuhlenbäumer, G., et al. (2017). Topography of essential tremor. *Parkinsonism & Related Disorders, 40*, 58–63. https://doi.org/10.1016/j.parkreldis.2017.04.012.

Cohen, O., Pullman, S., Jurewicz, E., Watner, D., & Louis, E. D. (2003). Rest tremor in patients with essential tremor: Prevalence, clinical correlates, and electrophysiologic characteristics. *Archives of Neurology, 60*(3), 405–410. https://doi.org/10.1001/archneur.60.3.405.

Constantinides, V. C., Paraskevas, G. P., Paraskevas, P. G., Stefanis, L., & Kapaki, E. (2019). Corticobasal degeneration and corticobasal syndrome: A review. *Clinical Parkinsonism & Related Disorders, 1*, 66–71. https://doi.org/10.1016/j.prdoa.2019.08.005.

Cury, R. G., Fraix, V., Castrioto, A., Pérez Fernández, M. A., Krack, P., Chabardes, S., et al. (2017). Thalamic deep brain stimulation for tremor in Parkinson disease, essential tremor, and dystonia. *Neurology, 89*(13), 1416–1423. https://doi.org/10.1212/WNL.0000000000004295.

Datta, A., Batra, N., & Pandey, S. (2021). Primary writing tremor: Current concepts. *Annals of Indian Academy of Neurology, 24*(3), 319–326. https://doi.org/10.4103/aian.AIAN_1264_20.

de Lima Xavier, L., & Simonyan, K. (2020). Neural representations of the voice tremor spectrum. *Movement Disorders: Official Journal of the Movement Disorder Society, 35*(12), 2290–2300. https://doi.org/10.1002/mds.28259.

Deng, H., Wu, S., & Jankovic, J. (2019). Essential tremor: Genetic update. *Expert Reviews in Molecular Medicine, 21*. https://doi.org/10.1017/erm.2019.7, e8.

Deuschl, G., Bain, P., & Brin, M. (1998). Consensus statement of the Movement Disorder Society on Tremor. Ad Hoc Scientific Committee. *Movement Disorders: Official Journal of the Movement Disorder Society, 13*(Suppl. 3), 2–23. https://doi.org/10.1002/mds.870131303.

Deuschl, G., Raethjen, J., Hellriegel, H., & Elble, R. (2011). Treatment of patients with essential tremor. *The Lancet. Neurology, 10*(2), 148–161. https://doi.org/10.1016/S1474-4422(10)70322-7.

Diez-Fairen, M., Bandres-Ciga, S., Houle, G., Nalls, M. A., Girard, S. L., Dion, P. A., et al. (2019). Genome-wide estimates of heritability and genetic correlations in essential tremor. *Parkinsonism & Related Disorders, 64*, 262–267. https://doi.org/10.1016/j.parkreldis.2019.05.002.

Dyke, J. P., Cameron, E., Hernandez, N., Dydak, U., & Louis, E. D. (2017). Gray matter density loss in essential tremor: A lobule by lobule analysis of the cerebellum. *Cerebellum & Ataxias, 4*, 10. https://doi.org/10.1186/s40673-017-0069-3.

Elble, R. J. (1986). Physiologic and essential tremor. *Neurology, 36*(2), 225–231. https://doi.org/10.1212/wnl.36.2.225.

Elble, R. J. (2000). Essential tremor frequency decreases with time. *Neurology, 55*(10), 1547–1551. https://doi.org/10.1212/wnl.55.10.1547.

Elble, R. J., Moody, C., & Higgins, C. (1990). Primary writing tremor. A form of focal dystonia? *Movement Disorders: Official Journal of the Movement Disorder Society, 5*(2), 118–126. https://doi.org/10.1002/mds.870050205.

Elias, W. J., Lipsman, N., Ondo, W. G., Ghanouni, P., Kim, Y. G., Lee, W., et al. (2016). A randomized trial of focused ultrasound thalamotomy for essential tremor. *The New England Journal of Medicine, 375*(8), 730–739. https://doi.org/10.1056/NEJMoa1600159.

Erickson-Davis, C. R., Faust, P. L., Vonsattel, J.-P. G., Gupta, S., Honig, L. S., & Louis, E. D. (2010). "Hairy baskets" associated with degenerative Purkinje cell changes in essential tremor. *Journal of Neuropathology and Experimental Neurology, 69*(3), 262–271. https://doi.org/10.1097/NEN.0b013e3181d1ad04.

Erickson-DiRenzo, E., Sung, C. K., Ho, A. L., & Halpern, C. H. (2020). Intraoperative evaluation of essential vocal tremor in deep brain stimulation surgery. *American Journal of Speech-Language Pathology, 29*(2), 851–863. https://doi.org/10.1044/2019_AJSLP-19-00079.

Fasano, A., Herzog, J., Raethjen, J., Rose, F. E. M., Muthuraman, M., Volkmann, J., et al. (2010). Gait ataxia in essential tremor is differentially modulated by thalamic stimulation. *Brain: A Journal of Neurology, 133*(Pt. 12), 3635–3648. https://doi.org/10.1093/brain/awq267.

Ferreira, J. J., Mestre, T. A., Lyons, K. E., Benito-León, J., Tan, E.-K., Abbruzzese, G., et al. (2019). MDS evidence-based review of treatments for essential tremor. *Movement Disorders: Official Journal of the Movement Disorder Society, 34*(7), 950–958. https://doi.org/10.1002/mds.27700.

Flora, E. D., Perera, C. L., Cameron, A. L., & Maddern, G. J. (2010). Deep brain stimulation for essential tremor: A systematic review. *Movement Disorders: Official Journal of the Movement Disorder Society, 25*(11), 1550–1559. https://doi.org/10.1002/mds.23195.

Frucht, S. J., & Riboldi, G. M. (2020). Alcohol-responsive hyperkinetic movement disorders—A mechanistic hypothesis. *Tremor and Other Hyperkinetic Movements (New York, N.Y.), 10*, 47. https://doi.org/10.5334/tohm.560.

Gironell, A., Kulisevsky, J., Barbanoj, M., López-Villegas, D., Hernández, G., & Pascual-Sedano, B. (1999). A randomized placebo-controlled comparative trial of gabapentin and propranolol in essential tremor. *Archives of Neurology*, *56*(4), 475–480. https://doi.org/10.1001/archneur.56.4.475.

Hachinski, V. C., Thomsen, I. V., & Buch, N. H. (1975). The nature of primary vocal tremor. *The Canadian Journal of Neurological Sciences*, *2*(3), 195–197. https://doi.org/10.1017/s0317167100020254.

Helmchen, C., Hagenow, A., Miesner, J., Sprenger, A., Rambold, H., Wenzelburger, R., et al. (2003). Eye movement abnormalities in essential tremor may indicate cerebellar dysfunction. *Brain: A Journal of Neurology*, *126*(Pt. 6), 1319–1332. https://doi.org/10.1093/brain/awg132.

Helmich, R. C., Toni, I., Deuschl, G., & Bloem, B. R. (2013). The pathophysiology of essential tremor and Parkinson's tremor. *Current Neurology and Neuroscience Reports*, *13*(9), 378. https://doi.org/10.1007/s11910-013-0378-8.

Holtbernd, F., & Shah, N. J. (2021). Imaging the pathophysiology of essential tremor—A systematic review. *Frontiers in Neurology*, *12*, 680254. https://doi.org/10.3389/fneur.2021.680254.

Hubble, J. P., Busenbark, K. L., Pahwa, R., Lyons, K., & Koller, W. C. (1997). Clinical expression of essential tremor: Effects of gender and age. *Movement Disorders: Official Journal of the Movement Disorder Society*, *12*(6), 969–972. https://doi.org/10.1002/mds.870120620.

Jimenez-Jimenez, F. J., Cabrera-Valdivia, F., Orti-Pareja, M., Gasalla, T., Tallon-Barranco, A., & Zurdo, M. (1998). Bilateral primary writing tremor. *European Journal of Neurology*, *5*(6), 613–614. https://doi.org/10.1046/j.1468-1331.1998.560613.x.

Justicz, N., Hapner, E. R., Josephs, J. S., Boone, B. C., Jinnah, H. A., & Johns, M. M. (2016). Comparative effectiveness of propranolol and botulinum for the treatment of essential voice tremor. *The Laryngoscope*, *126*(1), 113–117. https://doi.org/10.1002/lary.25485.

Kachi, T., Rothwell, J. C., Cowan, J. M., & Marsden, C. D. (1985). Writing tremor: Its relationship to benign essential tremor. *Journal of Neurology, Neurosurgery, and Psychiatry*, *48*(6), 545–550. https://doi.org/10.1136/jnnp.48.6.545.

Klawans, H. L., Glantz, R., Tanner, C. M., & Goetz, C. G. (1982). Primary writing tremor: A selective action tremor. *Neurology*, *32*(2), 203–206. https://doi.org/10.1212/wnl.32.2.203.

Koda, J., & Ludlow, C. L. (1992). An evaluation of laryngeal muscle activation in patients with voice tremor. *Otolaryngology—Head and Neck Surgery: Official Journal of American Academy of Otolaryngology-Head and Neck Surgery*, *107*(5), 684–696. https://doi.org/10.1177/019459989210700510.

Koller, W. C., & Martyn, B. (1986). Writing tremor: Its relationship to essential tremor. *Journal of Neurology, Neurosurgery, and Psychiatry*, *49*(2), 220. https://doi.org/10.1136/jnnp.49.2.220.

Kralic, J. E., Criswell, H. E., Osterman, J. L., O'Buckley, T. K., Wilkie, M. E., Matthews, D. B., et al. (2005). Genetic essential tremor in gamma-aminobutyric acidA receptor alpha1 subunit knockout mice. *The Journal of Clinical Investigation*, *115*(3), 774–779. https://doi.org/10.1172/JCI23625.

Krüger, M. T., Avecillas-Chasin, J. M., Tamber, M. S., Heran, M. K. S., Sandhu, M. K., Polyhronopoulos, N. E., et al. (2021). Tremor and quality of life in patients with advanced essential tremor before and after replacing their standard deep brain stimulation with a directional system. *Neuromodulation: Journal of the International Neuromodulation Society*, *24*(2), 353–360. https://doi.org/10.1111/ner.13301.

Kundu, B., Schrock, L., Davis, T., & House, P. A. (2018). Thalamic deep brain stimulation for essential tremor also reduces voice tremor. *Neuromodulation: Journal of the International Neuromodulation Society*, *21*(8), 748–754. https://doi.org/10.1111/ner.12739.

Kuo, S.-H., Wang, J., Tate, W. J., Pan, M.-K., Kelly, G. C., Gutierrez, J., et al. (2017). Cerebellar pathology in early onset and late onset essential tremor. *Cerebellum (London, England)*, *16*(2), 473–482. https://doi.org/10.1007/s12311-016-0826-5.

Lederman, R. J. (2012). Primary bowing tremor: A task-specific movement disorder of string instrumentalists. *Medical Problems of Performing Artists*, *27*(4), 219–223.

Lee, A., Chadde, M., Altenmüller, E., & Schoonderwaldt, E. (2014). Characteristics of task-specific tremor in string instrument players. *Tremor and Other Hyperkinetic Movements (New York, N.Y.)*, *4*, 198. https://doi.org/10.7916/D86Q1V9W.

Lee, A., Furuya, S., & Altenmüller, E. (2014). Epidemiology and treatment of 23 musicians with task specific tremor. *Journal of Clinical Movement Disorders*, *1*, 5. https://doi.org/10.1186/2054-7072-1-5.

Lee, A., Tominaga, K., Furuya, S., Miyazaki, F., & Altenmüller, E. (2013). Task-specific tremor in violinists: Evidence of coactivation in the 3 to 8 Hz frequency range. *Movement Disorders: Official Journal of the Movement Disorder Society*, *28*(13), 1890–1892. https://doi.org/10.1002/mds.25569.

Lee, A., Voget, J., Furuya, S., Morise, M., & Altenmüller, E. (2016). Quantification of sound instability in embouchure tremor based on the time-varying fundamental frequency. *Journal of Neural Transmission (Vienna, Austria: 1996)*, *123*(5), 515–521. https://doi.org/10.1007/s00702-016-1533-6.

Lenka, A., Bhalsing, K. S., Jhunjhunwala, K. R., Chandran, V., & Pal, P. K. (2015). Are patients with limb and head tremor a clinically distinct subtype of essential tremor? *The Canadian Journal of Neurological Sciences*, *42*(3), 181–186. https://doi.org/10.1017/cjn.2015.23.

Lenka, A., Jhunjhunwala, K. R., Panda, R., Saini, J., Bharath, R. D., Yadav, R., et al. (2017). Altered brain network measures in patients with primary writing tremor. *Neuroradiology*, *59*(10), 1021–1029. https://doi.org/10.1007/s00234-017-1895-y.

Lenka, A., & Louis, E. D. (2020). Do we belittle essential tremor by calling it a syndrome rather than a disease? Yes. *Frontiers in Neurology*, *11*, 522687. https://doi.org/10.3389/fneur.2020.522687.

Liao, C., Sarayloo, F., Rochefort, D., Houle, G., Akçimen, F., He, Q., et al. (2020). Multiomics analyses identify genes and pathways relevant to essential tremor. *Movement Disorders: Official Journal of the Movement Disorder Society*, *35*(7), 1153–1162. https://doi.org/10.1002/mds.28031.

Lin, C.-Y., Louis, E. D., Faust, P. L., Koeppen, A. H., Vonsattel, J.-P. G., & Kuo, S.-H. (2014). Abnormal climbing fibre-Purkinje cell synaptic connections in the essential tremor cerebellum. *Brain: A Journal of Neurology*, *137*(Pt. 12), 3149–3159. https://doi.org/10.1093/brain/awu281.

Llinás, R., & Volkind, R. A. (1973). The olivo-cerebellar system: Functional properties as revealed by harmaline-induced tremor. *Experimental Brain Research*, *18*(1), 69–87. https://doi.org/10.1007/BF00236557.

Lorenz, D., Frederiksen, H., Moises, H., Kopper, F., Deuschl, G., & Christensen, K. (2004). High concordance for essential tremor in monozygotic twins of old age. *Neurology*, *62*(2), 208–211. https://doi.org/10.1212/01.wnl.0000103236.26934.41.

Lou, J. S., & Jankovic, J. (1991). Essential tremor: Clinical correlates in 350 patients. *Neurology*, *41*(2 (Pt. 1)), 234–238. https://doi.org/10.1212/wnl.41.2_part_1.234.

Louis, E. D. (2001). Samuel Adams' tremor. *Neurology*, *56*(9), 1201–1205. https://doi.org/10.1212/wnl.56.9.1201.

Louis, E. D. (2011). Essential tremor. *Handbook of Clinical Neurology*, *100*, 433–448. https://doi.org/10.1016/B978-0-444-52014-2.00033-1.

Louis, E. D. (2013). When do essential tremor patients develop head tremor? Influences of age and duration and evidence of a biological clock. *Neuroepidemiology*, *41*(2), 110–115. https://doi.org/10.1159/000351698.

Louis, E. D., Babij, R., Cortés, E., Vonsattel, J.-P. G., & Faust, P. L. (2013). The inferior olivary nucleus: A postmortem study of essential tremor cases versus controls. *Movement Disorders: Official Journal of the Movement Disorder Society*, *28*(6), 779–786. https://doi.org/10.1002/mds.25400.

Louis, E. D., Bares, M., Benito-Leon, J., Fahn, S., Frucht, S. J., Jankovic, J., et al. (2020). Essential tremor-plus: A controversial new concept. *The Lancet. Neurology, 19*(3), 266–270. https://doi.org/10.1016/S1474-4422(19)30398-9.

Louis, E. D., & Faust, P. L. (2020). Essential tremor pathology: Neurodegeneration and reorganization of neuronal connections. *Nature Reviews. Neurology, 16*(2), 69–83. https://doi.org/10.1038/s41582-019-0302-1.

Louis, E. D., Faust, P. L., Vonsattel, J.-P. G., Honig, L. S., Henchcliffe, C., Pahwa, R., et al. (2009). Older onset essential tremor: More rapid progression and more degenerative pathology. *Movement Disorders: Official Journal of the Movement Disorder Society, 24*(11), 1606–1612. https://doi.org/10.1002/mds.22570.

Louis, E. D., Faust, P. L., Vonsattel, J.-P. G., Honig, L. S., Rajput, A., Robinson, C. A., et al. (2007). Neuropathological changes in essential tremor: 33 cases compared with 21 controls. *Brain: A Journal of Neurology, 130*(Pt. 12), 3297–3307. https://doi.org/10.1093/brain/awm266.

Louis, E. D., Ferrer, M., Eliasen, E. H., Gaini, S., & Petersen, M. S. (2019). Tremor in normal adults: A population-based study of 1158 adults in the Faroe Islands. *Journal of the Neurological Sciences, 400*, 169–174. https://doi.org/10.1016/j.jns.2019.03.019.

Louis, E. D., Ford, B., & Barnes, L. F. (2000). Clinical subtypes of essential tremor. *Archives of Neurology, 57*(8), 1194–1198. https://doi.org/10.1001/archneur.57.8.1194.

Louis, E. D., Ford, B., & Frucht, S. (2003). Factors associated with increased risk of head tremor in essential tremor: A community-based study in northern Manhattan. *Movement Disorders: Official Journal of the Movement Disorder Society, 18*(4), 432–436. https://doi.org/10.1002/mds.10395.

Louis, E. D., Ford, B., Frucht, S., & Ottman, R. (2001). Mild tremor in relatives of patients with essential tremor: What does this tell us about the penetrance of the disease? *Archives of Neurology, 58*(10), 1584–1589. https://doi.org/10.1001/archneur.58.10.1584.

Louis, E. D., Hafeman, D., Parvez, F., Liu, X., Alcalay, R. N., Islam, T., et al. (2011). Tremor severity and age: A cross-sectional, population-based study of 2,524 young and midlife normal adults. *Movement Disorders: Official Journal of the Movement Disorder Society, 26*(8), 1515–1520. https://doi.org/10.1002/mds.23674.

Louis, E. D., Hernandez, N., Clark, L. N., & Ottman, R. (2013). Familial aggregation of cranial tremor in familial essential tremor. *Neuroepidemiology, 41*(1), 48–53. https://doi.org/10.1159/000348553.

Louis, E. D., Hernandez, N., Rabinowitz, D., Ottman, R., & Clark, L. N. (2013). Predicting age of onset in familial essential tremor: How much does age of onset run in families? *Neuroepidemiology, 40*(4), 269–273. https://doi.org/10.1159/000345253.

Louis, E. D., Honig, L. S., Vonsattel, J. P. G., Maraganore, D. M., Borden, S., & Moskowitz, C. B. (2005). Essential tremor associated with focal nonnigral Lewy bodies: A clinicopathologic study. *Archives of Neurology, 62*(6), 1004–1007. https://doi.org/10.1001/archneur.62.6.1004.

Louis, E. D., Huey, E. D., & Cosentino, S. (2021). Features of "ET plus" correlate with age and tremor duration: "ET plus" may be a disease stage rather than a subtype of essential tremor. *Parkinsonism & Related Disorders, 91*, 42–47. https://doi.org/10.1016/j.parkreldis.2021.08.017.

Louis, E. D., Kuo, S.-H., Wang, J., Tate, W. J., Pan, M.-K., Kelly, G. C., et al. (2017). Cerebellar pathology in familial vs. sporadic essential tremor. *Cerebellum (London, England), 16*(4), 786–791. https://doi.org/10.1007/s12311-017-0853-x.

Louis, E. D., Lee, M., Babij, R., Ma, K., Cortés, E., Vonsattel, J.-P. G., et al. (2014). Reduced Purkinje cell dendritic arborization and loss of dendritic spines in essential tremor. *Brain: A Journal of Neurology, 137*(Pt. 12), 3142–3148. https://doi.org/10.1093/brain/awu314.

Louis, E. D., & McCreary, M. (2021). How common is essential tremor? Update on the worldwide prevalence of essential tremor. *Tremor and Other Hyperkinetic Movements (New York, N.Y.), 11*, 28. https://doi.org/10.5334/tohm.632.

Louis, E. D., & Vonsattel, J. P. G. (2008). The emerging neuropathology of essential tremor. *Movement Disorders: Official Journal of the Movement Disorder Society, 23*(2), 174–182. https://doi.org/10.1002/mds.21731.

Louis, E. D., Wendt, K. J., Pullman, S. L., & Ford, B. (1998). Is essential tremor symmetric? Observational data from a community-based study of essential tremor. *Archives of Neurology, 55*(12), 1553–1559. https://doi.org/10.1001/archneur.55.12.1553.

Lu, G., Luo, L., Liu, M., Zheng, Z., Zhang, B., Chen, X., et al. (2020). Outcomes and adverse effects of deep brain stimulation on the ventral intermediate nucleus in patients with essential tremor. *Neural Plasticity, 2020*, 2486065. https://doi.org/10.1155/2020/2486065.

Lyons, K. E., & Pahwa, R. (2008). Pharmacotherapy of essential tremor: An overview of existing and upcoming agents. *CNS Drugs, 22*(12), 1037–1045. https://doi.org/10.2165/0023210-200822120-00006.

Lyons, M., Shneyder, N., & Evidente, V. (2013). Primary writing tremor responds to unilateral thalamic deep brain stimulation. *Turkish Neurosurgery, 23*(1), 122–124. https://doi.org/10.5137/1019-5149.JTN.4425-11.0.

Massey, E. W., & Paulson, G. W. (1985). Essential vocal tremor: Clinical characteristics and response to therapy. *Southern Medical Journal, 78*(3), 316–317.

Miller, K. M., Okun, M. S., Fernandez, H. F., Jacobson, C. E., Rodriguez, R. L., & Bowers, D. (2007). Depression symptoms in movement disorders: Comparing Parkinson's disease, dystonia, and essential tremor. *Movement Disorders: Official Journal of the Movement Disorder Society, 22*(5), 666–672. https://doi.org/10.1002/mds.21376.

Minguez-Castellanos, A., Carnero-Pardo, C., Gómez-Camello, A., Ortega-Moreno, A., García-Gómez, T., Arjona, V., et al. (1999). Primary writing tremor treated by chronic thalamic stimulation. *Movement Disorders: Official Journal of the Movement Disorder Society, 14*(6), 1030–1033. https://doi.org/10.1002/1531-8257(199911)14:6<1030::aid-mds1021>3.0.co;2-o.

Modugno, N., Nakamura, Y., Bestmann, S., Curra, A., Berardelli, A., & Rothwell, J. (2002). Neurophysiological investigations in patients with primary writing tremor. *Movement Disorders: Official Journal of the Movement Disorder Society, 17*(6), 1336–1340. https://doi.org/10.1002/mds.10292.

Muthuraman, M., Raethjen, J., Koirala, N., Anwar, A. R., Mideksa, K. G., Elble, R., et al. (2018). Cerebello-cortical network fingerprints differ between essential, Parkinson's and mimicked tremors. *Brain: A Journal of Neurology, 141*(6), 1770–1781. https://doi.org/10.1093/brain/awy098.

Neely, K. A., Kurani, A. S., Shukla, P., Planetta, P. J., Wagle Shukla, A., Goldman, J. G., et al. (2015). Functional brain activity relates to 0-3 and 3-8 Hz force oscillations in essential tremor. *Cerebral Cortex (New York, N.Y.: 1991), 25*(11), 4191–4202. https://doi.org/10.1093/cercor/bhu142.

Ng, A. S. L., Lim, W. K., Xu, Z., Ong, H. L., Tan, Y. J., Sim, W. Y., et al. (2020). NOTCH2NLC GGC repeat expansions are associated with sporadic essential tremor: Variable disease expressivity on long-term follow-up. *Annals of Neurology, 88*(3), 614–618. https://doi.org/10.1002/ana.25803.

Nicoletti, G., Manners, D., Novellino, F., Condino, F., Malucelli, E., Barbiroli, B., et al. (2010). Diffusion tensor MRI changes in cerebellar structures of patients with familial essential tremor. *Neurology, 74*(12), 988–994. https://doi.org/10.1212/WNL.0b013e3181d5a460.

Nietz, A., Krook-Magnuson, C., Gutierrez, H., Klein, J., Sauve, C., Hoff, I., et al. (2020). Selective loss of the GABAAα1 subunit from Purkinje cells is sufficient to induce a tremor phenotype. *Journal of Neurophysiology, 124*(4), 1183–1197. https://doi.org/10.1152/jn.00100.2020.

Ondo, W. G., & Satija, P. (2012). Task-specific writing tremor: Clinical phenotypes, progression, treatment outcomes, and proposed nomenclature. *The International Journal of Neuroscience, 122*(2), 88–91. https://doi.org/10.3109/00207454.2011.630544.

Paige, C., Hopewell, B. L., Gamsarian, V., Myers, B., Patel, P., Garrett, C. G., et al. (2018). Characterizing the normative voice tremor frequency in essential vocal tremor. *JAMA Otolaryngology—Head & Neck Surgery*, *144*(12), 1169–1173. https://doi.org/10.1001/jamaoto.2018.2566.

Papapetropoulos, S., & Singer, C. (2006). Treatment of primary writing tremor with botulinum toxin type A injections: Report of a case series. *Clinical Neuropharmacology*, *29*(6), 364–367. https://doi.org/10.1097/01.WNF.0000236765.00785.9C.

Paris-Robidas, S., Brochu, E., Sintes, M., Emond, V., Bousquet, M., Vandal, M., et al. (2012). Defective dentate nucleus GABA receptors in essential tremor. *Brain: A Journal of Neurology*, *135*(Pt. 1), 105–116. https://doi.org/10.1093/brain/awr301.

Patel, A., & Frucht, S. J. (2015). Isolated vocal tremor as a focal phenotype of essential tremor: A retrospective case review. *Journal of Clinical Movement Disorders*, *2*, 4. https://doi.org/10.1186/s40734-015-0016-5.

Peng, J., Wang, L., Li, N., Li, J., Duan, L., & Peng, R. (2020). Distinct non-motor features of essential tremor with head tremor patients. *Acta Neurologica Scandinavica*, *142*(1), 74–82. https://doi.org/10.1111/ane.13242.

Prasad, S., Bhalsing, K. S., Jhunjhunwala, K., Lenka, A., Binu, V. S., & Pal, P. K. (2019). Phenotypic variability of essential tremor based on the age at onset. *The Canadian Journal of Neurological Sciences*, *46*(2), 192–198. https://doi.org/10.1017/cjn.2018.384.

Putzke, J. D., Wharen, R. E., Obwegeser, A. A., Wszolek, Z. K., Lucas, J. A., Turk, M. F., et al. (2004). Thalamic deep brain stimulation for essential tremor: Recommendations for long-term outcome analysis. *The Canadian Journal of Neurological Sciences*, *31*(3), 333–342. https://doi.org/10.1017/s0317167100003413.

Racette, B. A., Dowling, J., Randle, J., & Mink, J. W. (2001). Thalamic stimulation for primary writing tremor. *Journal of Neurology*, *248*(5), 380–382. https://doi.org/10.1007/s004150170177.

Ragothaman, M., Swaminath, P. V., Pal, P. K., & Muthane, U. B. (2007). Embouchure dystonia and tremor in a professional windpipe "Nadaswaram" player. *Movement Disorders: Official Journal of the Movement Disorder Society*, *22*(14), 2133–2135. https://doi.org/10.1002/mds.21709.

Rajput, A., Robinson, C. A., & Rajput, A. H. (2004). Essential tremor course and disability: A clinicopathologic study of 20 cases. *Neurology*, *62*(6), 932–936. https://doi.org/10.1212/01.wnl.0000115145.18830.1a.

Rosenbaum, F., & Jankovic, J. (1988). Focal task-specific tremor and dystonia: Categorization of occupational movement disorders. *Neurology*, *38*(4), 522–527. https://doi.org/10.1212/wnl.38.4.522.

Rothwell, J. C., Traub, M. M., & Marsden, C. D. (1979). Primary writing tremor. *Journal of Neurology, Neurosurgery, and Psychiatry*, *42*(12), 1106–1114. https://doi.org/10.1136/jnnp.42.12.1106.

Schneier, F. R., Barnes, L. F., Albert, S. M., & Louis, E. D. (2001). Characteristics of social phobia among persons with essential tremor. *The Journal of Clinical Psychiatry*, *62*(5), 367–372. https://doi.org/10.4088/jcp.v62n0511.

Schnitzler, A., Münks, C., Butz, M., Timmermann, L., & Gross, J. (2009). Synchronized brain network associated with essential tremor as revealed by magnetoencephalography. *Movement Disorders: Official Journal of the Movement Disorder Society*, *24*(11), 1629–1635. https://doi.org/10.1002/mds.22633.

Shill, H. A., Adler, C. H., Sabbagh, M. N., Connor, D. J., Caviness, J. N., Hentz, J. G., et al. (2008). Pathologic findings in prospectively ascertained essential tremor subjects. *Neurology*, *70*(16 Pt. 2), 1452–1455. https://doi.org/10.1212/01.wnl.0000310425.76205.02.

Siokas, V., Aloizou, A.-M., Tsouris, Z., Liampas, I., Aslanidou, P., Dastamani, M., et al. (2020). Genetic risk factors for essential tremor: A review. *Tremor and Other Hyperkinetic Movements (New York, N.Y.)*, *10*, 4. https://doi.org/10.5334/tohm.67.

Soland, V. L., Bhatia, K. P., Volonte, M. A., & Marsden, C. D. (1996). Focal task-specific tremors. *Movement Disorders: Official Journal of the Movement Disorder Society*, *11*(6), 665–670. https://doi.org/10.1002/mds.870110611.

Sulica, L., & Louis, E. D. (2010). Clinical characteristics of essential voice tremor: A study of 34 cases. *The Laryngoscope*, *120*(3), 516–528. https://doi.org/10.1002/lary.20702.

Sun, Q.-Y., Xu, Q., Tian, Y., Hu, Z.-M., Qin, L.-X., Yang, J.-X., et al. (2020). Expansion of GGC repeat in the human-specific NOTCH2NLC gene is associated with essential tremor. *Brain: A Journal of Neurology*, *143*(1), 222–233. https://doi.org/10.1093/brain/awz372.

Symanski, C., Shill, H. A., Dugger, B., Hentz, J. G., Adler, C. H., Jacobson, S. A., et al. (2014). Essential tremor is not associated with cerebellar Purkinje cell loss. *Movement Disorders: Official Journal of the Movement Disorder Society*, *29*(4), 496–500. https://doi.org/10.1002/mds.25845.

Tallón-Barranco, A., Vázquez, A., Javier Jiménez-Jiménez, F., Ortí-Pareja, M., Gasalla, T., Cabrera-Valdivia, F., et al. (1997). Clinical features of essential tremor seen in neurology practice: A study of 357 patients. *Parkinsonism & Related Disorders*, *3*(4), 187–190. https://doi.org/10.1016/s1353-8020(97)00031-x.

Tan, E.-K. (2010). LINGO1 and essential tremor: Linking the shakes. Linking LINGO1 to essential tremor. *European Journal of Human Genetics*, *18*(7), 739–740. https://doi.org/10.1038/ejhg.2010.25.

Tan, E. K., Fook-Chong, S., Lum, S. Y., Gabriel, C., Koh, K. K., Prakash, K. M., et al. (2005). Non-motor manifestations in essential tremor: Use of a validated instrument to evaluate a wide spectrum of symptoms. *Parkinsonism & Related Disorders*, *11*(6), 375–380. https://doi.org/10.1016/j.parkreldis.2005.04.007.

Tanner, C. M., Goldman, S. M., Lyons, K. E., Aston, D. A., Tetrud, J. W., Welsh, M. D., et al. (2001). Essential tremor in twins: An assessment of genetic vs environmental determinants of etiology. *Neurology*, *57*(8), 1389–1391. https://doi.org/10.1212/wnl.57.8.1389.

Tsuboi, T., Wong, J. K., Eisinger, R. S., Okromelidze, L., Burns, M. R., Ramirez-Zamora, A., et al. (2021). Comparative connectivity correlates of dystonic and essential tremor deep brain stimulation. *Brain: A Journal of Neurology*, *144*(6), 1774–1786. https://doi.org/10.1093/brain/awab074.

Valls-Solé, J., Tolosa, E. S., Nobbe, F., Diéguez, E., Muñoz, E., Sanz, P., et al. (1997). Neurophysiological investigations in patients with head tremor. *Movement Disorders: Official Journal of the Movement Disorder Society*, *12*(4), 576–584. https://doi.org/10.1002/mds.870120415.

Wade, P., Gresty, M. A., & Findley, L. J. (1982). A normative study of postural tremor of the hand. *Archives of Neurology*, *39*(6), 358–362. https://doi.org/10.1001/archneur.1982.00510180036009.

Wang, L., Lei, D., Suo, X., Li, N., Lu, Z., Li, J., et al. (2018). Resting-state fMRI study on drug-naive patients of essential tremor with and without head tremor. *Scientific Reports*, *8*(1), 10580. https://doi.org/10.1038/s41598-018-28778-z.

Wang, S., Wang, X., Zhao, M., Li, T., Zhang, C., Wang, M., et al. (2020). Long-term efficacy and cognitive effects of voltage-based deep brain stimulation for drug-resistant essential tremor. *Clinical Neurology and Neurosurgery*, *194*, 105940. https://doi.org/10.1016/j.clineuro.2020.105940.

Welton, T., Cardoso, F., Carr, J. A., Chan, L.-L., Deuschl, G., Jankovic, J., et al. (2021). Essential tremor. *Nature Reviews. Disease Primers*, 7(1), 83. https://doi.org/10.1038/s41572-021-00314-w.

CHAPTER TWO

Is essential tremor a family of diseases or a syndrome? A syndrome

Sanjay Pandey*
Department of Neurology, Govind Ballabh Pant Postgraduate Institute of Medical Education and Research, New Delhi, India
*Corresponding author: e-mail address: sanjaysgpgi2002@yahoo.co.in

Contents

Abstract

In a consensus statement, a task force of the "International Parkinson and Movement Disorder Society" (IPMDS) has recently proposed a two axes classification for tremor: axis I (clinical manifestations) and axis II (etiology). In the axis, I, the clinical features of tremor in a given patient are specified in terms of medical history, tremor characteristics, associated signs, and laboratory tests for some tremors leading to the discovery of axis 2 etiologies. Based on axis I sign and symptoms a specific clinical syndrome is diagnosed which have been categorized as isolated tremor syndrome (a syndrome consisting only of tremor) and combined tremor syndrome (consisting of tremor and other systemic or neurological signs). The IPMDS task force defined essential tremor as an isolated tremor syndrome of bilateral upper limb action tremor of at least 3 years duration with or without a tremor in other locations (e.g., head, voice or lower limbs) in absence of other neurological signs, such as dystonia, ataxia, or parkinsonism.

International Review of Neurobiology, Volume 163
ISSN 0074-7742
https://doi.org/10.1016/bs.irn.2022.02.002

Patients with neurological signs of uncertain significance (such as impaired tandem gait, questionable dystonic posturing, or memory impairment) are classified as essential tremor plus.

In this paper, the author will make the argument that essential tremor is a syndrome with multiple causes.

1. Introduction

The first consensus statement of the movement disorder society on tremor published in 1998 stated that "Essential tremor" (ET) is a tremor syndrome classically defined by a mostly hereditary, mainly postural tremor of the hands and sometimes of the head (Deuschl, Bain, & Brin, 1998). In the recent consensus classification, the tremor was classified into two axes: Axis 1 (Clinical manifestations) and Axis 2 (Etiology) (Bhatia et al., 2018). The task force broadly defined two groups of axes I tremor syndromes: isolated (tremor is the only abnormal sign) and combined (additional signs such as dystonia and ataxia). "Essential tremor" (ET) was defined as an isolated tremor syndrome of bilateral upper limb action tremor of at least 3 years duration, with or without a tremor in other locations (head, voice, or lower limbs) and absence of other neurological signs, such as dystonia, ataxia, and parkinsonism. The consensus group added a new terminology "Essential tremor plus" (ET plus), which was defined as a tremor with the characteristics of ET and additional neurological signs of insignificance such as questionable dystonia, impaired tandem gait, memory impairment, or other mild neurologic signs. Other clearly defined syndromes like dystonic tremor (DT) and task-specific tremor are not included in the ET plus syndrome. The exclusionary criteria for ET and ET plus require ruling out isolated focal tremor (voice and head), orthostatic tremor (frequency >12 Hz), task, and position-specific tremors, and sudden onset and step-wise deterioration. The consensus group acknowledged that the status of an ET patient may change over a period of time as new symptoms (dystonia, parkinsonism, and gait problem) may evolve and there may be a change in the diagnosis of the patient from ET to ET plus provided they do not meet the criteria for alternate neurological diagnosis. Further, the longstanding ET patients may ultimately develop other neurologic conditions such as dystonia and Parkinson's disease (PD). This heterogeneous presentation of ET and evolution of signs and symptoms over a period of time has sparked a new debate whether ET should be classified as "Syndrome" or "Family of diseases" (Louis, 2020; Louis et al., 2020).

2. Definition of a syndrome

The term "Syndrome" is derived from Greek ("syn" together and "dromus" a course) meaning a running together or concurrence (Jablonski, 1991). It is one of the oldest terminologies in the medical literature and is said to be used by Hippocrates and Galen (Murray, Bradley, Craigie, & Onions, 1919). The syndrome is also defined as a "complex of symptoms, a set of symptoms that occur together; a sum of signs of any morbid state" (Dorland & Taylor, 1988). In most of the medical literature, Dorland's definition is still being used. Throughout the 19th century and during the early 20th centuries the term syndrome was used to name disorders where multiple symptoms occurred together and involved various unrelated organs or systems with no explainable cause (Jablonski, 1991). Sometimes "Syndrome" was also used for disorders that baffled physicians, or whenever enough was not known about the etiology. Occasionally the term "Syndrome" was named after the physician who described it (The Gilles de la Tourette syndrome) and sometimes after the phenomenology (Stiff-person syndrome) or the etiology (Fragile X syndrome) or expressions (Latah syndrome) or sometimes imaginative syndrome designations (Jumping Frenchman of Maine syndrome) (Jablonski, 1991). In recent literature, the syndrome is defined as a recognizable complex of symptoms and physical findings which indicate a specific condition for which a direct cause is not necessarily understood (Calvo, Karras, Phillips, Kimball, & Wolf, 2003; Lenka & Louis, 2020).

The concept of the disease has been defined differently. Historically doctors defined a disease according to a cluster of symptoms. There are other views defining disease based on phenomenology such as schizophrenia or etiology such as genetic causes. With dramatic advancements in the field of diagnostics including genetics, the prevailing view is that a disease state should be defined by the etiology and pathogenesis, however, there is a clear overlap in how we define or approach the syndrome or the disease. Based on the historical perspectives and the current discussion we can conclude that "Syndrome" refers to a set of signs and symptoms and there is no knowledge about pathogenesis or disease course whereas a disease is one in which there is some understanding of the latter. In practice, clinicians use "Syndrome" for a situation where there is uncertainty regarding the causative agent like "Viral syndrome." Once a cause, agent, or process with a high degree of certainty is known, clinicians may refer them to a disease and not syndrome (Albert, Munson, & Resnik, 1988; Calvo et al., 2003).

3. Essential tremor: "Syndrome" or "family of diseases"?

With this background regarding the basic concepts of syndrome vs disease, we will review the clinical, etiological, pathophysiological, and therapeutic data to show that ET better fits the definition of a syndrome than the disease designation (Fig. 1). Considering the marked heterogeneity ET has also been referred to as a "family of diseases" or "the essential tremors" with an "evolving definition" (Louis, 2018, 2021). But it is important to note that the "family of diseases" and syndrome are not the same.

3.1 Clinical spectrum

3.1.1 Essential tremor: Historical perspectives

The term "Essential tremor" (ET) was for the first time used in 1874 by, Pietro Burresi from the University of Siena, Italy to describe an 18-year-old man with severe, isolated action tremor of the arms with head tremor (Burresi, 1874) (Table 1). Perhaps he used this term to indicate that the tremor was the only symptom detected in these patients (Louis et al., 2008). Later, many authors reported ET under different nomenclatures (Dana, 1887; Maragliano, 1879; Nagy, 1890; Raymond, 1892). The term ET appeared more regularly since the early years of the 20th century when several authors did a neurophysiologic study to demonstrate a tremor frequency ranging from 3 to 12 Hz (Flatau, 1908; Van Bogaert & De Savitsch, 1937). In 1949, Macdonald Critchley provided the first detailed review of ET and reported that patients may have cranial involvement including head, neck, jaw, eyelids, eyeballs, larynx, tongue, and palate (Buechele, Baumann, Lees, & Deuschl, 2021; Critchley, 1949). Rarely the diaphragm and truncal muscles may also be involved in the late stages. Based on the site of involvement he proposed various subgroups (hands only, hands and lips, etc.) of ET. Critchley also reported that occasionally ET patients may have other movement disorders such as choreiform movements of the face or head, facial twitching, and spasmodic torticollis (Critchley, 1949). But, the tone, reflexes, and sensory examination are usually normal, and signs of parkinsonism and ataxia are absent. Later, in 1983 Marsden and colleagues reasoned that "Essential tremor is not a single entity" and indeed they included DT and postural tremor of PD in the definition of ET (Marsden et al., 1983). Other contemporary studies also supported the view that ET is frequently not a stable phenotype but may evolve into a more complex syndrome with parkinsonism, dementia, ataxia,

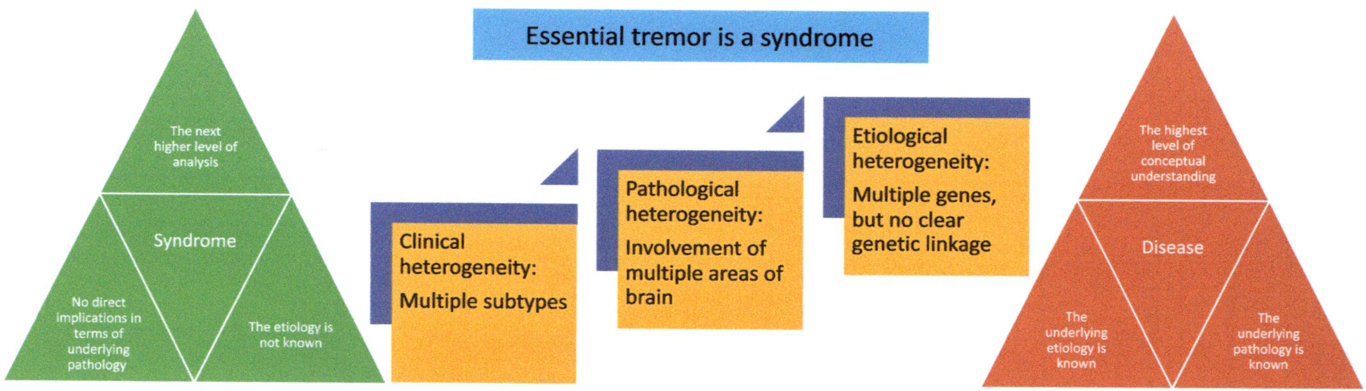

Fig. 1 Core features of "Syndrome" and "Disease" and evidence supporting that "Essential tremor" is a syndrome rather than a disease.

Table 1 Historical evolution of the definition of "Essential tremor."

Author/year	Term	Salient features
Burresi, 1874	"Tremore semplice essenziale" (i.e., simple essential tremor)	• Isolated action tremor of the arms with head tremor and absence of paralysis or other central nervous signs
Maragliano, 1879	Idiopathic action tremor	• Isolated action tremor of all four limbs, mild gait unsteadiness, early onset, and heritable nature (present in mother)
Dana, 1887	"Hereditary" tremor	• A fine tremor affecting voluntary muscle
Nagy, 1890	"Essentieller Tremor" (i.e., essential tremor)	• Severe tremor in arm, head, leg and trunk • Alcohol responsiveness • Heritable nature • Early onset in all members • No other neurological signs except mild gait disorder
Raymond, 1892	"Tremblement essentiel he're'ditaire" (i.e., hereditary essential tremor)	• Postural tremor in arm • 4–5 Hz tremor • Alcohol responsiveness • Early onset • Heritable nature (father, mother and brother)
Van Bogaert & De Savitsch, 1937	Heredofamilial tremor	• Families with head tremor, nystagmus, dystonia of neck, epilepsy, psychopathy and mental defect
Critchley, 1949	"Essential (heredofamilial) tremor"	• Common groupings of musculature involved by tremor • Occasionally other movement disorders may coexist (chorea of face/head, facial twitching/ spasmodic torticollis) • Parkinsonism, ataxia, increased tone and sensory disturbances are absent

Table 1 Historical evolution of the definition of "Essential tremor."—cont'd

Author/year	Term	Salient features
Critchley, 1972	Essential tremor	• Hereditary nature • Hyperhidrosis, cramps, dyskinetic movements, and ataxia • Apart from minor alterations in tone, expression, and arm swing, features of Parkinsonism were notably absent
Marsden, Obeso, & Rothwell, 1983	Essential tremor (ET)	Four types: (1) Enhanced physiologic tremor, (2) Benign ET, (3) Severe ET, and (4) Symptomatic ET
Martinelli, Gabellini, Gulli, & Lugaresi, 1987	Typical essential tremor	• Buccolingual dyskinesia, dystonia and mild extrapyramidal signs may be seen
Deuschl, Zimmermann, Genger, & Lücking, 1994 (The Tremor Investigation Group)	Definite essential tremor	• 5-year duration • Bilateral postural tremor with or without kinetic tremor, involving hands and forearms • Exclusion criteria: Other neurological signs
	Probable essential tremor	• 3-year duration • Bilateral postural tremor with or without kinetic tremor, involving hands and forearms • Exclusion criteria: Other neurological signs
	Possible essential tremor	• Type I: Exhibit other recognizable neurologic disorders such as parkinsonism, dystonia, myoclonus • Type II: Monosymptomatic and isolated tremors of uncertain relation to essential tremor
Deuschl et al., 1998 (Consensus Statement of the Movement Disorder Society on Tremor)	Classic ET	• Inclusion: Bilateral, largely symmetric postural or kinetic tremor involving hands and forearms that is visible and persistent • Head tremor may occur in the absence of abnormal posturing • Exclusion: Other abnormal neurologic signs. Especially dystonia

Continued

Table 1 Historical evolution of the definition of "Essential tremor."—cont'd

Author/year	Term	Salient features
	Possible ET	Type I: 1. Parkinsonian tremor syndromes 2. Dystonic tremor syndromes 3. Neuropathic tremor syndromes
		Type II: Monosymptomatic and isolated tremors
Bhatia et al., 2018 (Consensus Statement on the Classification of Tremors, From the Task Force on Tremor of the International Parkinson and Movement Disorder Society)	Essential tremor	• Isolated tremor syndrome of bilateral upper limb action tremor of at least 3 years duration with absence of other neurological signs
	Essential tremor plus	• Tremor with the characteristics of ET and additional neurological signs of uncertain significance such as questionable dystonic posturing, impaired tandem gait, and memory impairment. • ET with rest tremor

and autonomic dysfunctions (Critchley, 1972; Martinelli et al., 1987). Overall, the lack of a clear definition of ET was a major limitation in the clinical research.

3.1.2 Essential tremor: Clinical variants

ET may have different types of presentations, like early-onset and late-onset, familial and sporadic, responsive to alcohol or not, and with or without head/voice tremor dominant (Gövert, Becktepe, & Deuschl, 2016). The early-onset and late-onset distinctions are based on studies data reporting that the prevalence of ET significantly increases after the age of 50 (Louis & Ferreira, 2010; Louis & McCreary, 2021). Early-onset tremor is also labeled as "Classical type ET" and they are likely to have better motor and cognitive functioning (Hopfner, Ahlf, et al., 2016). They are more likely to have a family history, response to alcohol, and longer survival. In contrast, late-onset ET is associated with shorter life expectancy and worse age-related parameters including motor and cognitive scores (Gövert et al., 2016). The late-onset

ET is also labeled as "Aging-Related Tremor" indicating specific premature aging. The pattern of progression has also been used to classify ET patients as faster progressors and slower progressors (Putzke, Whaley, Baba, Wszolek, & Uitti, 2006). In ET families, there is a fourfold difference in the rate of progression, with some families being faster progressors than others (Louis, Hernandez, Ionita-Laza, Ottman, & Clark, 2013). Several other factors have been considered to influence the rate of progressions such as older age of onset and asymmetrical tremor.

3.1.3 Essential tremor: Definition

One of the major problems in ET definition was perhaps it was defined too broadly. In 1998, the first consensus statement of the "Movement Disorder Society on Tremor" defined ET and included the "The Tremor Investigation Group" (TRIG) criteria for the diagnosis (Deuschl et al., 1998, 1994). The term "Classic ET" was used for definite and probable ET in the TRIG classification and the remaining entities were classified as "possible ET." "Classic ET" was defined as bilateral, largely symmetric postural or kinetic tremor involving hands and forearm with additional or isolated head tremor but in the absence of abnormal posturing. Patients with additional neurological signs (especially dystonia), primary orthostatic tremor, isolated tremor (voice, leg, tongue or chin, position-specific or tasks-specific), enhanced physiological tremor, drug-induced tremor, and psychogenic tremor were excluded. However, the consensus group did not include tremor duration to classify ET as proposed by the TRIG classification (5 years for definite ET and 3 years for probable ET). The consensus definition was certainly an important advancement in highlighting the heterogeneous character of ET, but, there were some major problems also. First, in the absence of minimum time duration and lack of clear definition regarding head posturing the so-called "Classic ET" remained an unstable syndrome as many patients were likely to develop additional neurological signs with time leading to some alternate diagnosis (Hopfner, Haubenberger, et al., 2016). Second, there was a great degree of variation among clinicians, whether ET should be considered as isolated or combined tremor syndrome. Finally, due to the lack of disease-specific diagnostic markers, a very high prevalence of misdiagnosis was still being reported ranging from 37% to 50%, mainly with dystonia, parkinsonism, and rarely myoclonus (Elble, 2013; Jain, Lo, & Louis, 2006; Schrag, Münchau, Bhatia, Quinn, & Marsden, 2000).

3.1.4 Essential tremor plus

To facilitate deeper and consensual phenotyping of ET patients, an updated version of the consensus statement of the "International Parkinson Movement Disorder Society" (IPMDS) was published in 2018 (Bhatia et al., 2018; Pandey, Benito-León, & Kuo, 2021). ET was defined as an isolated tremor syndrome of bilateral upper limb action tremor of at least 3 years duration with or without a tremor in other locations such as head, voice, or upper limbs. In the new definition, the tremor is the only permissible neurologic sign with the absence of other signs such as dystonia, ataxia, or parkinsonism. ET patients frequently have a family history, response to alcohol, and onset of tremor in the upper limb. However, these features were not included in the definition as they are not consistent and convincing data are lacking to include these features in the diagnostic criterion. A new terminology "ET plus" was added to address a major problem in the diagnosis of ET as specialists had a different threshold for identifying dystonia, ataxia, parkinsonism, and other neurological signs (Elble, 2020). ET plus is defined as a tremor with the characteristics of ET and additional neurological signs of uncertain significance such as questionable dystonic posturing, impaired tandem gait, memory impairment, or other mild neurological signs. ET with rest tremor is also defined as ET plus. The ET plus syndrome does not include other clearly defined syndromes such as a DT. The task force concluded that clinicians should be encouraged consistently to document these questionable signs.

3.1.4.1 Essential tremor plus is likely to be more common than essential tremor

Following the new classification, a significant number of patients from the previous ET cohort have been reclassified as ET plus due to the presence of questionable or soft signs (Pandey, Bhattad, & Hallett, 2020). Rajalingam and colleagues reclassified 283 patients with lower limb tremor and 110 patients (38.87%) were classified as ET plus (Rajalingam, Breen, Lang, & Fasano, 2018). ET plus patients were significantly older at age of tremor onset and they had a higher "Fahn-Tolosa-Martin Tremor Rating Scale" (TRS) score. Prasad and colleagues also reclassified a preexisting ET cohort of 252 patients and 54% ($n = 99$) of them were labeled as ET plus including 21 patients who had mild cervical dystonia (Prasad & Pal, 2019). Pandey et al. published a video-based assessment of questionable dystonia in their ET cohort (Pandey & Bhattad, 2019). First, they completed a retrospective chart review of 121 ET patients who were diagnosed based on the previous (1998) criteria. After applying the new criteria (2018) 42 patients were excluded (duration <3 years in 38 patients and isolated head tremors in 4 patients). Of the remaining 79 patients, 34 (43.03%) were reclassified as ET plus due to additional soft signs including questionable dystonia present in 19 patients.

Huang et al. also reviewed the medical records of 280 ET syndrome patients and reclassified them using the recent criteria (Huang et al., 2020). Twenty-five patients were reclassified as indeterminate tremor (<3 years duration) and 17 were diagnosed as focal tremor. Concerning the remaining 238 patients, a total of 117 patients (49.15%) were reclassified as ET plus and 121 (50.84%) as ET. Older age at tremor onset, greater upper limb action tremor, and tremor spread to the cranial region were significantly associated with ET plus. In another study, Amlang et al. reevaluated 104 previously diagnosed ET patients (Amlang, Trujillo Diaz, & Louis, 2020). A postevaluation diagnoses of ET was made in 45.2% ($n=47$); dystonia in 27.9% ($n=29$); and 26.9% ($n=28$) of patients had other diagnosis including PD. Factors associated with alternate postevaluation diagnoses other than ET were diagnoses made by a nonneurologist, shorter tremor duration, irregular tremor, and abnormal limb postures. In another study, Louis et al. evaluated a cohort of 241 ET patients and they observed that numerous ET plus features were significantly correlated with action tremor duration and age (Louis, Huey, & Cosentino, 2021). We can make some important conclusions from these studies. First, following the new tremor classification, ET plus is likely to be more common than ET. Second, ET plus patients are likely to be older with a longer duration of disease than ET patients. Third, ET patients are likely to develop other neurologic signs (dystonia, parkinsonism, ataxia, or dementia) over a period of time. Fourth, several of these papers also concluded that "ET plus" is likely to be just a later stage of what has been labeled as "ET" (Iglesias-Hernandez et al., 2021; Louis et al., 2021). So, it is important to determine whether these new neurological signs are coincidental or pathologically/etiologically related.

3.1.4.2 Essential tremor plus: Soft signs

In the recent consensus statement, the detection of soft signs has become an integral part of tremor classification (Bhatia et al., 2018). Soft signs are neurological signs of uncertain relationship to tremor such as mild memory impairment, impaired tandem gait, and subtle body posturing that could be dystonic. However, we do not know the exact prevalence and significance of these signs. In a prospective study of 45 patients, ET plus diagnosis was based on the presence of soft signs in 43 patients and rest tremor in 2 patients (Pandey & Bhattad, 2021). The majority of the patients (66.66%; $n=30$) had only one soft sign, 15.55% ($n=7$) had two soft signs and 13.33% ($n=6$) had three soft signs. The most common soft sign present was questionable dystonia ($n=42$), followed by impaired tandem gait ($n=11$) and mild cognitive impairment ($n=9$). In a retrospective study where ET patients ($n=238$) were reclassified as ET plus ($n=117$) and ET ($n=121$), 61.53% ($n=72$) of ET plus patients had a single soft sign, and the remaining (38.46%; $n=45$) patients had multiple soft signs (Huang et al., 2020). The presence of soft signs suggests that ET is no longer a monosymptomatic illness. But, soft signs are not sufficient to diagnose another tremor syndrome or movement disorder. During follow-up, soft signs may become hard signs leading to a more definitive diagnosis of combined tremor syndrome.

3.1.4.2.1 Questionable dystonia Questionable dystonia (subtle body posturing that could be dystonic) is the most common soft sign present in ET plus patients. In two of our published studies, we observed that questionable dystonia was more likely to be present in the upper limb followed by the neck (Pandey & Bhattad, 2019, 2021). Dystonia was labeled as questionable if there was discordance between the two examiners regarding its presence. In the upper limb, questionable dystonia was more common in the thumb followed by the second digit and wrist (Pandey & Bhattad, 2019). The diagnosis of questionable dystonia in the neck and upper limb is difficult and observer-dependent as in some patients it may point toward DT (Rajalingam et al., 2018). Rajan and colleagues published the interrater and intrarater agreements where two independent raters reviewed the video recordings of tremor patients and applied the new criteria to classify the tremors into the axis I syndromes (Rajan, Pandey, Anandapadmanabhan, & Srivastava, 2018). They concluded that in a clinical setting interpretation of questionable dystonic posturing is subjective with a high rate of discordance and is one of the major challenges faced in differentiating ET plus from a DT syndrome. Regarding the questionable dystonia, certain clinical features such as sensory trick, null point, and irregular head tremor with directional quality persisting during sleep favors DT and help in differentiating it from ET (Albanese et al., 2013). But they are not always present. Surface electromyography (EMG) and other electrophysiological techniques (blink reflex recovery curve and somatosensory temporal discrimination threshold) have been used to differentiate DT from ET, but they have not been validated and are not available easily (Pandey & Sarma, 2016). So, there is a need for better biomarkers or objective measures of dystonia to aid in accurate classification. Also, certain terminologies such as jerkiness and asymmetry need to be defined in a better way.

3.1.4.2.2 Parkinsonism ET and PD are common in the aging population and they share many overlapping features that can make it difficult to differentiate between the two. Also, there is a possible pathophysiological overlap that has been advocated in the past. The hallmark of ET is action tremor, however, approximately 30% of patients may also develop rest tremor (Louis, Hernandez, & Michalec, 2015). Long duration of disease, greater disease severity, and widespread tremor distribution are more likely to be associated with ET and rest tremor. In the recent classification, ET patients with rest tremor have been classified as ET plus (Bhatia et al., 2018). In a clinical setting, PD poses less diagnostic challenge considering its clear diagnostic criteria. Rest tremor in PD has certain characteristic features such as the reemergent phenomenon. However, there may be some diagnostic difficulties when ET patients develop rest tremor or PD patients develop a postural tremor (Algarni & Fasano, 2018). Positive family history and response to alcohol are more likely to favor ET. Tremor stability index has also been used to differentiate tremor dominant PD from ET patients, but it is not always feasible (di Biase et al., 2017). Bradykinesia is usually not observed in ET patients, but in one study, clinical examinations of ET patients documented more bradykinesia than controls (Montgomery, Baker, Lyons, & Koller, 2000). Dopamine transporter (DAT) scan findings have also been used to differentiate ET from PD. Isaias et al. reported mildly decreased DAT binding in ET patients compared to controls (Isaias et al., 2010). Studies have reported up to fourfold increased risks for developing PD in ET patients (Benito-León, Louis, Bermejo-Pareja, & Neurological Disorders in Central Spain Study Group, 2009; Sepúlveda Soto & Fasano, 2019). ET has been reported more frequently in relatives of PD patients and also, ET and PD have been reported from the same family (Algarni & Fasano, 2018). The controversial link between ET and PD is supported by several lines of other evidence including overlapping motor and nonmotor features, increase prevalence of PD in patients with longstanding antecedent ET, high prevalence of rapid eye movement sleep behavior disorder in ET patients, and the presence of "Lewy bodies" in the brains of some ET patients (Tarakad & Jankovic, 2019). The similarities between PD and ET in terms of total nonmotor scores also raise the question of whether there is an association between the two (Shalash et al., 2021).

So, the relationship between ET and PD is complex (Rajput, Rajput, Bocking, Auer, & Rajput, 2019).

3.1.4.2.3 Impaired tandem gait Balance and gait impairments have been commonly reported in ET patients, but questions remain regarding their prevalence and the precise degree to which they change during the normal aging process (Arkadir & Louis, 2013). In an older patient, 3 missteps in a 10-step tandem walk may not be excluded from the definition of ET, but a younger patient not able to perform tandem walk will certainly be classified as tremor ataxia syndrome and not ET or ET plus (Elble, 2020; Singer, Sanchez-Ramos, & Weiner, 1994). In a systematic review on ataxic gait in ET, Rao and Louis reported that factors associated with balance and gait impairment in ET included age, presence of tremor in midline structures, and cognitive dysfunction (Rao & Louis, 2019). The pattern of gait impairment in ET was similar to what is reported in cerebellar ataxias characterized by a wide base of support, short step length, slow gait speed, increased step-to-step variability, and deviations in the walking path. However, the presentation in ET was less severe. The reported findings indicated the role of the cerebellum in the pathogenesis of ET.

3.1.4.2.4 Mild cognitive impairment ET patients exhibit heterogeneous cognitive functioning. Though it is normal in the majority of the patients, studies have documented deficits primarily in attention, memory, and visuospatial abilities (Louis, 2021). An increased risk for mild cognitive impairment and dementia is also observed in ET patients (Louis, Benito-León, Vega-Quiroga, Bermejo-Pareja, & Neurological Disorders in Central Spain (NEDICES) Study Group, 2010). Predictive factors likely to develop MCI or dementia in ET patients have been reported (Cersonsky et al., 2020). The basis for the cognitive changes in ET is likely to be multifactorial and further longitudinal studies are needed to better characterize the progression of these changes.

3.1.4.2.5 Validation and standardization of soft signs The concept of dichotomizing examinations by documenting and validating the soft signs has important implications in the new tremor classification system (Bhatia et al., 2018; Fearon et al., 2019). However, the retrospective data from the different cohorts have highlighted that the interrater reliability of these soft signs is variable, so their relevance may be questionable (Pandey et al., 2020). Also, clinicians have different thresholds regarding the consideration of neurological signs as soft vs significant (Rajan et al., 2018). To assess the presence or absence of clinical signs in axis I classification, Becktepe et al. recently designed a new clinical tool: the "Standardized Tremor Elements Assessment" (STEA) (Becktepe et al., 2021). Four movement disorder specialists (3 tremor focused and 1 dystonia focused) evaluated 59 videos of patients with upper limb tremor by using STEA. A greater agreement was observed for items characterizing tremor than signs of dystonia. Regarding the final diagnosis, an interrater agreement was higher within the tremor specialists and poor between dystonia and tremor specialists. The findings of this study suggest that tremor and dystonia diagnoses are strongly influenced by a clinician's research interest, training, or clinical practice and better interrater agreement can be achieved by training and teaching video libraries. Also, future studies are needed to determine whether the identification of some of the soft signs is useful or needed. The validity of soft signs and their normal limits also require additional studies.

3.1.5 Essential tremor-like tremors

Tremor associated with dystonia (TAD) and tremor with a dystonia gene (i.e., endophenotype of dystonia gene) may present with ET-like tremor (Pandey, Bhattad, & Dinesh, 2021; Pandey & Sarma, 2017; Shaikh et al., 2021). In the 1998 MDS consensus criteria, DT syndrome was categorized as DT (tremor in dystonic body segment) and TAD (tremor in nondystonic body segment) (Deuschl et al., 1998). DT and TAD were considered different because postural tremors resembling ET can occur at extremities not involved in dystonia. In the recent MDS task force classification, DT syndromes were defined as tremor syndromes combining tremor and dystonia as the leading neurological signs (Bhatia et al., 2018). The task force choose to retain the term DT and TAD, but they concluded that the pathophysiology of both these types is likely to be the same. TAD would be regarded as a form of ET if dystonia were not present. Also, a patient with longstanding isolated bibrachial tremor may develop additional diagnostic signs such as PD, dystonia, or ataxia, and their axis I diagnosis may change from one tremor syndrome to another (Bhatia et al., 2018; Pandey & Bhattad, 2020) (Fig. 2). The consensus group recommended that if a patient with axis I diagnosis of ET or ET plus develop dystonia, such patients should be reclassified as DT with antecedent ET, and assumptions regarding axis II etiology should be avoided as the etiology of ET may not be the same as the etiology of subsequently combined dystonia syndrome with additional neurological signs of significance (Bhatia et al., 2018).

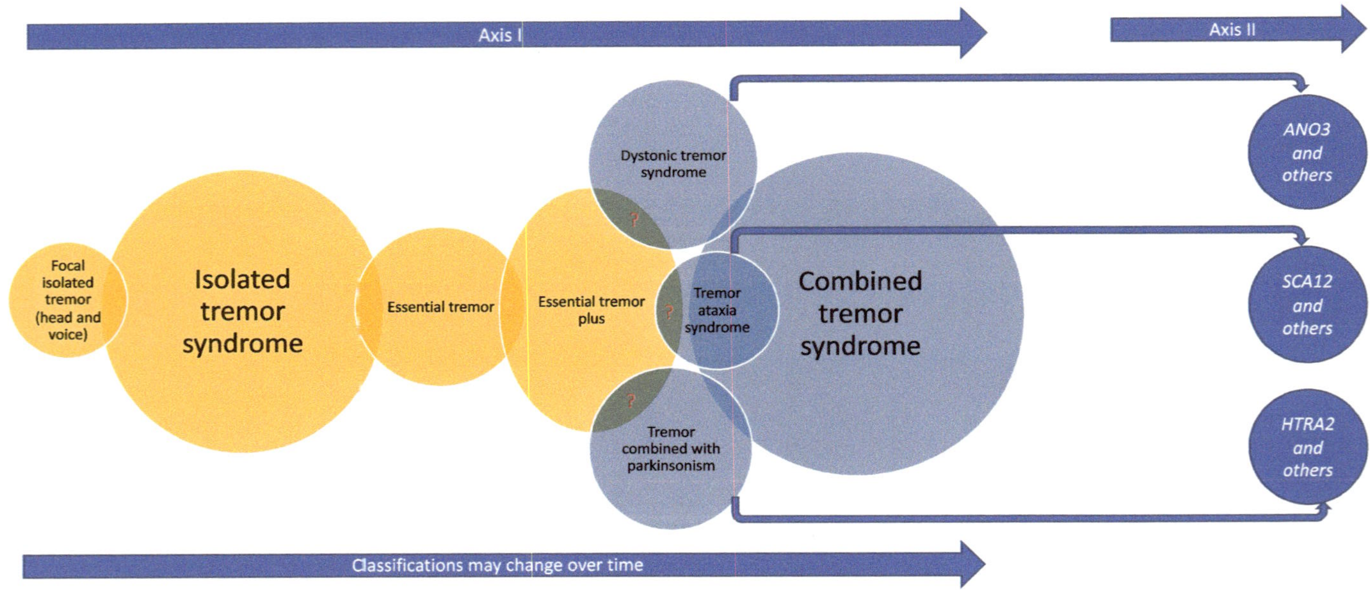

Fig. 2 The new tremor classification broadly categorizes tremor patients as "Isolated" and "Combined" tremor syndrome in axis I. Essential tremor plus patients have questionable signs (?) which may be overlapping with clinical features of combined tremor syndrome patents. The axis I diagnosis of a patient may change over time from one tremor syndrome to another. In axis II multiple etiologies may be seen in tremor syndrome patients.

3.2 Essential tremor: Lack of consensus on pathophysiology

The pathophysiology of ET is still being carefully studied with multiple overlapping hypotheses and many unanswered questions (Welton et al., 2021). Different areas of the brain have been implicated in the pathophysiology of ET (Hopfner, Ahlf, et al., 2016). Originally an olivary model was proposed suggesting a tremor pacemaker in the inferior olivary nucleus (Buijink et al., 2015). This hypothesis is one of the oldest ones and is partially based on the harmaline animal model of ET (Handforth, 2012). However, this model has several problems, and its relevance to ET has been increasingly questioned (Louis & Lenka, 2017). Recent studies have provided more evidence of the involvement of the cerebello-thalamo-cortical circuit and supplementary motor, visual and parietal areas (Archer et al., 2018; Nieuwhof, Panyakaew, van de Warrenburg, Gallea, & Helmich, 2018). Interest in cerebellar research is primarily supported by the neuroimaging studies which strongly implicate the role of this brain region in the pathogenesis of ET (Passamonti, Cerasa, & Quattrone, 2012). Both altered dentate nucleus function and pathological changes in the cerebellar cortex have been described in ET patients, however, it is not clear whether these changes are primary or secondary (Boecker et al., 2010; Buijink et al., 2015; Paris-Robidas et al., 2012). Clinical studies have consistently reported cerebellar signs in ET patients (Singer et al., 1994). Also, several cerebellar disorders such as spinocerebellar ataxias (SCAs) and fragile X-related tremor/ataxia syndrome (FXTAS) are known to cause tremor, and some of the features seen in ET patients such as gait, nystagmus, and intention tremor also point toward the cerebellar involvement (Schmahmann & Sherman, 1998). Pathophysiological differences based on ET subtypes have also been linked to the cerebellum (Muthuraman et al., 2015). Cerebellar involvement in ET is also supported by neurochemical studies highlighting the role of GABA in the pathophysiology of ET (Gironell, 2014; Louis, 1999). GABA inhibits the dentate nucleus and other deep cerebellar output neurons enabling Purkinje cells (PC) to play an important role in controlling the motor movements (Bowery et al., 2002). Some pathological studies have documented the PC loss in ET and decrease of GABA levels in the cerebral spinal fluid of ET patients (Málly, Baranyi, & Vizi, 1996). Studies have also demonstrated decreased levels of GABAa and GABAb receptors in patients with ET (Paris-Robidas et al., 2012). Also, most drugs (gabapentin, pregabalin, benzodiazepines, topiramate, and alcohol) used in the treatment of ET targets the GABA system by enhancing its activity indicating the dysfunction of the GABA system in the pathogenesis of ET (Gironell et al., 2012; Louis, 1999). The notion that ET is a neurodegenerative disease linked to the cerebellum, has gained increasing momentum in recent years (Louis, 2010). The strongest argument for neurodegeneration is PC loss and increase in the number of torpedoes (rounded swellings of the proximal portion of the PC axon) as well as other degenerative changes in PC and neighboring neuronal populations (Louis & Faust, 2020; Louis, Faust, & Vonsattel, 2011, 2012; Louis et al., 2019). Gionco and colleagues recently reported a detailed postmortem study of 50 ET cases (24 ET and 26 ET plus) using a set of 14 quantitative metrics of cerebellar pathology including histologic and immunohistochemical methods (Gionco et al., 2021). Similar findings were reported in ET and ET plus patients with respect to 13 of 14 cerebellar metrics. The authors concluded that there were no pathological differences in the cerebellar cortex between ET and ET plus cases. Contradictory findings have been reported by two studies that did not reproduce the findings of PC loss in individuals with ET (Rajput, Robinson, Rajput, Robinson, & Rajput, 2012; Symanski et al., 2014). However, the observations reported by these studies had some inherent methodological flaws (Louis & Faust, 2020). PC loss has also been reported in diseases not causing tremor such as Huntington's disease and Alzheimer's disease (Mavroudis et al., 2019; Singh-Bains et al., 2019). Further, PC degeneration also does not always result in ET manifestations (Rajput, Robinson, & Rajput, 2013). Considering the lack of consistent and specific findings in different studies we cannot conclude with certainty that PC pathology is tremorgenic.

3.3 Essential tremor: A polygenic disorder

ET is an isolated tremor syndrome with the age of onset ranging from childhood to adulthood. ET has an additive heritability of at least 75%, and large families with the autosomal dominant pattern are common (Diez-Fairen et al., 2019; Hopfner & Deuschl, 2018; Hopfner & Helmich, 2018; Magrinelli et al., 2020). After decades of extensive research, many candidate loci and a gene associated with tremor have been identified like fused in sarcoma gene (*FUS*) (Merner et al., 2012), Htra Serine Peptidase 2 gene (*HTRA2*) (Sun et al., 2020), GGC repeat expansion in the Notch 2 N-terminal like C gene (*NOTCH2NLC*) (Unal Gulsuner et al., 2014) and teneurin transmembrane protein 4 genes (*TENM4*) (Hor et al., 2015). However, these findings have not been replicated in future studies and the etiology of ET remains controversial (Balint & Bhatia, 2015). Also, families with these mutations have illustrated phenotypes other than ET, like parkinsonism, ataxia, and autonomic dysfunction. There are other factors also hampering the genetic research of ET. First, in the absence of robust criteria, previous studies on ET recruited many tremor disorders misdiagnosed with ET such as FXTAS and SCAs (Emamikhah et al., 2021; O'Hearn, Holmes, Calvert, Ross, & Margolis, 2001). Also, sometimes tremor may be the initial manifestation of combined tremor syndrome-like DT (e.g., *DYT-ANO3*) (Stamelou et al., 2014). This complexity is likely to be observed more in ET plus patients with soft neurological signs such as questionable dystonia. Second, in many families with ET reduced penetrance is observed. Third, the possible contribution of multiple risk loci has also been a possible reason for the conflicting data on genetic research related to ET. Fourth, the possible environmental nongenetic risk factor may be responsible for identical phenotypes being reported due to different genetic mutations. Finally, there is a lack of good animal models that fully recapitulate the disease.

In summary, ET is clearly a polygenic disorder.

3.4 Essential tremor: Inconsistent therapeutic response

In ET, there is marked variability in response to the medical and surgical treatments. Several trials have indicated that the response to medical treatment is effective only in approximately 50% of ET patients (Sharma & Pandey, 2020). Further, the therapeutic response is not uniform across the patients, and there is a sizeable number of nonresponders as well (Gironell et al., 1999). Patients having low amplitude tremor often have a better response than individuals with high amplitude tremor (Hopfner & Deuschl, 2018). Response to medical treatment is also worse in patients with intention tremor compared with postural tremor. Based on the electrophysiological evidence Deuschl and colleagues proposed two forms of ET (Deuschl, Lücking, & Schenck, 1987). One (type A) with preponderant synchronous activity in the antagonists with a tremor frequency of 5.5–10 Hz and normal long latency reflex (LLR) and other (type B) with preponderant alternating activity in antagonist muscles with a tremor frequency of 5.5–8 Hz and abnormal LLR. The authors hypothesized that type A patients respond to propranolol whereas type B patients do not, but may respond to primidone. In a neuroimaging (MRI-based) study, where 32 patients (propranolol responder group = 18; nonresponder group = 14) with ET were recruited, the nonresponder group showed severe cortical atrophy in the right temporal and left orbitofrontal cortex (Chung et al., 2013). Botulinum toxin treatment used in ET patients also has variable response based on the phenotypes with axial symptoms responding better than the appendicular symptoms (Hertegård, Granqvist, & Lindestad, 2000; Wissel, Masuhr, Schelosky, Ebersbach, & Poewe, 1997). The surgical outcome of ET patients is not uniform and the right target is still a matter of debate considering the heterogeneity in ET patients (Kremer et al., 2021; Martínez-Fernández & Pineda-Pardo, 2020). The treatment of ET plus has not been discussed much but it has been speculated that they may have a different target and different outcome primarily depending upon the presence of soft signs such as questionable dystonia (Patel, Deeb, & Okun, 2018). In a retrospective study, 44 patients with ET (diagnosed according to the previous MDS tremor classification), who had received thalamic ventral intermediate nucleus or the posterior subthalamic area deep brain stimulation were retrospectively reclassified as either ET ($n=13$; 30%) or ET plus ($n=31$; 70%) (Steffen et al., 2020). Rest tremor ($n=14$) was the most frequent plus sign followed by mild cognitive impairment ($n=13$), dystonic posturing ($n=9$), and gait ataxia ($n=5$). ET plus patients had higher TRS scores at baseline. Comparable tremor reduction was seen in ET plus patients, albeit they required higher stimulation parameters than ET patients which were probably related to a more dorsal location of active electrode contacts. Future prospective studies are urgently needed using the newly developed and standardized classification into ET, ET plus, and DT.

4. Discussion

Considering the marked clinical variability and heterogeneity in terms of etiology, pathophysiology, and pharmacological responses there is a broad consensus that ET is not one entity. Two divergent views are emerging, whether ET should be regarded as a "syndrome" or a "family of diseases." Syndromes are useful as they promote detailed phenotyping and facilitate the discovery of specific etiologies and useful treatments (Elble, 2020). Also, a syndrome may have multiple etiologies, and one particular etiology may produce multiple clinical syndromes (Bhatia et al., 2018). So, ET is better classified as a "syndrome" that facilitates deeper phenotyping of patients which is needed to uncover likely differences among patients with various etiologies of ET. In the new classification, a new terminology ET plus has been added to address the clinical variability of ET (Bhatia et al., 2018). It encourages clinicians to document all deviations from the ET syndrome that are of questionable significance (Elble, 2020). But, the concept of ET plus has raised more questions than addressing the current knowledge gap (Lenka & Pandey, 2021; Louis, 2020). ET plus is likely to represent a state condition (representing patients with ET who might develop soft neurological signs when the disease is at a more advanced stage) rather than a trait condition (a distinct and completely separate subtype of ET) (Louis, 2020; Vidailhet, 2020). Another major area of concern remains is the definition of soft neurological signs. Expressions such as "questionable dystonia" indicate uncertainty in the mind of clinicians. Minimum criteria for the definition of dystonia are an urgent need and probably the recognition of possible, probable and definite may be useful for diagnostic gradation (Albanese, 2018). The new classification of tremor is a welcome step forward providing a much-needed improvement of the old classification. But, in some aspects, the new classification is backward trying to accommodate discarded terminology like "plus" with no clear pathophysiological or diagnostic markers. While a change in nomenclature from ET to any other designation is welcomed, however, it should be supported by the biological data. Considering the syndromic nature of the ET, future research should be focused on facilitating its etiologic diagnoses and a molecular approach to future therapeutic endeavors.

5. Conclusion

The recent MDS classification puts ET in the proper perspective of being a clinical syndrome. A new terminology ET plus has also been added to address the mild neurological abnormalities commonly missed or dismissed in the evaluation of possible ET patients. But, in absence of any clear guidelines or definitions, the diagnosis of "ET plus" remains questionable and unstable. Classifying ET as a syndrome would certainly be clinically useful in finding specific etiology and this has certainly been an important step forward. Future longitudinal studies of carefully defined subtypes of ET are needed to understand the natural history of ET and ET plus.

References

Albanese, A. (2018). Classifying tremor: Language matters. *Movement Disorders: Official Journal of the Movement Disorder Society*, *33*(1), 3–4. https://doi.org/10.1002/mds.27178.

Albanese, A., Bhatia, K., Bressman, S. B., Delong, M. R., Fahn, S., Fung, V. S., et al. (2013). Phenomenology and classification of dystonia: A consensus update. *Movement Disorders: Official Journal of the Movement Disorder Society*, *28*(7), 863–873. https://doi.org/10.1002/mds.25475.

Albert, D. A., Munson, R., & Resnik, M. D. (1988). *Reasoning in medicine*. Baltimore: Johns Hopkins University Press.

Algarni, M., & Fasano, A. (2018). The overlap between essential tremor and Parkinson disease. *Parkinsonism & Related Disorders, 46*(Suppl. 1), S101–S104. https://doi.org/10.1016/j.parkreldis.2017.07.006.

Amlang, C. J., Trujillo Diaz, D., & Louis, E. D. (2020). Essential tremor as a "Waste Basket" diagnosis: Diagnosing essential tremor remains a challenge. *Frontiers in Neurology, 11*, 172. https://doi.org/10.3389/fneur.2020.00172.

Archer, D. B., Coombes, S. A., Chu, W. T., Chung, J. W., Burciu, R. G., Okun, M. S., et al. (2018). A widespread visually-sensitive functional network relates to symptoms in essential tremor. *Brain: A Journal of Neurology, 141*(2), 472–485.

Arkadir, D., & Louis, E. D. (2013). The balance and gait disorder of essential tremor: What does this mean for patients? *Therapeutic Advances in Neurological Disorders, 6*(4), 229–236. https://doi.org/10.1177/1756285612471415.

Balint, B., & Bhatia, K. P. (2015). Isolated and combined dystonia syndromes—An update on new genes and their phenotypes. *European Journal of Neurology, 22*(4), 610–617. https://doi.org/10.1111/ene.12650.

Becktepe, J., Gövert, F., Balint, B., Schlenstedt, C., Bhatia, K., Elble, R., et al. (2021). Exploring interrater disagreement on essential tremor using a standardized tremor elements assessment. *Movement Disorders Clinical Practice, 8*(3), 371–376. https://doi.org/10.1002/mdc3.13150.

Benito-León, J., Louis, E. D., Bermejo-Pareja, F., & Neurological Disorders in Central Spain Study Group. (2009). Risk of incident Parkinson's disease and parkinsonism in essential tremor: A population based study. *Journal of Neurology, Neurosurgery, and Psychiatry, 80*(4), 423–425. https://doi.org/10.1136/jnnp.2008.147223.

Bhatia, K. P., Bain, P., Bajaj, N., Elble, R. J., Hallett, M., Louis, E. D., et al. (2018). Consensus statement on the classification of tremors. From the task force on tremor of the International Parkinson and Movement Disorder Society. *Movement Disorders: Official Journal of the Movement Disorder Society, 33*(1), 75–87. https://doi.org/10.1002/mds.27121.

Boecker, H., Weindl, A., Brooks, D. J., Ceballos-Baumann, A. O., Liedtke, C., Miederer, M., et al. (2010). GABAergic dysfunction in essential tremor: An 11C-flumazenil PET study. *Journal of Nuclear Medicine: Official Publication, Society of Nuclear Medicine, 51*(7), 1030–1035.

Bowery, N. G., Bettler, B., Froestl, W., Gallagher, J. P., Marshall, F., Raiteri, M., et al. (2002). International Union of Pharmacology. XXXIII. Mammalian gamma-aminobutyric acid(B) receptors: Structure and function. *Pharmacological Reviews, 54*(2), 247–264. https://doi.org/10.1124/pr.54.2.247.

Buechele, F., Baumann, C. R., Lees, A., & Deuschl, G. (2021). Encouraging a generation of tremor researchers: Macdonald Critchley's paper on essential tremor. *Movement Disorders Clinical Practice*. https://doi.org/10.1002/mdc3.13333.

Buijink, A. W., Broersma, M., van der Stouwe, A. M., van Wingen, G. A., Groot, P. F., Speelman, J. D., et al. (2015). Rhythmic finger tapping reveals cerebellar dysfunction in essential tremor. *Parkinsonism & Related Disorders, 21*(4), 383–388. https://doi.org/10.1016/j.parkreldis.2015.02.003.

Burresi, P. (1874). Sopra un caso di tremore essenziale. Memore originali. Conferenza raccolta dallo studente Alfredo Rubini (22 febbraio 1874, Siena). *Lo Sperimentale, 33*, 475–481.

Calvo, F., Karras, B. T., Phillips, R., Kimball, A. M., & Wolf, F. (2003). Diagnoses, syndromes, and diseases: A knowledge representation problem. *AMIA ... Annual Symposium proceedings. AMIA Symposium, 2003*, 802.

Critchley, M. (1949). Observations on essential (heredofamial) tremor. *Brain: A Journal of Neurology, 72*(Pt. 2), 113–139.

Cersonsky, T., Kellner, S., Chapman, S., Huey, E. D., Louis, E. D., & Cosentino, S. (2020). Profiles of normal cognition in essential tremor. *Journal of the International Neuropsychological Society (JINS), 26*(2), 197–209. https://doi.org/10.1017/S1355617719001140.

Chung, S. J., Kwon, H., Lee, D. K., Hong, J. Y., Sunwoo, M. K., Sohn, Y. H., et al. (2013). Neuroanatomical heterogeneity of essential tremor according to propranolol response. *PLoS One, 8*(12), e84054. https://doi.org/10.1371/journal.pone.0084054.

Critchley, E. (1972). Clinical manifestations of essential tremor. *Journal of Neurology, Neurosurgery, and Psychiatry, 35*(3), 365–372. https://doi.org/10.1136/jnnp.35.3.365.

Dana, C. L. (1887). Hereditary tremor, a hitherto undescribed form of motor neurosis. 1. *The American Journal of the Medical Sciences (1827-1924), 94*, 386.

Deuschl, G., Bain, P., & Brin, M. (1998). Consensus statement of the Movement Disorder Society on Tremor. Ad Hoc Scientific Committee. *Movement Disorders: Official Journal of the Movement Disorder Society, 13*(Suppl. 3), 2–23. https://doi.org/10.1002/mds.870131303.

Deuschl, G., Lücking, C. H., & Schenck, E. (1987). Essential tremor: Electrophysiological and pharmacological evidence for a subdivision. *Journal of Neurology, Neurosurgery, and Psychiatry, 50*(11), 1435–1441. https://doi.org/10.1136/jnnp.50.11.1435.

Deuschl, G., Zimmermann, R., Genger, H., & Lücking, C. H. (1994). Physiological classification of essential tremor. *Neurological Disease and Therapy, 30*, 195.

di Biase, L., Brittain, J. S., Shah, S. A., Pedrosa, D. J., Cagnan, H., Mathy, A., et al. (2017). Tremor stability index: A new tool for differential diagnosis in tremor syndromes. *Brain: A Journal of Neurology, 140*(7), 1977–1986. https://doi.org/10.1093/brain/awx104.

Diez-Fairen, M., Bandres-Ciga, S., Houle, G., Nalls, M. A., Girard, S. L., Dion, P. A., et al. (2019). Genome-wide estimates of heritability and genetic correlations in essential tremor. *Parkinsonism & Related Disorders, 64*, 262–267.

Dorland, W. A. N., & Taylor, E. J. (1988). *Dorland's illustrated medical dictionary*. Philadelphia, PA: W.B. Saunders.

Elble, R. J. (2013). What is essential tremor? *Current Neurology and Neuroscience Reports, 13*(6), 353. https://doi.org/10.1007/s11910-013-0353-4.

Elble, R. J. (2020). Do we belittle essential tremor by calling it a syndrome rather than a disease? No. *Frontiers in Neurology, 11*, 586606. https://doi.org/10.3389/fneur.2020.586606.

Emamikhah, M., Aghavali, S., Moghadas, F., Munhoz, R. P., Lang, A. E., Alavi, A., et al. (2021). Spinocerebellar Ataxia 40: Another etiology underlying essential tremor syndrome. *Movement Disorders Clinical Practice, 8*(6), 944–946. https://doi.org/10.1002/mdc3.13251.

Fearon, C., Espay, A. J., Lang, A. E., Lynch, T., Martino, D., Morgante, F., et al. (2019). Soft signs in movement disorders: Friends or foes? *Journal of Neurology, Neurosurgery, and Psychiatry, 90*(8), 961–962. https://doi.org/10.1136/jnnp-2018-318455.

Flatau, G. (1908). Ü ber hereditären essentiellen Tremor. *Archiv für Psychiatrie und Nervenkrankheiten, 44*, 306–340.

Gionco, J. T., Hartstone, W. G., Martuscello, R. T., Kuo, S. H., Faust, P. L., & Louis, E. D. (2021). Essential tremor versus "ET-plus": A detailed postmortem study of cerebellar pathology. *Cerebellum (London, England)*. https://doi.org/10.1007/s12311-021-01263-6.

Gironell, A. (2014). The GABA hypothesis in essential tremor: Lights and shadows. *Tremor and Other Hyperkinetic Movements (New York, N.Y.), 4*, 254. https://doi.org/10.7916/D8SF2T9C.

Gironell, A., Figueiras, F. P., Pagonabarraga, J., Herance, J. R., Pascual-Sedano, B., Trampal, C., et al. (2012). Gaba and serotonin molecular neuroimaging in essential tremor: A clinical correlation study. *Parkinsonism & Related Disorders*, *18*(7), 876–880. https://doi.org/10.1016/j.parkreldis.2012.04.024.

Gironell, A., Kulisevsky, J., Barbanoj, M., López-Villegas, D., Hernández, G., & Pascual-Sedano, B. (1999). A randomized placebo-controlled comparative trial of gabapentin and propranolol in essential tremor. *Archives of Neurology*, *56*(4), 475–480. https://doi.org/10.1001/archneur.56.4.475.

Gövert, F., Becktepe, J. S., & Deuschl, G. (2016). Current concepts of essential tremor. *Revue Neurologique*, *172*(8–9), 416–422. https://doi.org/10.1016/j.neurol.2016.07.010.

Handforth, A. (2012). Harmaline tremor: Underlying mechanisms in a potential animal model of essential tremor. *Tremor and Other Hyperkinetic Movements (New York, N.Y.)*, *2*. https://doi.org/10.7916/D8TD9W2P. 02-92-769-1.

Hertegård, S., Granqvist, S., & Lindestad, P. A. (2000). Botulinum toxin injections for essential voice tremor. *The Annals of Otology, Rhinology, and Laryngology*, *109*(2), 204–209. https://doi.org/10.1177/000348940010900216.

Hopfner, F., Ahlf, A., Lorenz, D., Klebe, S., Zeuner, K. E., Kuhlenbäumer, G., et al. (2016). Early- and late-onset essential tremor patients represent clinically distinct subgroups. *Movement Disorders: Official Journal of the Movement Disorder Society*, *31*(10), 1560–1566.

Hopfner, F., & Deuschl, G. (2018). Is essential tremor a single entity? *European Journal of Neurology*, *25*(1), 71–82. https://doi.org/10.1111/ene.13454.

Hopfner, F., Haubenberger, D., Galpern, W. R., Gwinn, K., Van't Veer, A., White, S., et al. (2016). Knowledge gaps and research recommendations for essential tremor. *Parkinsonism & Related Disorders*, *33*, 27–35. https://doi.org/10.1016/j.parkreldis.2016.10.002.

Hopfner, F., & Helmich, R. C. (2018). The etiology of essential tremor: Genes versus environment. *Parkinsonism & Related Disorders*, *46*(Suppl. 1), S92–S96. https://doi.org/10.1016/j.parkreldis.2017.07.014.

Hor, H., Francescatto, L., Bartesaghi, L., Ortega-Cubero, S., Kousi, M., Lorenzo-Betancor, O., et al. (2015). Missense mutations in TENM4, a regulator of axon guidance and central myelination, cause essential tremor. *Human Molecular Genetics*, *24*(20), 5677–5686. https://doi.org/10.1093/hmg/ddv281.

Huang, H., Yang, X., Zhao, Q., Ning, P., Shen, Q., Wang, H., et al. (2020). Clinical characteristics of patients with essential tremor or essential tremor plus. *Acta Neurologica Scandinavica*, *141*(4), 335–341. https://doi.org/10.1111/ane.13209.

Iglesias-Hernandez, D., Delgado, N., McGurn, M., Huey, E. D., Cosentino, S., & Louis, E. D. (2021). "ET Plus": Instability of the diagnosis during prospective longitudinal follow-up of essential tremor cases. *Frontiers in Neurology*, *12*, 782694. https://doi.org/10.3389/fneur.2021.782694.

Isaias, I. U., Marotta, G., Hirano, S., Canesi, M., Benti, R., Righini, A., et al. (2010). Imaging essential tremor. *Movement Disorders: Official Journal of the Movement Disorder Society*, *25*(6), 679–686. https://doi.org/10.1002/mds.22870.

Jablonski, S. (1991). Syndrome: le mot de jour. *American Journal of Medical Genetics*, *39*(3), 342–346. https://doi.org/10.1002/ajmg.1320390319.

Jain, S., Lo, S. E., & Louis, E. D. (2006). Common misdiagnosis of a common neurological disorder: How are we misdiagnosing essential tremor? *Archives of Neurology*, *63*(8), 1100–1104. https://doi.org/10.1001/archneur.63.8.1100.

Kremer, N. I., Pauwels, R., Pozzi, N. G., Lange, F., Roothans, J., Volkmann, J., et al. (2021). Deep brain stimulation for tremor: Update on long-term outcomes, target considerations and future directions. *Journal of Clinical Medicine*, *10*(16), 3468. https://doi.org/10.3390/jcm10163468.

Lenka, A., & Louis, E. D. (2020). Do we belittle essential tremor by calling it a syndrome rather than a disease? Yes. *Frontiers in Neurology*, *11*, 522687. https://doi.org/10.3389/fneur.2020.522687.

Lenka, A., & Pandey, S. (2021). Essential tremor: Five new things. *Neurology: Clinical Practice*. https://doi.org/10.1212/CPJ.0000000000001145.

Louis, E. D. (1999). A new twist for stopping the shakes? Revisiting GABAergic therapy for essential tremor. *Archives of Neurology*, *56*(7), 807–808. https://doi.org/10.1001/archneur.56.7.807.

Louis, E. D. (2010). Essential tremor: Evolving clinicopathological concepts in an era of intensive post-mortem enquiry. *The Lancet. Neurology*, *9*(6), 613–622. https://doi.org/10.1016/S1474-4422(10)70090-9.

Louis, E. D. (2018). The evolving definition of essential tremor: What are we dealing with? *Parkinsonism & Related Disorders*, *46*(Suppl. 1), S87–S91. https://doi.org/10.1016/j.parkreldis.2017.07.004.

Louis, E. D. (2020). "Essential Tremor Plus": A problematic concept: Implications for clinical and epidemiological studies of essential tremor. *Neuroepidemiology*, *54*(2), 180–184. https://doi.org/10.1159/000502862.

Louis, E. D. (2021). The essential tremors: Evolving concepts of a family of diseases. *Frontiers in Neurology*, *12*, 650601. https://doi.org/10.3389/fneur.2021.650601.

Louis, E. D., Bares, M., Benito-Leon, J., Fahn, S., Frucht, S. J., Jankovic, J., et al. (2020). Essential tremor-plus: A controversial new concept. *The Lancet. Neurology*, *19*(3), 266–270. https://doi.org/10.1016/S1474-4422(19)30398-9.

Louis, E. D., Benito-León, J., Vega-Quiroga, S., Bermejo-Pareja, F., & Neurological Disorders in Central Spain (NEDICES) Study Group. (2010). Faster rate of cognitive decline in essential tremor cases than controls: A prospective study. *European Journal of Neurology*, *17*(10), 1291–1297. https://doi.org/10.1111/j.1468-1331.2010.03122.x.

Louis, E. D., Broussolle, E., Goetz, C. G., Krack, P., Kaufmann, P., & Mazzoni, P. (2008). Historical underpinnings of the term essential tremor in the late 19th century. *Neurology*, *71*(11), 856–859. https://doi.org/10.1212/01.wnl.0000325564.38165.d1.

Louis, E. D., & Faust, P. L. (2020). Essential tremor pathology: Neurodegeneration and reorganization of neuronal connections. *Nature Reviews. Neurology*, *16*(2), 69–83. https://doi.org/10.1038/s41582-019-0302-1.

Louis, E. D., Faust, P. L., & Vonsattel, J. P. (2011). Purkinje cell loss is a characteristic of essential tremor. *Parkinsonism & Related Disorders*, *17*(6), 406–409. https://doi.org/10.1016/j.parkreldis.2011.05.004.

Louis, E. D., Faust, P. L., & Vonsattel, J. P. (2012). Purkinje cell loss is a characteristic of essential tremor: Towards a more mature understanding of pathogenesis. *Parkinsonism & Related Disorders*, *18*(8), 1003–1004. https://doi.org/10.1016/j.parkreldis.2012.06.017.

Louis, E. D., & Ferreira, J. J. (2010). How common is the most common adult movement disorder? Update on the worldwide prevalence of essential tremor. *Movement Disorders: Official Journal of the Movement Disorder Society*, *25*(5), 534–541. https://doi.org/10.1002/mds.22838.

Louis, E. D., Hernandez, N., Ionita-Laza, I., Ottman, R., & Clark, L. N. (2013). Does rate of progression run in essential tremor families? Slower vs. faster progressors. *Parkinsonism & Related Disorders*, *19*(3), 363–366. https://doi.org/10.1016/j.parkreldis.2012.10.005.

Louis, E. D., Hernandez, N., & Michalec, M. (2015). Prevalence and correlates of rest tremor in essential tremor: Cross-sectional survey of 831 patients across four distinct cohorts. *European Journal of Neurology*, *22*(6), 927–932. https://doi.org/10.1111/ene.12683.

Louis, E. D., Huey, E. D., & Cosentino, S. (2021). Features of "ET plus" correlate with age and tremor duration: "ET plus" may be a disease stage rather than a subtype of essential tremor. *Parkinsonism & Related Disorders*, *91*, 42–47. https://doi.org/10.1016/j.parkreldis.2021.08.017.

Louis, E. D., Kerridge, C. A., Chatterjee, D., Martuscello, R. T., Diaz, D. T., Koeppen, A. H., et al. (2019). Contextualizing the pathology in the essential tremor cerebellar cortex: A patholog-omics approach. *Acta Neuropathologica*, *138*(5), 859–876. https://doi.org/10.1007/s00401-019-02043-7.

Louis, E. D., & Lenka, A. (2017). The olivary hypothesis of essential tremor: Time to lay this model to rest? *Tremor and Other Hyperkinetic Movements (New York, N.Y.)*, 7, 473. https://doi.org/10.7916/D8FF40RX.

Louis, E. D., & McCreary, M. (2021). How common is essential tremor? Update on the worldwide prevalence of essential tremor. *Tremor and Other Hyperkinetic Movements (New York, N.Y.)*, *11*, 28. https://doi.org/10.5334/tohm.632.

Magrinelli, F., Latorre, A., Balint, B., Mackenzie, M., Mulroy, E., Stamelou, M., et al. (2020). Isolated and combined genetic tremor syndromes: A critical appraisal based on the 2018 MDS criteria. *Parkinsonism & Related Disorders*, 77, 121–140. https://doi.org/10.1016/j.parkreldis.2020.04.010.

Málly, J., Baranyi, M., & Vizi, E. S. (1996). Change in the concentrations of amino acids in CSF and serum of patients with essential tremor. *Journal of Neural Transmission (Vienna, Austria: 1996)*, *103*(5), 555–560. https://doi.org/10.1007/BF01273153.

Maragliano, E. (1879). Tremore essenziale congenito. In *Note di Clinica Medica* (pp. 7–12). Genova: Tipografia Del R. Istituto Sordo-Muti.

Marsden, C. D., Obeso, J. A., & Rothwell, J. C. (1983). Benign essential tremor is not a single entity. In M. D. Yahr (Ed.), *Current concepts of Parkinson's disease and related disorders* (pp. 31–46). Amsterdam, the Netherlands: Excerpta Medica.

Martinelli, P., Gabellini, A. S., Gulli, M. R., & Lugaresi, E. (1987). Different clinical features of essential tremor: A 200-patient study. *Acta Neurologica Scandinavica*, *75*(2), 106–111. https://doi.org/10.1111/j.1600-0404.1987.tb07903.x.

Martínez-Fernández, R., & Pineda-Pardo, J. A. (2020). Magnetic resonance-guided focused ultrasound for movement disorders: Clinical and neuroimaging advances. *Current Opinion in Neurology*, *33*(4), 488–497. https://doi.org/10.1097/WCO.0000000000000840.

Mavroudis, I., Petridis, F., Kazis, D., Njau, S. N., Costa, V., & Baloyannis, S. J. (2019). Purkinje cells pathology in Alzheimer's disease. *American Journal of Alzheimer's Disease and Other Dementias*, *34*(7–8), 439–449. https://doi.org/10.1177/1533317519859200.

Merner, N. D., Girard, S. L., Catoire, H., Bourassa, C. V., Belzil, V. V., Rivière, J. B., et al. (2012). Exome sequencing identifies FUS mutations as a cause of essential tremor. *American Journal of Human Genetics*, *91*(2), 313–319. https://doi.org/10.1016/j.ajhg.2012.07.002.

Montgomery, E. B., Jr., Baker, K. B., Lyons, K., & Koller, W. C. (2000). Motor initiation and execution in essential tremor and Parkinson's disease. *Movement Disorders: Official Journal of the Movement Disorder Society*, *15*(3), 511–515. https://doi.org/10.1002/1531-8257(200005)15:3<511::AID-MDS1014>3.0.CO;2-R.

Murray, J. A., Bradley, N., Craigie, W. A., & Onions, C. T. (1919). *English dictionary on historical principles*. Oxford: Clarendon Press.

Muthuraman, M., Deuschl, G., Anwar, A. R., Mideksa, K. G., von Helmolt, F., & Schneider, S. A. (2015). Essential and aging-related tremor: Differences of central control. *Movement Disorders: Official Journal of the Movement Disorder Society*, *30*(12), 1673–1680. https://doi.org/10.1002/mds.26410.

Nagy, A. (1890). Ueber hereditären juvenilen Tremor. *Neurologisches Zentralblatt*, *9*, 557–559.

Nieuwhof, F., Panyakaew, P., van de Warrenburg, B. P., Gallea, C., & Helmich, R. C. (2018). The patchy tremor landscape: Recent advances in pathophysiology. *Current Opinion in Neurology*, *31*(4), 455–461. https://doi.org/10.1097/WCO.0000000000000582.

O'Hearn, E., Holmes, S. E., Calvert, P. C., Ross, C. A., & Margolis, R. L. (2001). SCA-12: Tremor with cerebellar and cortical atrophy is associated with a CAG repeat expansion. *Neurology*, *56*(3), 299–303. https://doi.org/10.1212/wnl.56.3.299.

Pandey, S., Benito-León, J., & Kuo, S.-H. (2021). Editorial: Tremor syndromes: Current concepts and future perspectives. *Frontiers in Neurology, 12*, 752463. https://doi.org/10.3389/fneur.2021.752463.

Pandey, S., & Bhattad, S. (2019). Questionable dystonia in essential tremor plus: A video-based assessment of 19 patients. *Movement Disorders Clinical Practice, 6*(8), 722–723. https://doi.org/10.1002/mdc3.12838.

Pandey, S., & Bhattad, S. (2020). Diagnostic delay in cervical dystonia: Dystonia with antecedent essential tremor? *Movement Disorders: Official Journal of the Movement Disorder Society, 35*(6), 1085–1086. https://doi.org/10.1002/mds.28103.

Pandey, S., & Bhattad, S. (2021). Soft signs in essential tremor plus: A prospective study. *Movement Disorders Clinical Practice, 8*(8), 1275–1277. https://doi.org/10.1002/mdc3.13350.

Pandey, S., Bhattad, S., & Dinesh, S. (2021). Tremor in primary monogenic dystonia. *Current Neurology and Neuroscience Reports, 21*(9), 48. https://doi.org/10.1007/s11910-021-01135-w.

Pandey, S., Bhattad, S., & Hallett, M. (2020). The problem of questionable dystonia in the diagnosis of 'Essential Tremor-Plus'. *Tremor and Other Hyperkinetic Movements (New York, N.Y.), 10*, 27. https://doi.org/10.5334/tohm.539.

Pandey, S., & Sarma, N. (2016). Tremor in dystonia. *Parkinsonism & Related Disorders, 29*, 3–9. https://doi.org/10.1016/j.parkreldis.2016.03.024.

Pandey, S., & Sarma, N. (2017). Tremor in dystonia: A cross-sectional study from India. *Movement Disorders Clinical Practice, 4*(6), 858–863. https://doi.org/10.1002/mdc3.12546.

Paris-Robidas, S., Brochu, E., Sintes, M., Emond, V., Bousquet, M., Vandal, M., et al. (2012). Defective dentate nucleus GABA receptors in essential tremor. *Brain: A Journal of Neurology, 135*(Pt. 1), 105–116. https://doi.org/10.1093/brain/awr301.

Passamonti, L., Cerasa, A., & Quattrone, A. (2012). Neuroimaging of essential tremor: What is the evidence for cerebellar involvement? *Tremor and Other Hyperkinetic Movements (New York, N.Y.), 2*. https://doi.org/10.7916/D8F76B8G. 02-67-421-3.

Patel, A., Deeb, W., & Okun, M. S. (2018). Deep brain stimulation management of essential tremor with dystonic features. *Tremor and Other Hyperkinetic Movements (New York, N.Y.), 8*, 557. https://doi.org/10.7916/D8P85VBQ.

Prasad, S., & Pal, P. K. (2019). Reclassifying essential tremor: Implications for the future of past research. *Movement Disorders: Official Journal of the Movement Disorder Society, 34*(3), 437. https://doi.org/10.1002/mds.27615.

Putzke, J. D., Whaley, N. R., Baba, Y., Wszolek, Z. K., & Uitti, R. J. (2006). Essential tremor: Predictors of disease progression in a clinical cohort. *Journal of Neurology, Neurosurgery, and Psychiatry, 77*(11), 1235–1237. https://doi.org/10.1136/jnnp.2006.086579.

Rajalingam, R., Breen, D. P., Lang, A. E., & Fasano, A. (2018). Essential tremor plus is more common than essential tremor: Insights from the reclassification of a cohort of patients with lower limb tremor. *Parkinsonism & Related Disorders, 56*, 109–110. https://doi.org/10.1016/j.parkreldis.2018.06.029.

Rajan, R., Pandey, S., Anandapadmanabhan, R., & Srivastava, A. K. (2018). Interrater and intrarater agreement on the 2018 consensus statement on classification of tremors. *Movement Disorders: Official Journal of the Movement Disorder Society, 33*(12), 1966–1967. https://doi.org/10.1002/mds.27513.

Rajput, A. H., Rajput, E. F., Bocking, S. M., Auer, R. N., & Rajput, A. (2019). Parkinsonism in essential tremor cases: A clinicopathological study. *Movement Disorders: Official Journal of the Movement Disorder Society, 34*(7), 1031–1040. https://doi.org/10.1002/mds.27729.

Rajput, A. H., Robinson, C. A., & Rajput, A. (2013). Purkinje cell loss is neither pathological basis nor characteristic of essential tremor. *Parkinsonism & Related Disorders, 19*(4), 490–491. https://doi.org/10.1016/j.parkreldis.2012.11.019.

Rajput, A. H., Robinson, C. A., Rajput, M. L., Robinson, S. L., & Rajput, A. (2012). Essential tremor is not dependent upon cerebellar Purkinje cell loss. *Parkinsonism & Related Disorders*, *18*(5), 626–628. https://doi.org/10.1016/j.parkreldis.2012.01.013.

Rao, A. K., & Louis, E. D. (2019). Ataxic gait in essential tremor: A disease-associated feature? *Tremor and Other Hyperkinetic Movements (New York, N.Y.)*, *9*. https://doi.org/10.7916/d8-28jq-8t52.

Raymond, F. (1892). Sur le tremblement essentiel he're'ditaire. *Le Bulletin Médical*, *6*, 205–209.

Schmahmann, J. D., & Sherman, J. C. (1998). The cerebellar cognitive affective syndrome. *Brain: A Journal of Neurology*, *121*(Pt. 4), 561–579. https://doi.org/10.1093/brain/121.4.561.

Schrag, A., Münchau, A., Bhatia, K. P., Quinn, N. P., & Marsden, C. D. (2000). Essential tremor: An overdiagnosed condition? *Journal of Neurology*, *247*(12), 955–959. https://doi.org/10.1007/s004150070053.

Sepúlveda Soto, M. C., & Fasano, A. (2019). Essential tremor: New advances. *Clinical Parkinsonism & Related Disorders*, *3*, 100031. https://doi.org/10.1016/j.prdoa.2019.100031.

Shaikh, A. G., Beylergil, S. B., Scorr, L., Kilic-Berkmen, G., Freeman, A., Klein, C., et al. (2021). Dystonia and tremor: A cross-sectional study of the dystonia coalition cohort. *Neurology*, *96*(4), e563–e574. https://doi.org/10.1212/WNL.0000000000011049.

Shalash, A. S., Hamid, E., Elrassas, H., Bahbah, E. I., Mansour, A. H., Mohamed, H., et al. (2021). Non-motor symptoms in essential tremor, akinetic rigid and tremor-dominant subtypes of Parkinson's disease. *PLoS One*, *16*(1), e0245918. https://doi.org/10.1371/journal.pone.0245918.

Sharma, S., & Pandey, S. (2020). Treatment of essential tremor: Current status. *Postgraduate Medical Journal*, *96*(1132), 84–93. https://doi.org/10.1136/postgradmedj-2019-136647.

Singer, C., Sanchez-Ramos, J., & Weiner, W. J. (1994). Gait abnormality in essential tremor. *Movement Disorders: Official Journal of the Movement Disorder Society*, *9*(2), 193–196. https://doi.org/10.1002/mds.870090212.

Singh-Bains, M. K., Mehrabi, N. F., Sehji, T., Austria, M., Tan, A., Tippett, L. J., et al. (2019). Cerebellar degeneration correlates with motor symptoms in Huntington disease. *Annals of Neurology*, *85*(3), 396–405. https://doi.org/10.1002/ana.25413.

Stamelou, M., Charlesworth, G., Cordivari, C., Schneider, S. A., Kägi, G., Sheerin, U. M., et al. (2014). The phenotypic spectrum of DYT24 due to ANO3 mutations. *Movement Disorders: Official Journal of the Movement Disorder Society*, *29*(7), 928–934. https://doi.org/10.1002/mds.25802.

Steffen, J. K., Jergas, H., Petry-Schmelzer, J. N., Dembek, T. A., Thies, T., Jost, S. T., et al. (2020). Thalamic deep brain stimulation in essential tremor plus is as effective as in essential tremor. *Brain Sciences*, *10*(12), 970. https://doi.org/10.3390/brainsci10120970.

Sun, Q. Y., Xu, Q., Tian, Y., Hu, Z. M., Qin, L. X., Yang, J. X., et al. (2020). Expansion of GGC repeat in the human-specific NOTCH2NLC gene is associated with essential tremor. *Brain: A Journal of Neurology*, *143*(1), 222–233. https://doi.org/10.1093/brain/awz372.

Symanski, C., Shill, H. A., Dugger, B., Hentz, J. G., Adler, C. H., Jacobson, S. A., et al. (2014). Essential tremor is not associated with cerebellar Purkinje cell loss. *Movement Disorders: Official Journal of the Movement Disorder Society*, *29*(4), 496–500. https://doi.org/10.1002/mds.25845.

Tarakad, A., & Jankovic, J. (2019). Essential tremor and Parkinson's disease: Exploring the relationship. *Tremor and Other Hyperkinetic Movements (New York, N.Y.)*, *8*, 589. https://doi.org/10.7916/D8MD0GVR.

Unal Gulsuner, H., Gulsuner, S., Mercan, F. N., Onat, O. E., Walsh, T., Shahin, H., et al. (2014). Mitochondrial serine protease HTRA2 p.G399S in a kindred with essential tremor and Parkinson disease. *Proceedings of the National Academy of Sciences of the United States of America*, *111*(51), 18285–18290.

Van Bogaert, L., & De Savitsch, E. (1937). Sur une maladie congénitale et hérédofamiliale comportant un tremblement rythmique de la tête, des globes oculaires et des membres supérieurs. *L'Encéphale*, *32*, 133–139.

Vidailhet, M. (2020). Essential tremor-plus: A temporary label. *The Lancet. Neurology*, *19*(3), 202–203. https://doi.org/10.1016/S1474-4422(19)30442-9.

Welton, T., Cardoso, F., Carr, J. A., Chan, L. L., Deuschl, G., Jankovic, J., et al. (2021). Essential tremor. *Nature Reviews. Disease Primers*, 7(1), 83. https://doi.org/10.1038/s41572-021-00314-w.

Wissel, J., Masuhr, F., Schelosky, L., Ebersbach, G., & Poewe, W. (1997). Quantitative assessment of botulinum toxin treatment in 43 patients with head tremor. *Movement Disorders: Official Journal of the Movement Disorder Society*, *12*(5), 722–726. https://doi.org/10.1002/mds.870120516.

SECTION II

Essential tremor: Is this an electrical or a degenerative disorder? Or is it both?

Elan D. Louis[a] and Sheng-Han Kuo[b,c]

[a]Department of Neurology, University of Texas Southwestern Medical Center, Dallas, TX, United States
[b]Department of Neurology, College of Physicians and Surgeons, Columbia University, New York, NY, United States
[c]Initiative for Columbia Ataxia and Tremor, Columbia University, New York, NY, United States

As part of the current series, which highlights current controversies in the essential tremor (ET) field, we commissioned authors to submit papers with opposing viewpoints, one using available data to put forth the argument that ET is degenerative (Faust, 2022) and the other, using available data to put forth the argument that ET is an electric disturbance (Cho, 2022). We are grateful to both sets of authors for thoughtfully making their cases and here we provide additional commentary.

The traditional pathophysiological model of ET, the olivary model of ET (Louis & Lenka, 2017), arose in the 1970s and was grounded in a series of insightful and elegant physiological studies of animals on the presence tremor generators in the central nervous system as well as an animal model of tremor involving the neurotoxin harmaline (Llinas & Volkind, 1973; Llinas & Yarom, 1986). Because these observations were physiological in nature, model building was physiological in its approach. As such, action tremors could arise from oscillations and oscillatory disturbances in motor networks involving the brainstem, cerebellum, thalamus and motor cortex. Models for the generation of action tremor *in general* and *in animals* were then applied to a *specific type of tremor* that occurs *in humans*, namely ET. Hence, there was an attempt to extrapolate from the general to the specific. In parallel with this was the virtual absence of postmortem studies of patients with ET; indeed, a scanty postmortem literature was superficial and uncontrolled (Louis & Vonsattel, 2007). These studies failed to observe what subsequent more systematic work easily detected—a plethora of tissue based changes in the ET cerebellum, with a number of these consistent with neurodegeneration (Louis & Faust, 2020a; Louis, Faust, Vonsattel, et al., 2007). The initial apparent lack of tissue based changes served to further reinforce the notion that ET is an electrical disorder. In addition, observations from human physiology studies on patients with ET were used to support the notion of a synchronized network of brain regions that were responsible for propagating tremor (Muthuraman, Raethjen, Koirala, et al., 2018).

Yet more recent work, as outlined by Faust (2022), involving detailed, systematic and controlled studies of human brain material, has resulted in the discovery of a plethora of postmortem changes in ET (Louis & Faust, 2020a, 2020b) and, with these, for the first time a disease model that was grounded in tissue based changes, including those that are degenerative (Louis & Faust, 2020b). The degenerative model has gained increasing traction with the addition of a myriad of studies over time (Louis & Faust, 2020b).

With the growth of studies, two schools of thought have arisen. One, that ET is an electrical disorder, with the fundamental problem being a physiological disarrangement of a normal set of brain circuits. The second is that ET is a degenerative disorder, with the primary problem being a set of molecular events that leads to neuronal death. In the current series, authors have been asked to argue each side and each provides convincing data to support their model. Yet at the same time, it is of interest that each author acknowledges the merits of the data on the other side. This is important because it provides the grounds for more integrated thought.

Rather than an *either/or* situation, it is possible that *both* are occurring in ET; that is, that ET is both a degenerative disorder and one that is electrical in nature. As noted above, both Dr. Faust and Dr. Cho recognize this (Cho, 2022; Faust, 2022) and Drs. Lang and Handforth, in another similar piece is this series do as well (Lang & Handforth, 2022). One possibility is that a primary degenerative process in the Purkinje cells leads to a subsequent change in downstream (i.e., deep cerebellar nuclei) physiology, as has been argued previously (Louis & Faust, 2020b) and is argued by Dr. Faust (Faust, 2022). As such, an initial degenerative process leads to an abnormal Purkinje cell output to the deep cerebellar nuclei, resulting in an abnormal physiology (i.e., an electrical disarrangement) along the remainder of the tremor loop. The other possibility (see paper by Lang and Handforth, 2022) is that a both degenerative and non-degenerative mechanisms lead to changes in the function of tremor pathways. Those authors note that traditionally, ET research has focused on the hypothesis of increased coupling of inferior olivary neurons as the cause of hypersynchronous complex spike activity underlying tremor (Lang & Handforth, 2022; Lin et al., 2014; Wu et al., 2021). They note, however, that recent postmortem studies of humans with ET and evidence from mutant mice, particularly the *hotfoot17* mouse, that largely replicate the pathology and genetics of ET, suggest that the abnormal innervation of multiple Purkinje cells by climbing fibers could be related to tremor (Lang & Handforth, 2022). They further note that ET brains show partial Purkinje cell loss and axon terminal sprouting by surviving Purkinje cells, and that this may provide another mechanism for tremor (Lang & Handforth, 2022). As such, they proposed that in ET, these three mechanisms could each promote tremor (Lang & Handforth, 2022). Regardless of what happens in ET, it is apparent that groups of authors are increasingly recognizing that both degenerative as well as electrical process likely underlie ET. As such, groups are incorporating into their work the concept of *pathophysiology*—tissue based changes with physiological changes. Such approaches are more nuanced and more integrated and move beyond simple uni-dimensional explanations for what is likely to be a complex process.

References

Cho, H. J. (2022). Is essential tremor a degenerative or an electrical disorder? Electrical disorder. *International Review of Neurobiology, 163*, 103–128.

Faust, P. L. (2022). Is essential tremor a degenerative disorder or an electric disorder? Degenerative disorder. *International Review of Neurobiology, 163*, 65–101.

Lang, E. J. & Handforth, A. (2022). Is the inferior olive central to essential tremor? Yes. *International Review of Neurobiology, 163*, 133–165.

Lin, C. Y., Louis, E. D., Faust, P. L., Koeppen, A. H., Vonsattel, J. P., & Kuo, S. H. (2014). Abnormal climbing fibre-Purkinje cell synaptic connections in the essential tremor cerebellum. *Brain, 137*, 3149–3159.

Llinas, R., & Volkind, R. A. (1973). The olivo-cerebellar system: Functional properties as revealed by harmaline-induced tremor. *Experimental Brain Research, 18*, 69–87.

Llinas, R., & Yarom, Y. (1986). Oscillatory properties of guinea-pig inferior olivary neurones and their pharmacological modulation: An in vitro study. *The Journal of Physiology, 376*, 163–182.

Louis, E. D., & Faust, P. L. (2020a). Essential tremor within the broader context of other forms of cerebellar degeneration. *Cerebellum, 19*, 879–896.

Louis, E. D., & Faust, P. L. (2020b). Essential tremor pathology: Neurodegeneration and reorganization of neuronal connections. *Nature Reviews. Neurology, 16*, 69–83.

Louis, E. D., Faust, P. L., Vonsattel, J. P., et al. (2007). Neuropathological changes in essential tremor: 33 cases compared with 21 controls. *Brain, 130*, 3297–3307.

Louis, E. D., & Lenka, A. (2017). The olivary hypothesis of essential tremor: Time to lay this model to rest? *Tremor and Other Hyperkinetic Movements*, 7, 473.

Louis, E. D., & Vonsattel, J. P. (2007). The emerging neuropathology of essential tremor. *Movement Disorders, 23*, 174–182.

Muthuraman, M., Raethjen, J., Koirala, N., et al. (2018). Cerebello-cortical network fingerprints differ between essential, Parkinson's and mimicked tremors. *Brain, 141*, 1770–1781.

Wu, Y. C., Louis, E. D., Gionco, J., Pan, M. K., Faust, P. L., & Kuo, S. H. (2021). Increased climbing fiber lateral crossings on Purkinje cell dendrites in the cerebellar hemisphere in essential tremor. *Movement Disorders, 36*, 1440–1445.

CHAPTER THREE

Is essential tremor a degenerative disorder or an electric disorder? Degenerative disorder

Phyllis L. Faust*
Department of Pathology and Cell Biology, Columbia University Vagelos College of Physicians and Surgeons and the New York Presbyterian Hospital, New York, NY, United States
*Corresponding author: e-mail address: plf3@cumc.columbia.edu

Contents

Abstract

Essential tremor (ET) is a highly prevalent neurologic disease and is the most common of the many tremor disorders. ET is a progressive condition with marked clinical heterogeneity, associated with a spectrum of both motor and non-motor features. However, its disease mechanisms remain poorly understood. Much debate has centered on whether ET should be considered a degenerative disorder, with underlying pathological changes in brain causing progressive disease manifestations, or an electric disorder, with overactivity of intrinsically oscillatory motor networks that occur without underlying structural brain abnormalities. Converging data from clinical, neuroimaging and pathological studies in ET now provide considerable evidence for the neurodegenerative hypothesis. A major turning point in this debate is that rigorous tissue-based studies have recently identified a series of structural changes in the ET cerebellum. Most of these pathological changes are centered on the Purkinje cell and connected neuronal populations, which can result in partial loss of Purkinje cells and circuitry reorganizations that would disturb cerebellar function. There is significant overlap in clinical and

International Review of Neurobiology, Volume 163
ISSN 0074-7742
https://doi.org/10.1016/bs.irn.2022.02.003

pathological features of ET with other disorders of cerebellar degeneration, and an increased risk of developing other degenerative diseases in ET. The combined implication of these studies is that ET could be degenerative. The evidence in support of the degenerative hypothesis is presented.

Abbreviations

AD	Alzheimer's disease
CF	climbing fiber
DTI	diffusion tensor imagingfiber
EAAT2	excitatory amino acid transporter type 2
ET	essential tremor
fMRI	functional magnetic resonance imaging
ION	inferior olivary nucleus
MR	magnetic resonance
MRS	magnetic resonance spectroscopy
PC	Purkinje cell
PD	Parkinson's disease
PET	positron emission tomography
PSP	progressive supranuclear palsy
SCA	spinocerebellar ataxia

1. Introduction

Essential tremor (ET) is a highly prevalent and progressive tremor disorder, with its most recognizable feature being a 4–12 Hz kinetic tremor (i.e., tremor occurring during volitional movement) of the arms. The clinical phenotype of ET has expanded from a monosymptomatic entity focusing on its kinetic arm tremor to one with several other types of tremors, other motor features, and a range of non-motor features (Louis, 2021). Thus, there is considerable clinical heterogeneity in ET, leading to the concept that it may well be a family of diseases. While it is currently accepted that motor circuits involving the cerebello-thalamo-cortical loop and brainstem circuits including olivo-cerebellar networks are involved in ET, the underlying pathobiology for how tremor is generated in the brain in ET remains controversial.

There are currently two main postulates for tremor generation in ET: (1) it is a degenerative disorder, (2) it is an electric disorder. The degenerative hypothesis was previously discarded, as early studies on ET claimed there was no definable pathology (Rajput, Robinson, & Rajput, 2004). Indeed, during the 100-year period from 1903 to 2003, there were only 15 published postmortems, mainly comprising isolated case reports with no conclusive findings (Louis & Vonsattel, 2008). These studies lacked rigor, as none used quantitative or immunohistochemical approaches nor compared ET to control brains, and clinical definitions lacked precision as many of the cases had dystonia or chorea rather than ET. This thinking persisted in the literature, with many textbook chapters and review articles simply omitting any discussion of ET pathology well into the 2000s.

In the early 1970s an alternative model took favor that ET is an electric/electrophysiologic disorder, much like epilepsy, due to overactivity of synchronous firing from pacemaking neurons in the inferior olivary nucleus (ION). This electric hypothesis of ET was fueled by studies on an animal model of action tremor induced by injection of harmaline into rats (Louis & Lenka, 2017). Harmaline primarily targets the ION where it enhances gap-junction mediated synchronous firing of olivary neurons. Overactivity in the ION could then induce rhythmical burst firing in cerebellar Purkinje cells (PCs) and cerebellar nuclei through excitatory climbing fiber (CF) input, leading to tremor by propagation through cerebello-thalamo-cortical output channels. There remain many problematic issues with the harmaline model as an accurate representation of human ET, as (1) it is an acute model of action tremor, lacking the chronicity of ET, (2) it does not respond to all the same pharmacological agents as ET, and (3) harmaline injection in rats causes massive destruction of PCs in parasagittal bands, a morphologic feature not seen in human autopsies of ET brains (Louis & Lenka, 2017). Furthermore, there are neurons with pacemaking properties in several other brain regions, including PCs, cerebellar nuclei, thalamus and locus coeruleus, among others. More recent studies in ET have not shown any consistent neuroimaging (Wills, Jenkins, Thompson, Findley, & Brooks, 1994) or pathological abnormalities in the ION (Louis, Babij, Cortés, Vonsattel, & Faust, 2013), raising issues with this single ION oscillator model. Pathology case studies of ET patients that developed olivary hypertrophy (Elkouzi, Kattah, & Elble, 2016) or olivary degeneration (Louis et al., 2018) demonstrated no change in their tremor characteristics, arguing against the hypothesis that the ION alone plays a crucial role in genesis of tremor in ET. Considering these findings, the focus of the electrical model of ET shifted to encompass network properties. It is proposed there is an oscillatory network comprising the ION, cerebellum, thalamus and other brainstem regions, and where all network components may act as oscillators that dynamically entrain each other (Helmich, Toni, Deuschl, & Bloem, 2013; Raethjen & Deuschl, 2012). However, this model is largely a theoretic construct with little empirical support.

Since 2003, increasing support for the neurodegenerative hypothesis of ET has grown through consideration of the clinical features of ET and its similarities with other degenerative disorders, advanced neuroimaging techniques demonstrating structural changes consistent with neuronal loss, and an increasing number of pathological studies demonstrating degenerative changes centered in the cerebellum in ET. Indeed, the idea that ET could be neurodegenerative was proposed long ago. In the landmark paper by Critchley and Greenfield, the authors wrote: "Although anatomical proof is as yet lacking, there are at least a number of clinical points to make question whether "essential tremor" may not, at times any rate, represent an incomplete or a premature variant of one of the cerebellar atrophies" (Critchley & Greenfield, 1948). In this chapter, we review this growing body of evidence from these three perspectives, which support the conclusion that ET is a degenerative disorder. We present evidence that these findings in ET fulfill central tenets for the definition of a degenerative disease: a *chronic* and *progressive* disorder, often with an *insidious onset*, with an intrinsic *selective vulnerability* of specific cell populations leading to *cell loss* and *loss of structure and function* in the nervous system.

2. Clinical features of ET as a degenerative disorder

ET is increasingly recognized as a clinically heterogeneous condition, with differences in age of onset, anatomic distribution of tremor, rate of disease progression and other degenerative disease associations. Many of these clinical features suggest that ET is a degenerative disorder (Table 1).

First, ET has an insidious onset (Critchley, 1949; Larsson & Sjogren, 1960). Age of symptom onset can be difficult to precisely recall, and individuals with ET may initially dismiss the tremor as just "nervousness." In contrast, electric disorders such as seizures, often have an abrupt onset.

As in many degenerative diseases, the prevalence of ET increases markedly with age and, furthermore, increases in an exponential manner with advanced age. While prevalence estimates for ET differ across studies due to a variety of differences in methodology and population demographics, reproducible patterns of increasing prevalence with age are still obtained. In the largest meta-analysis to date, based on 42 studies across 6 continents and 23 countries reported between 1960 and 2020, the pooled prevalence of

Table 1 Clinical features suggesting that ET is neurodegenerative.

Evidence	**Comment**
Insidious onset	Tremor may initially be dismissed as just "nervousness" due to slow onset
Prevalence increases with age, and exponentially with advanced age	Prevalence across all ages ~1.33%, increases to 5.79% in individuals ≥65 years, over 20% in oldest old (≥95 years)
Older age of onset associated with more rapid disease progression	ET associated with aging processes
Somatotopic spread of tremor to cranium (neck, voice, jaw) with disease progression	Suggests progressive destruction of different brain regions, as seen in AD and PD
Intention tremor, gait ataxia, oculomotor abnormalities	Clinical resemblance to other cerebellar atrophies, suggesting shared disease mechanisms
Isolated rest tremor in arm only with longstanding disease	Pathological involvement of extra-cerebellar circuits or abnormal interactions with basal ganglia
Associated non-motor features Cognitive deficits Psychiatric features (depression, anxiety, social phobia) Mild olfactory dysfunction, hearing impairment, sleep dysregulation	Suggests more widespread involvement of different functional brain regions Increased prevalence or incidence of MCI and dementia in ET Many neurodegenerative diseases are neuropsychiatric disorders. Depression in ET may be primary Increased in some ET patients. Not specific to neurodegeneration
Increased risk of AD	Increased conversion from MCI to AD-type dementia. Late onset-ET increases risk of developing AD by ~2-fold
Increased risk of PD	Four- to fivefold increased risk of developing PD. Suggests overlap in disease mechanisms
Hereditability factors lead to earlier onset disease	Similar to that seen in other neurodegenerative diseases

AD, Alzheimer's disease; ET, essential tremor; MCI, mild cognitive impairment; PD, Parkinson's disease.

ET (all ages) was 1.33% (95% CI = 0.88%–2.02%) (Louis & McCreary, 2021). In individuals age ≥ 65 years, prevalence increases to 5.79% (95% CI = 4.14%–8.05%) and is even higher in oldest age groups (80s, 90s and older) ranging from 1.2% to 42.9% (mean = 11.4%, median = 9.3%).

ET prevalence increases by 74% for every decade increase in age ($P < 0.00001$) (Louis & McCreary, 2021). In another meta-analysis based on 29 studies reported between 2000 and 2019, the estimated global prevalence of ET was 0.32% (95% CI = 0.12%–0.91%) (Song et al., 2021). The prevalence of ET increased dramatically with advancing age, where the prevalence estimate in people aged under 20 years was 0.04% (95% confidence interval = 0.00%–0.29%) and that in elderly aged 80 years and above was 2.87% (95% confidence interval = 1.07%–7.49%). Thus, aging is a clear risk factor for ET, and similar exponential increases in prevalence are found in patients with Parkinson's disease (PD), Alzheimer's disease (AD), and other degenerative diseases of the nervous system (Louis, 2019).

ET is a *chronic* disorder with a gradual yet *progressive* clinical course that proceeds over many years (Louis, Ford, & Barnes, 2000; Putzke, Whaley, Baba, Wszolek, & Uitti, 2006). A few longitudinal studies have attempted to quantify the clinical rate of tremor progression from baseline in ET patients, estimated as 12% per year (Putzke et al., 2006) or a 3.1%–5.3% average annual increase in tremor severity (Louis, Agnew, Gillman, Gerbin, & Viner, 2011). In a prospective study of ET patients, disease duration and age were independently associated with tremor severity in ET (Louis, Jurewicz, & Watner, 2003). These findings raise the possibilities that reported increases in tremor severity in ET may be related not only to inherent worsening of disease pathophysiology with increase duration but also with age and age-related processes such as neuronal attrition. In a large clinical cohort of 348 ET cases, older age of onset was associated with more rapid disease progression, indicating an intimate relationship between disease progression and aging (Louis, Faust, Vonsattel, Honig, Henchcliffe, et al., 2009a). With disease progression in ET, there is a somatotopic spread to other body parts, including head (neck), voice and jaws (Louis, 2021), which may reflect spreading of disease to different brain regions as seen in other degenerative disorders. Last, once the disease begins there are no reported cases of spontaneous remission of ET, which might occur if the pathophysiology of ET was based on a central oscillating pacemaker.

A wide spectrum of cerebellar symptoms may be observed with disease progression, supporting the notion that the cerebellum might be centrally involved in ET. These include intentional tremor, gait ataxia, dysarthria, oculomotor abnormalities, and deficits in hand-eye coordination (Benito-León & Labiano-Fontcuberta, 2016; Louis & Faust, 2020a, 2020b). For instance, as many as 50% of patients with ET have intention tremor, which may involve arms as well as head and legs, and this is usually more evident in advanced ET cases. Similarly, gait disturbances such as ataxia typically occur with longer disease duration in ET, and those patients with head tremors have the greatest gait impairment. These ET-related gait abnormalities may reflect onset of disease in the medial cerebellum that controls balance and locomotion, consistent with prominent atrophy of the cerebellar vermis in a magnetic resonance imaging (MRI) study (Quattrone et al., 2008) and greater cerebellar pathology in a postmortem study of ET vermis (Louis et al., 2011) among patients with cranial tremors. A variety of oculomotor abnormalities occur in ET, and they are particularly evident among those ET patients also exhibiting intention tremor (Helmchen et al., 2003). In sum, these other cerebellar abnormalities in ET resemble that seen in cerebellar atrophies, suggesting the possibility of shared disease mechanisms (Louis & Faust, 2020a, 2020b).

Rest tremor involving only the arms may occur in as many as 20%–30% of ET patients, associated with a kinetic tremor that is more severe, more disseminated and of longer duration (Cohen, Pullman, Jurewicz, Watner, & Louis, 2003). In a postmortem analysis of 9 ET cases with rest tremor, Lewy body pathology was not present in the basal ganglia (Louis et al., 2011). Rest tremor in PD, and more broadly in other disorders (e.g., ET, dystonia), is postulated to be induced by abnormal basal ganglia activity, generated by the thalamus, and modulated or reinforced by the cerebellum (Duval, Daneault, Hutchison, & Sadikot, 2016). Thus, both ET and PD have circuits involving cerebellar-thalamo-cortical networks associated with rest and kinetic tremor generation.

As with many neurodegenerative movement disorders (e.g., PD, Huntington's disease), the clinical features in ET extend beyond the motor system. These non-motor features may affect several distinct domains, including cognitive, psychiatric (depression, anxiety, social phobia), sensory (olfactory dysfunction, hearing impairment) and sleep disturbances (Louis, 2016, 2021). Many studies have demonstrated cognitive deficits in ET, ranging from mild cognitive impairment to frank dementia, and a greater rate of cognitive decline occurs in older ET patients, all more than seen in age-matched controls (Benito-León, 2014; Radler et al., 2020). Some of this impairment might reflect degenerative pathology in the cerebral cortex (Farrell et al., 2019; Pan et al., 2014) although there are some biochemical distinctions in neocortical tau isoforms in ET versus other tauopathies with an increase in tau without N-terminal inserts (Kim et al., 2021). A cerebellar contribution is also possible, as a cerebellar cognitive affective syndrome is well-known to occur and includes hallmark deficits in executive function (Janicki, Cosentino, & Louis, 2013). Mild olfactory dysfunction occurs in a subset of ET patients, and does not correlate with disease duration or severity, indicating it may occur early in the disease process. Furthermore, emerging evidence suggests that some of these features (e.g., cognition, depression) could be primary and may even pre-date motor features in ET (Benito-León, 2014; Louis, 2016).

ET itself has been associated with other degenerative disorders. There is a longstanding relationship between ET and PD from a clinical, epidemiologic, and genetic perspective (Clark & Louis, 2018; Thenganatt & Jankovic, 2016), including a four- to fivefold increased risk of developing PD in ET patients (Benito-León, 2014). In older onset ET, the risk of developing AD is increased nearly twofold (Louis, 2019). Family history is also a strong risk factor for ET. The disease is familial in as many as 30%–70% of ET patients and this genetic predisposition lowers age of disease onset (Clark & Louis, 2018; Louis, 2019), as seen in other degenerative disorders with a hereditary basis. These various associations suggest that ET may share pathogenic mechanisms with these disorders.

While many of the clinical features discussed in this section are not specific to degenerative diseases when occurring in isolation, the constellation of findings, all present in the same disease, is more compelling (Table 1).

3. Neuroimaging findings suggest cerebellar degeneration in ET

Neuroimaging studies provide an in vivo modality to localize disease related changes in the brain. Several recent reviews have compiled the ever-increasing number of neuroimaging studies in ET that use a broad array of methods, including magnetic resonance (MR) volumetry, MR spectroscopy, diffusion tensor imaging (DTI), functional MR imaging (fMRI), other MR imaging, and positron emission tomography (PET) (Cerasa & Quattrone, 2016; Louis, Huang, Dyke, Long, & Dydak, 2014; Pietracupa, Bologna, Tommasin, Berardelli, & Pantano, 2021). Several fMRI studies and a considerable number of PET studies have identified neuronal activity changes in the cerebellar-thalamic network in ET, clearly demonstrating involvement of these brain regions for tremor generation. As fMRI and PET do not necessarily assess neuronal loss or degeneration, they will not be discussed further here. Although there is still significant variability in results across these various neuroimaging studies, overall, the combined structural data provide the greatest support for a cerebellar neuronal degeneration in ET (Table 2).

Table 2 Neuroimaging studies in ET demonstrating structural cerebellar changes.

Method	Authors (year)	# Subjects	Main findings in ET patients versus controls
MRI volumetry	Quattrone et al. (2008)	50 ET, 32 controls	Marked atrophy of cerebellar vermis in ET patients with head tremor ($n = 20$)
	Cerasa et al. (2009)	46 ET, 28 controls	Reduced cerebellar volume in ET patients with head tremor ($n = 19$)
	Benito-León et al. (2009)	19 ET, 20 controls	Gray and white matter losses in bilateral cerebellum and other regions (parietal lobes; right frontal, insular, limbic lobes; left medulla)
	Bagepally et al. (2012)	20 ET, 17 controls	Gray matter losses in bilateral cerebellum, vermis, bilateral frontal and occipital lobes
	Bhalsing et al. (2014)	25 ET, 25 controls	Gray matter loss in anterior and posterior cerebellar lobes, right medial frontal gyrus in ET patients with cognitive impairment
	Choi et al. (2015)	45 ET, 45 PD, 45 controls	Reduced cerebellar volume in ET patients with head tremor (n = 19); no change in PD versus controls
	Shin et al. (2016)	39 ET, 36 controls	Reduced cerebellar volume in several vermal areas, most evident in ET patients with (n = 20) versus without (n = 19) cerebellar signs
	Dyke, Cameron, Hernandez, Dydak, and Louis (2017)	47 ET, 36 controls	Decreased cerebellar volume in the vermis and several cerebellar lobules in ET patients with cranial tremors
DTI	Shin et al. (2016)	10 ET, 8 controls	Reduced FA in bilateral cerebellum as well as pons, midbrain and cerebral cortex (orbitofrontal, lateral frontal, parietal, temporal)
	Nicoletti et al. (2010)	25 ET, 15 PD, 15 controls	Reduced FA in dentate nucleus and superior cerebellar peduncle correlating with ET disease duration; not seen in PD patients
	Klein et al. (2011)	14 ET, 20 controls	Reduced FA in right inferior cerebellar peduncle; Increased MD of inferior cerebellar peduncle bilaterally and left parietal white matter

Continued

Table 2 Neuroimaging studies in ET demonstrating structural cerebellar changes.—cont'd

Method	Authors (year)	# Subjects	Main findings in ET patients versus controls
	Jia, Jia-Lin, Qin, Qing, and Yan (2011)	15 ET, 15 controls	Increased diffusion values in red nucleus; only limited regions examined that did not include cerebellum
	Saini et al. (2012)	20 ET, 17 controls	Increased diffusivity in cerebellar hemispheric white matter, thalamus, brainstem and bilateral cerebral hemispheres
	Novellino et al. (2016)	67 ET, 39 controls	Increased diffusivity in cerebellum
	Pietracupa et al. (2019)	19 ET, 15 controls	White abnormalities in corticospinal tract, cerebellar peduncles, corpus callosum, several associative white matter bundles
MRS	Louis et al. (2002)	16 ET, 11 controls	Reduced mean cerebellar cortical NAA/Cr ratio in ET patients
	Pagan, Butman, Dambrosia, and Hallett (2003)	10 ET, 10 controls	Reduced NAA/Cr ratio in bilateral cerebellar hemispheres
Other	Novellino et al. (2013)	24 ET, 25 controls	Whole brain voxel-based analysis consistent with increased iron deposits in globus pallidus, substantia nigra, right dentate nucleus
PET	Boecker et al. (2010)	8 ET, 11 controls	Increased cerebellar dentate, ventrolateral thalamus and premotor cerebral cortex uptake of ^{11}C-flumazenil
	Gironell et al. (2012)	10 ET, no controls	Significant correlation of ^{11}C-flumazenil uptake and tremor rating scales

CR, creatinine; DTI, diffusion tensor imaging; ET, essential tremor; FA, fractional anisotropy; MRI, magnetic resonance imaging; MRS, magnetic resonance spectroscopy; NAA, N-actetylaspartic acid; PD, Parkinson's disease; PET, positron emission tomography.

In a comprehensive review published in 2014 (Louis et al., 2014), four of six studies using voxel-based morphometry (i.e., MR volumetry) reported a reduction in cerebellar volume in ET (Bagepally et al., 2012; Benito-León et al., 2009;Cerasa et al., 2009; Quattrone et al., 2008), and this was particularly prominent in patients with head tremor in two studies (Cerasa et al., 2009; Quattrone et al., 2008). In another recent review, cerebellar gray matter volume reductions were the most common structural finding in ET, seen in 7 of 12 studies that examined the cerebellum (Pietracupa et al., 2021). Two additional MR volumetry studies confirmed greater cerebellar volume losses in ET patients with head tremors, either more globally present and distinguishing patients with ET from those with PD (Choi et al., 2015) or present within vermis and particular cerebellar lobules (Dyke et al., 2017). ET patients with clinical cerebellar signs (e.g., ataxia) had atrophy in several contiguous regions of the cerebellar vermis versus those lacking cerebellar signs (Shin et al., 2016). Last, ET patients with cognitive impairment had gray matter loss in anterior and posterior lobes of the cerebellum, as well as medial frontal gyrus, in comparison to controls (Bhalsing et al., 2014); this is of interest as posterior cerebellar lobes are implicated in high-level cognitive functions.

In addition to gray matter volume reductions, most studies using diffusion imaging have demonstrated microstructural changes in cerebellar white matter between ET cases and controls. Five of seven studies reported as of 2014 demonstrated differences between ET cases and controls (Table 2), and the two studies with null results had important methodologic limitations (Louis, Lee, Babij, et al., 2014). A subsequent study also showed diffusion changes within the cerebellum (Novellino et al., 2016). More widespread white matter changes have been detected in cerebellar peduncles as well as corticospinal tracts, corpus callosum and several associative white matter bundles (Pietracupa et al., 2019). In sum, these diffusion imaging studies demonstrate that some orientation dependent aspect of tissue microstructure is abnormal in ET. The bulk of studies show evidence of axonal changes within the cerebellum itself along with abnormalities in other brain regions, many of which are part of the cerebellum's tremor outflow pathways.

Some diffusion imaging studies have highlighted a relationship between areas of cerebellar degeneration and specific disease features in ET. Increased diffusion changes in the dentate nucleus region completely distinguished familial ET patients from PD patients and controls, and patients with longer ET disease duration had more severe degeneration of white matter within/around the dentate (Nicoletti et al., 2010). In patients with late-onset compared to early-onset ET, more degenerative changes were seen in the superior cerebellar peduncle, which may relate to faster rate of tremor progression in late-onset ET (Louis, Faust, Vonsattel, Honig, Henchcliffe, et al., 2009a; Saini et al., 2012).

MR spectroscopy will detect in vivo concentrations of certain cellular metabolites such as *N*-acetyl aspartate (NAA), creatine (Cr) or choline (Cho). NAA is a marker of neuronal integrity, thus reduced NAA is suggestive of processes that cause neuronal damage, neuronal dysfunction, and neuronal loss. There are two currently available MRS studies that have examined the cerebellum and provide evidence of cerebellar involvement in ET based on reduced NAA (Louis et al., 2002; Pagan et al., 2003). The reduced NAA/cr ratio inversely correlated with tremor severity in ET (Louis et al., 2002).

Brain iron accumulation is a common phenomenon in neurodegenerative processes, as previously demonstrated in AD, PD and other degenerative disorders. In a study of 24 ET compared with 25 controls, a whole brain voxel-based analysis demonstrated significant differences consistent with increased iron deposition in bilateral globus pallidus, substantia nigra and right dentate nucleus (Novellino et al., 2013). This study provides indirect evidence supportive of degeneration. Iron accumulation has yet to be studied in the postmortem ET brain.

There is considerable evidence that defects in the GABAergic system underly disease pathogenesis in the ET brain (Gironell, 2014). For instance, the majority of drugs used to treat ET act via GABAergic pathways. One ET case-control PET study examined ^{11}C-flumazenil binding to the $GABA_A$ complex, demonstrating increased ligand uptake in cerebellar dentate, ventrolateral thalamus and premotor cerebral cortex in ET patients, raising the possibility of a receptor upregulation due to either GABA deficiency or abnormal/dysfunctional $GABA_A$ receptors in ET (Boecker et al., 2010). Further evidence for GABAergic dysfunction in ET was provided by tissue-based studies of ET cerebellum demonstrating a localized GABAergic defect due to cerebellar PC loss and reduced GABA receptors in cerebellar dentate nucleus (Louis & Faust, 2020c; Paris-Robidas, Brochu, Sintes, et al., 2012), supporting the presence of cerebellar degeneration in ET. A subsequent study in ET patients only demonstrated correlation between ^{11}C-flumazenil uptake and tremor rating scales, again suggesting an abnormality at the level of $GABA_A$ receptors in ET (Gironell et al., 2012).

In summary, neuroimaging studies in ET have provided convincing evidence that the cerebellum plays a key role in disease pathophysiology, with many of these studies supporting a degenerative process. More widespread changes are also seen in other brain regions, many of which are part of the cerebellum's tremor outflow pathway. Some of the changes outside the cerebellum may correlate with cognitive impairments in ET, as detailed elsewhere (Pietracupa et al., 2021). Distinct challenges remain for these neuroimaging methods, particularly with respect to issues of both spatial and temporal resolution and to fully capture the clinical heterogeneity of ET.

4. Pathological findings in ET demonstrate cerebellar degeneration

A fundamental question about the degenerative nature of a neurological disorder is whether there are histopathological changes in the brain that provide evidence of neuronal loss. In this section we discuss the expanding literature demonstrating that there is indeed *selective vulnerability* of specific cell populations (e.g., cerebellar PCs) leading to *neuronal loss* and *loss of structure and function* in the nervous system.

Considerable accumulating evidence now suggests that ET is a disease of cerebellar degeneration. A chronology of major pathological findings identified in the ET cerebellum since the turn of the 21st century is listed in Table 3, with many studies that now include modern methodologies as well as comparisons with both neurologically normal controls and various degenerative disease controls. These studies summarize growing evidence that there are indeed structural abnormalities centered on the cerebellar PC along with altered PC synaptic contacts by neighboring basket neurons and olivary CFs in cerebellar cortex. Indeed, it has been proposed that ET may be a "Purkinjopathy" (Grimaldi & Manto, 2013), where dysfunction of PCs results in dis-inhibition of the cerebellar nuclei, which then impacts functioning of cerebello-thalamo-cortical pathways and cerebellar-brainstem loops. In this section, a "best guess" anatomical-physiologic model is presented, based on findings identified cross-sectionally at the time of death. This model places the pathological findings in ET cerebellum within a putative cascade of "early," "middle" and "late" changes (Louis & Faust, 2020c), and highlights how these changes provide support for a degenerative disease process (Fig. 1).

Table 3 Chronology of Neuropathology studies in ET during the 21st century.

Authors (year)	# Subjects	Main findings
Rajput et al. (2004)	20 ET, no controls	Mild PC loss in 2 of 20 ET cases. *Qualitative methods only*
Ross et al. (2004)	11 ET, 11 control	Some ET cases with more pathology in brainstem and cerebellum. *Qualitative methods only*
Louis et al. (2006)	10 ET, 12 control	Two disease patterns in ET: increased torpedoes and Bergmann glia (4/10), brainstem Lewy bodies often with atypical distribution (6/10)
Louis et al. (2007)	33 ET, 21 control	PC loss (25%, $P<0.01$), $\sim 7\times$ increase in torpedoes ($P<0.001$). Qualitative dentate degeneration in 2 cases. Lewy bodies in 8 ET cases, which did not have significant cerebellar changes
Axelrad et al. (2008)	14 ET, 11 control	PC loss as high as 38%. PC loss correlated with age and number of torpedoes
Shill et al. (2008)	26 ET, 21 control	Increased cerebellar Bergmann gliosis and PC loss in 7/26 ET cases. *Qualitative methods only*
Louis, Faust, Vonsattel, Honig, Henchcliffe, et al. (2009a)	9 ET, no controls	Older age of ET onset associated with more degenerative pathology (PC loss, torpedoes) and more rapid disease progression
Louis, Faust, Vonsattel, Honig, Rajput, et al. (2009b)	40 ET, 25 control, 21 AD, 14 PD/DLBD	Increased torpedoes in ET versus controls, intermediate number of torpedoes in AD and PD
Louis, Yi, Erickson-Davis, Vonsattel, and Faust (2009)	4 ET, no controls	Torpedoes contain disorganized neurofilaments and organelles in unmyelinated PC axon segment
Erickson-Davis et al. (2010)	37 ET, 21 control	Structural changes in basket cell-PC contacts ("Hairy Baskets"); correlate with torpedo and PC loss pathology
Louis, Asabere, et al. (2011)	9 ET, no controls	PC loss and increased torpedoes, but no Lewy body pathology in basal ganglia, in ET with isolated rest tremor
Louis, Faust, et al. (2011)	24 ET, 10 controls	Increase torpedoes in ET vermis versus controls (2.7-fold), particularly in ET with cranial tremors

Kuo et al. (2011)	35 ET, 21 control, 11 PSP	Increased heterotopic PCs in ET cerebellar molecular layer, not seen in controls or PSP cases
Kuo, Faust, Vonsattel, Ma, and Louis (2011)	20 ET, 19 control	Similar parallel fiber counts and density in ET and controls
Rajput, Robinson, Rajput, and Rajput (2011)	7 ET, 2 control, 6 PD	No significant difference in PC counts among ET, PD-tremor and control groups. *Underpowered study, type II statistical error*
Yu et al. (2012)	20 ET, 19 control	Increased number of PC dendritic swellings in ET
Paris-Robidas et al. (2012)	10 ET, 16 control, 10 PD	Reduction in GABA receptors in the dentate nucleus in ET, not in PD
Louis, Babij, Cortés, et al. (2013)	14 ET, 15 control	No morphological differences in inferior olivary nucleus in ET
Kuo et al. (2013)	11 ET, 12 control	Increased Lingo-1 expression in ET cerebellum and elongated basket cell pinceau
Babij et al. (2013)	49 ET, 39 control	Abnormal PC axonal anatomy in ET with changes in shape and connectivity
Louis, Babij, Lee, Cortés, and Vonsattel (2013)	32 ET, 16 control	Reduced PC linear density in 100 micrometer thick sections, inversely correlated with torpedo counts
Louis, Lee, Cortés, Vonsattel, and Faust (2014)	25 ET, no controls	Clinical laterality or symmetry of tremor correlates with laterality of axonal pathological changes
Lin et al. (2014)	12 ET, 13 control	Abnormal climbing fiber-PC synaptic connections in ET
Louis, Lee, Babij, et al. (2014)	27 ET, 27 control	Reduced PC dendritic arborization and loss of dendritic spines in ET using Golgi silver staining
Lee et al. (2014)	16 ET, 13 control	Decreased EAAT2 expression in ET cerebellar cortex

Continued

Table 3 Chronology of Neuropathology studies in ET during the 21st century.—cont'd

Authors (year)	# Subjects	Main findings
Delay et al. (2014)	9 ET, 16 control, 10 PD	Increased LINGO-1 expression in ET cerebellum, not seen in PD
Symanski et al. (2014)	56 ET, 62 control	ET is not associated with cerebellar PC loss. *Potential problems with study design*
Louis, Lin, Faust, Koeppen, and Kuo (2015)	37 ET, no controls	Climbing fiber synaptic changes correlate with clinical features in ET
Béliveau et al. (2015)	9 ET, 16 control, 10 PD	Insoluble Aβ42 accumulates in the cerebellar cortex of ET, but not in age-matched PD
Choe et al. (2016)	50 ET, 25 control	PC loss in ET, assessed by random sampling method and nearest neighbor analysis
Kuo et al. (2016)	39 ET, no controls	DBS treatment changes pattern of abnormal CF-PC connections in ET cases
Wang et al. (2016)	61 ET, 25 control	Increased EAAT2 expression in ET dentate nucleus, decreased expression ET cerebellar cortex
Kuo et al. (2017)	20 ET, 25 control, 30 disease controls	Degenerative movement disorders differ in CF-PC synaptic pathology, extent of PC loss and torpedo formation
Kuo et al. (2017)	60 ET, 30 control	Early onset and late onset ET cases share similar cerebellar postmortem features
Louis, Kuo, Tate, Kelly, and Faust (2017)	60 ET, 30 control	Familial and sporadic ET cases share similar cerebellar postmortem features
Louis et al. (2018)	43 ET, 22 control, 31 SCAs	Heterotopic PCs and PC loss in ET are along spectrum of changes seen in SCAs

Lee, Gan, Faust, Louis, and Kuo (2018)	60 ET, 30 control	Similar climbing fiber-PC synaptic changes across clinical subtypes of ET
Lee et al. (2019)	50 ET, 25 control, 27 SCA, 25 PD	"Empty baskets" provides an indirect measure of PC loss, correlates with PC loss and torpedoes
Louis et al. (2019)	50 ET, 25 control, 81 disease controls	Distinctive patterns of pathologic changes in ET compared to primary cerebellar degenerations and other disorders involving cerebellum
Pan et al. (2020)	15 ET, 15 control	75% reduction in cerebellar GluRδ2 protein levels, inverse correlation with metrics of CF rewiring
Hartstone et al. (2021)	25 ET, 25 control	No difference in dentate nucleus neuronal density or total neuronal number in ET versus control
Wu et al. (2021)	15 ET, 15 control	Increased climbing fiber-PC crossings in ET, correlates with tremor severity
Gionco et al. (2021)	50 ET, no controls	No pathological differences in cerebellar cortex between ET versus "ET-plus" cases

Aβ, amyloid beta; AD, Alzheimer's disease; DBS, deep brain stimulation; EAAT2, excitatory amino acid transporter type 2; ET, essential tremor; GABA, gamma-amino--butyric acid; GluRδ2, glutamate receptor delta 2; PC, Purkinje cell; PD/DLBD, Parkinson's disease/Diffuse Lewy Body disease; PSP, progressive supranuclear palsy; SCA, spinocerebellar ataxia.

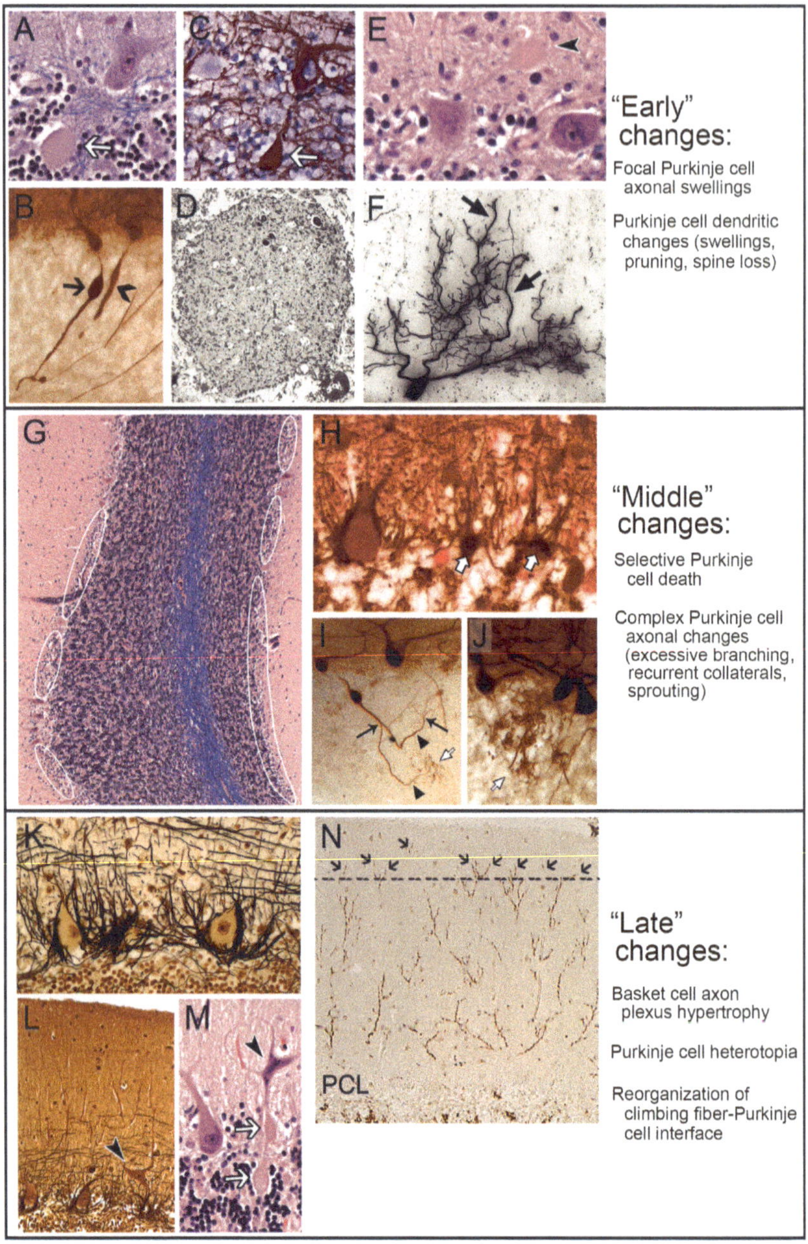

Fig. 1 Pathological changes in ET cerebellum. A putative cascade of early (A–F), middle (G–J) and late (K–N) changes is presented. *Early changes:* (A–D) Focal PC axonal swellings ("torpedoes") identified in LH&E stain (A, arrow), calbindinD28k immunostain (B, arrow), phosphorylated-neurofilament immunostain (C, arrow), and electron microscopy (D). There is abnormal accumulation of phosphorylated-neurofilaments in the cell body of a PC harboring a torpedo (C). In addition to rounded swellings, PC axons may be diffusely thickened (B, arrowhead). Dendritic abnormalities include focal dendrite

4.1 Putative "early" events: Purkinje cell axonal swellings and dendritic changes

Pathological events that indicate stress and underlying compromise in PCs are likely to occur early in the disease, prior to attempted compensatory responses or cell death in susceptible cells (Fig. 1A–F). This involves both axonal and dendritic compartments of PCs in ET.

4.1.1 Purkinje cell axonal torpedoes

One of the first pathological changes identified in cerebellum of postmortem ET cases was an increased number of PC torpedoes, which are round or ovoid swellings of the proximal portion of the PC axon (Louis et al., 2006, 2007; Louis, Vonsattel, Honig, Ross, et al., 2006) (Fig. 1A). Subsequent studies have demonstrated a 2- to 12-fold increase in number of torpedoes in ET versus age-matched controls, in both the motor region of cerebellar cortex (Babij et al., 2013; Kuo, Wang, et al., 2017; Louis et al., 2007, 2017; Louis, Kuo, Tate, et al., 2017) and cerebellar vermis (Louis, Faust, et al., 2011). Torpedoes are also increased in ET over that seen in other neurodegenerative disorders such as PD, AD, and dystonia (Louis et al., 2019; Louis, Faust, Vonsattel, Honig, Rajput, et al., 2009b). Ultrastructurally, torpedoes contain massive accumulations of disorganized neurofilaments admixed with mitochondria and smooth endoplasmic reticulum and occur along unmyelinated segments of the PC axon (Fig. 1D) (Louis, Yi, Erickson-Davis, Vonsattel, & Faust, 2009). A quantitative Western blot analysis in postmortem cerebellar cortex showed significantly higher levels of both phosphorylated and non-phosphorylated forms of neurofilament heavy chain subunit in ET compared to controls, suggesting these molecular differences could underly torpedo formation in ET

swellings (E, arrowhead, LH&E stain) and excessive pruning of PC dendrites (F, arrows, Golgi-Kopsh stain). *Middle changes:* Decreased PC linear density and Bergmann gliosis in LH&E stain (G, ovals). PC loss detected in calbindinD28K-GAD dual immunostain by increased empty baskets (H, arrows). Complex PC axonal changes revealed by calbindinD28k immunostain, including excessive branching (I, black arrows), recurrent collaterals (I, arrowheads) and terminal axonal sprouting (J, K, white arrows). *Late changes:* Basket cell axonal plexus hypertrophy (K, Bielschowsky stain). Heterotopic PCs (L, M, arrowheads) typically lack a surrounding basket cell axon plexus (L, Bielschowsky stain) and may have torpedoes along their axon (M, arrows, LH&E stain). Abnormal CF-PC synapses (N, arrows, VGlut2 immunostain) in the outer molecular layer (demarcated by dotted line). GAD, glutamic acid decarboxylase; LH&E, luxol fast blue-hematoxylin & eosin stain.

(Louis et al., 2012). Abnormal phosphorylation of neuronal cytoskeletal proteins is one of the major pathological hallmarks of degenerative disorders such as AD, PD and amyotrophic lateral sclerosis (Petzold, 2005).

Torpedoes are not specific to ET, being seen in a wide range of conditions with cerebellar degeneration, such as spinocerebellar ataxia (SCA), multiple system atrophy, progressive supranuclear palsy and paraneoplastic cerebellar ataxia (Louis et al., 2019; Louis, Kuo, Vonsattel, & Faust, 2014). They do not prominently accumulate with normal advanced aging (Louis, Faust, Vonsattel, & Erickson-Davis, 2009). There is general agreement that torpedo formation is an injury associated alteration in PCs with resultant inhibition of both anterograde and retrograde axonal transport, consistent with the presence of phosphorylated neurofilaments in the cell body of PCs harboring a torpedo (Fig. 1B; Louis et al., 2012). This axonal injury might ultimately lead to "neuronal strangulation" (Louis & Faust, 2020c). While it remains unclear whether torpedoes represent a preterminal cellular response or a regenerative response, the robust inverse correlation between number of torpedoes and PC counts in ET and other cerebellar diseases suggests that torpedoes are part of a degenerative cascade (Axelrad et al., 2008; Choe et al., 2016; Louis, Kuo, Vonsattel, & Faust, 2014).

4.1.2 Purkinje cell dendritic changes

Dendritic abnormalities traditionally occur in degenerative diseases where they reflect derangements in cellular homeostatic processes that often precede neuronal death. An increased number of PC focal dendritic swellings in ET compared to control brains has been demonstrated in two studies (Louis et al., 2019; Yu et al., 2012) (Fig. 1E). The robust correlation between number of PC dendritic swellings and number of torpedoes in a brain ($r = 0.60$, P-0.005) (Yu et al., 2012) suggests a cascade of interconnected events in the degenerative process involving PCs.

Other changes in the PC dendritic compartment in ET also support a degenerative process. In a detailed quantitative study of PC dendrite arbors in 27 ET and 27 age-matched controls using the Golgi-Kopsh method, there were significant reductions in spine density (19% reduced, $P = 0.03$) and several measures of dendritic complexity (21%–47% reduced, $P = 0.001$–0.01) in ET (Louis, Lee, Babij, et al., 2014) (Fig. 1F). The dendritic arbor and spine density variables were all highly intercorrelated, indicating a system-wide set of regressive changes. In addition, tremor of greater severity was correlated with more pruning of the PC dendritic arbor. Regressive dendritic changes are not specific to ET, being seen in patients with hereditary ataxias (Shintaku & Kaneda, 2009) as well as chronic alcoholics (Ferrer, Fabregues, Pineda, Gracia, & Ribalta, 1984), but may be considered a marker of neurons in extremis. Structural perturbations of the PC dendrite, such as swellings and changes in complexity, would likely alter the interface with the multiple synaptic inputs to PCs. These are likely to have physiological implications.

4.2 Putative "middle" events: Complex PC axonal changes and selective PC death

Although symptomatic loss of function in degenerative diseases is frequently attributed to the loss of neurons, events before this neuronal loss might well result in disease manifestations. Following early cellular changes, middle events represent ongoing responses in PCs that lead to neuronal remodeling, and failed attempts leading to eventual PC death (Fig. 1G–I).

4.2.1 Remodeling of PC axons

The cellular response of PCs to injury is not typical of that seen in most CNS neurons, as they survive for prolonged periods after axonal injury despite the inability to regenerate their severed axon (Rossi, Gianola, & Corvetti, 2006). An initial formation of torpedoes and hypertrophy of recurrent collateral branches (which form connections with neighboring PCs) is followed by a vigorous inclination towards axonal sprouting along the intracortical segment of their axon. These PC axonal reorganizations might enable PCs to access trophic factors by establishing additional connections with other PCs or granule layer neurons in their attempt to preserve connections and prevent cell death (Carulli, Buffo, & Strata, 2004). However, the extent to which these plasticity-type changes in PC axons remain neuroprotective versus inducing further degeneration is not known.

Morphologic analysis of the PC axonal compartment was performed in 49 ET and 39 control brains using calbindinD28k immunohistochemistry in 100-μm thick cerebellar cortical tissue sections (Babij et al., 2013), enabling detailed visualization of the trajectory of PC axons as they traverse the granule cell layer. In addition to axonal torpedoes, other axonal alterations were all present to an increased degree in ET cases versus controls, including changes in axonal shape (diffusely thickened axonal profiles, 17% increase, $P = 0.006$) and axonal connectivity (axonal recurrent collaterals, 2.3× increase; axonal branching, 2.2× increase; terminal axonal sprouting, 2.5× increase; all $P < 0.001$) (Fig. 1C, G, H). These changes in axonal shape and connectivity were significantly intercorrelated and were three- to fivefold more frequently seen on axons of PCs with torpedoes, indicating a set of biologically related processes. These findings indicate there is a broader rewiring of PCs in ET, with potential for formation of aberrant synapses and abnormal circuits that might precede or coincide with neuronal loss.

In a subsequent clinicopathological study of ET patients, the extent of these complex PC axonal changes was quantified in relation to laterality or symmetry of tremor on neurological examination (Louis, Lee, Cortés, et al., 2014). Thus, if these pathological changes were tremor-producing, they should be more prominent in the cerebellar hemisphere ipsilateral to the arm with more severe action tremor. There was a high concordance between clinical and pathological features in 18 of 25 (72%) ET cases ($P = 0.007$). A positive correlation between the degree of clinical asymmetry and pathological asymmetry ($r = 0.52$, $P = 0.01$) was even more robust in the subset of ET patients that had the greatest clinical asymmetry ($r = 0.78$, $P = 0.39$). This study provides important construct validation that the pathological changes observed in the cerebellum in ET are of patho-mechanistic importance.

4.2.2 Purkinje cell loss

Many pathology series, but not all, have identified a significant decrease in PCs in ET compared to controls. The first postmortem study to examine PC loss in the largest number of ET specimens available at that time compared changes in 33 ET and 21 age-matched controls, demonstrating 25% lower mean PC counts per 100× microscopic field in all patients with ET ($P < 0.01$) and 31% lower in a subset of 25 ET cases that did not have Lewy bodies ($P < 0.01$) (Louis et al., 2007).

Among the current pathological findings in ET, the issue of PC loss remains controversial. Some of the earlier studies still lacked ET-control comparisons (Table 3) (Rajput et al., 2004) or did not use quantitative methods (Ross et al., 2004; Shill et al., 2008). Two subsequent studies did not reproduce the finding of PC loss in ET (Rajput et al., 2011; Symanski et al., 2014). As previously described in detail (Louis & Faust, 2020c), these studies were hampered by several methodologic limitations. A follow-up study on a new cohort of brains from 50 ET versus 25 age-matched controls used a more sophisticated random sampling scheme and a nearest neighbor analysis and replicated the finding of significant PC loss in individuals with ET (Choe et al., 2016). In this study, 94% of individuals with ET had a PC count below the central tendency (the mean for the norm) for the controls, and the mean distance between PC bodies was greater in ET, further consistent with PC loss.

A novel assay was subsequently developed to further examine the question of PC loss in ET by quantifying "empty baskets" (Lee et al., 2019). Basket cells are interneurons in the cerebellar molecular layer whose axons form a plexus around the PC soma. With PC loss, the basket plexus appears empty (Fig. 1J). In a study of 50 ET and 25 controls, the median percentage of empty baskets in ET patients was 1.5 times higher than controls (48.8% versus 33.5%, $P < 0.001$). Even higher percentage of empty baskets was found in 22 SCA cases, including SCA1 (59.7%, $P = 0.011$), SCA2 (77.5%, $P < 0.003$), and SCA6 (87.0%, $P < 0.001$). The presence of empty baskets, particularly in SCA cases where PC loss is often marked, indicates that the basket cell plexus persists even after the loss of PCs. Thus, quantification of empty baskets provides an indirect and alternative method to quantify PC loss. Furthermore, the strong correlation between the percentage of empty baskets, PC loss and torpedo counts suggests that they are all part of the same pathophysiological cascade.

In sum, these studies suggest that some PCs become overwhelmed in ET, resulting in death of these cells. The reason for the selective nature of PC loss is not known.

4.3 Putative "late" events: Further remodeling of basket cell plexus and PCs and reorganization of climbing fiber-PC synapses

Several pathological changes have been observed in ET that might be a response to PC death, including remodeling of neuronal populations that normally synapse with PCs (basket cells, CFs) or involving the PCs themselves (Fig. 1K–N).

4.3.1 Basket cell axonal remodeling

Basket cells are GABAergic inhibitory interneurons whose axonal processes form a pericellular basket around the PC body and in a "pinceau" structure around the PC axon initial segment (AIS). The density of the basket cell plexus around the PC soma on Bielschowsky-stained sections of neocerebellum was examined in 37 ET, 48 disease control brains (patients with AD, PD, progressive supranuclear palsy or dystonia) and 21 normal controls using a semiquantitative hypertrophy rating scale (Erickson-Davis et al., 2010). A dense and tangled appearance of the perisomal basket cell plexus (termed "hairy baskets") was seen in 25% of ET brains as compared with 4% of other brains ($P = 0.001$) (Fig. 1K). This was the first observation, subsequently confirmed in additional studies (Kuo, Wang, et al., 2017; Louis, Kuo, Tate, et al., 2017; Louis, Kuo, Wang, et al., 2017), indicating that structural changes in ET are not restricted to the PC but also involve their functional network. It is postulated that in the context of PC loss in ET, relative preservation of basket cells can lead to basket cell axon sprouting onto remaining PCs, resulting in increased process density and hypertrophic appearance of the basket cell plexus.

A second finding involved the basket cell plexus pinceau in ET. Genomic association studies in ET identified a sequence variation in the gene encoding leucine rich repeat and immunoglobulin-like domain-containing nogo receptor-interacting protein 1 (Lingo-1) (Clark & Louis, 2018). Lingo-1 immunohistochemistry labeled the basket cell pinceau, and ET cases had abnormally long pinceau processes targeting more distal segments of the PC axon over that seen in control brains (Kuo et al., 2013). Elongated Lingo-1-positive pinceau strongly correlated with number of PC axonal torpedoes and the basket cell hypertrophy rating, features more commonly seen in ET brains. This basket cell pinceau remodeling in ET could represent a disturbance in the AIS and neuronal polarity, as accumulations of nonphosphorylated neurofilaments in the PC AIS (Louis et al., 2012), and phosphorylated neurofilaments in PC soma (Erickson-Davis et al., 2010) have also been observed in ET cases (Fig. 1B).

Two studies have demonstrated increased cerebellar expression of Lingo-1 in ET versus control brains (Delay et al., 2014; Kuo et al., 2013). As Lingo-1 is a negative regulator of axonal regeneration and neurite outgrowth, its upregulation might be a response to the increased axonal sprouting and branching of PC axons observed in ET, and possibly in other neuronal populations yet to be explored. Lingo-1 upregulation is seen in several disorders, such as PD, multiple sclerosis and spinal cord injury (see Delay et al., 2014), and would be consistent with a degenerative process in ET cerebellum.

4.3.2 Heterotopic Purkinje cells

PCs are normally located as a monolayer between the molecular layer and granule cell layers of the cerebellar cortex. Heterotopic PCs, whose cell body is mislocalized in the molecular layer (Fig. 1L–M), were quantified in the ET cerebellum in two studies, comparing ET brains to both neurologically normal controls as well those from individuals with progressive supranuclear palsy (PSP) and SCAs (Kuo, Erickson-Davis, et al., 2011; Louis, Kuo, et al., 2018). The mean or median number of heterotopic PCs was twofold higher in ET than in normal controls ($P < 0.05$), and heterotopic PCs in ET were seen singly rather than in clusters, most consistent with a degenerative process (Yang et al., 2000). Heterotopic PCs were not increased in PSP brains but were even more marked in several of the SCAs. In this regard, ET could represent an intermediate or less advanced state of spinocerebellar ataxia (Louis et al., 2019; Louis & Faust, 2020b, 2020c). As the number of heterotopic PCs and number of PCs are inversely related in ET cerebellum (Louis, Diaz, et al., 2018; Louis, Kuo, et al., 2018), these heterotopic PCs may reflect an untethering associated with anatomical rearrangement in the molecular layer, perhaps due to the PC dendritic pruning and PC loss observed in ET. Combinations of pathological changes are also seen in heterotopic PCs, with a loss of the surrounding inhibitory basket cell axonal plexus and axonal torpedoes along their axon (Fig. 1L–M), further suggesting these PCs are functionally abnormal.

4.3.3 Reorganization of climbing fibers

CFs from olivary neurons provide one of the major excitatory inputs to PCs. These synaptic connections are essential for normal cerebellar-mediated motor control and are dynamically regulated by activity and disease states (Watanabe, 2008). Normally, CFs form synapses mainly on the thick, proximal PC dendrites in the inner portion of the molecular layer, whereas parallel fibers from granule cells form synapse in the outer molecular layer and on the thin, distal PC spiny branchlets. In a study of ET cerebellum versus age-matched controls, three main abnormalities in CF distribution were detected in ET, including decreased CF-PC synaptic density (33% reduction, $P< 0.001$), an abnormal extension of CFs to the outer portion of the molecular layer (1.9-fold increase, $P < 0.001$) and more CF-PC synapses on the thin PC distal branchlets (2.2-fold increase, $P < 0.003$) (Lin et al., 2014). The abnormal extension of CFs into domains normally occupied by parallel fiber synapses indicates a redistribution of synaptic wiring in the molecular layer in ET. Correlations between CF extension into the outer molecular layer, number of torpedoes and extent of PC loss in ET suggest that this change is a downstream response to PC loss (Lin et al., 2014). There was also an unexpected strong negative correlation between CF synaptic density on PC distal branchlets and total tremor scores in ET (Lin et al., 2014; Louis et al., 2015). Thus, the pattern of CF-PC innervation may change with disease severity in ET, where PC degeneration and/or persistent long-standing tremor may cause preferential pruning of distal PC branches and CFs, obscuring and/or reversing the pattern that is seen in mild ET.

Another important feature of CFs and PCs is a 1:1 pattern of innervation, where one CF innervates one PC (Kano, Watanabe, Uesaka, & Watanabe, 2018). Using a novel immunohistochemical staining technique that labeled a subset of cerebellar CFs, CFs in ET were found to cross a greater number of PC somas (19% increase, $P = 0.023$) and PC dendrites (39% increased, $P = 0.014$) than in control cases, and these increases in CF-PC crossings positively correlated with tremor severity (Wu et al., 2021). While the pathophysiologic implication of these changes is uncertain, it might indicate a rewiring, with increased synaptic contacts between CFs and PCs and/or potentially multiple PC innervations by CFs.

The redistribution of CF-PC synapses shows some specificity to ET, as this pathological change is enriched in ET cases compared to other cerebellar degenerations and movement disorders, including SCAs, multiple system atrophy, PD, and dystonia (Kuo, Lin, et al., 2017; Louis et al., 2019). They are consistently observed in ET with varying clinical features (Lee et al., 2018), suggesting this synaptic pathology, among others, might be a core feature of ET.

Similar redistribution of CF synapses is also seen in a mouse model with defective glutamate receptor δ2 (GluRδ2), a key protein controlling the territorial distribution of CF and parallel fiber synapses on PC dendrites (Pan et al., 2020). In an analysis of 15 ET and 15 controls, patients with ET had approximately 75% reduction in mean GluRδ2 protein expression level, and GluRδ2 levels inversely correlated with the percentage of CFs extending into parallel fiber synaptic territory (Pan et al., 2020). Interestingly, GluRδ2-deficient mice develop an ET-like tremor, and CF-PC synaptic pruning deficits drive excessive cerebellar oscillatory activity that is necessary for tremor generation (Pan et al., 2020). This study was also the first to use cerebellar electroencephalography to confirm the presence of excessive cerebellar oscillations in ET patients compared to controls, and that the extent of these oscillations correlate with clinical tremor scores. Thus, the combination of CF structural alterations identified in ET across these multiple pathological studies might result in entrainment of PC activity to a more synchronous and rhythmic mode, leading to tremor. It is also possible that the abnormal CF-PC synaptic connections in ET could represent a compensatory change, resulting from longstanding rhythmic firing of CFs onto PCs.

4.3.4 Changes in astroglia

Pathological studies have identified an increased presence of Bergmann glia in ET, a stereotypic reaction in cerebellum associated with PC loss

(Louis et al., 2007; Shill et al., 2008). Genetic polymorphisms in Solute carrier family 1 (glial high affinity glutamate transporter), member 2 (*SLC1A2*) have been linked with ET (Clark & Louis, 2018). *SLC1A2* encodes excitatory amino acid transporter type 2 (EAAT2), which clears glutamate from the synaptic cleft. Using both immunohistochemistry and Western blot analyses, EAAT2 expression was decreased by 40%–65% in cerebellar cortex of ET compared to controls (Lee et al., 2014). This finding raises the possibility that ET cerebellar cortex might be more vulnerable to excitotoxic damage than those of controls, resulting from its numerous glutamatergic synapses.

EAAT2 expression was also found to be 1.3- to 1.5-fold higher in the dentate nucleus of ET patients compared to controls (Wang et al., 2016). The mechanism behind this differential expression pattern of EAAT2 in the ET cerebellar cortex versus dentate nucleus remains to be investigated. The balance between inhibitory inputs from multiple PCs and excitatory inhibitory inputs from mossy fiber and CF collaterals on dentate neurons is critical for modulating their spike frequency and thus important for cerebellar function and proper motor control (Person & Raman, 2012). It is possible that lower levels of inhibition in the dentate nucleus, due to PC loss in ET, could trigger a compensatory increase in EAAT2 to maintain excitation-inhibition balance in cerebellar nuclei.

4.3.5 Changes in the dentate nucleus

PCs send dense inhibitory GABAergic inputs to cerebellar nuclei that modulate dentate neuron spike frequency. In a study of 10 ET and 16 control brains, reductions in $GABA_A$ (35%) and $GABA_B$ (22%–31%) receptors was detected in ET ($P < 0.05$) (Paris-Robidas et al., 2012). Concentrations of $GABA_B$ receptors in dentate inversely correlated with duration of tremor symptoms in ET, suggesting this receptor loss follows disease progression, consistent with a degenerative process. A diminished GABA receptor efficacy in ET might lead to overactivity of cerebellar nuclei neurons and spread up through the cerebellar-thalamo-cortical circuit, contributing to the generation of tremors.

Current evidence does not support the presence of extensive neuronal loss in the dentate nucleus in ET. While neuronal loss was qualitatively noted in 2 of 33 (6.1%) ET patients in one case series (Louis et al., 2007), a larger quantitative study in 25 ET compared to 25 age-matched controls did not find differences in dentate nucleus neuronal density (Hartstone et al., 2021). A stereological analysis in a subset of ET and control brains in this study also did not show differences in total number of dentate neurons. These findings do not exclude that there are subtle anatomical changes, other than neuronal cell death, or neurochemical changes in the dentate that might contribute to tremor generation in ET.

5. Cerebellar pathology and clinical heterogeneity in ET

Considering the considerable clinical heterogeneity in ET, case definition has been an area of increasing focus. The term "ET-plus" was introduced in a recent consensus statement to distinguish and separate ET cases with signs of dystonia, cerebellar dysfunction, and rest tremor, among others, from those without such signs (Bhatia et al., 2018). This diagnostic concept has raised considerable controversy and its validity is not yet established. Indeed, "ET-plus" has not been distinguished from ET based on differences in genetics or prognosis. To address whether ET cases differ from "ET-plus" cases in underlying pathological changes in the postmortem brain, 50 ET cases (24 ET and 26 ET-plus) were examined using a set of 14 quantitative metrics of cerebellar pathology reflecting changes across the PC body, dendrites, and axon, and basket cell-PC and CF-PC synaptic alterations (Gionco et al., 2021). ET and ET-plus were similar with respect to 13 of 14 cerebellar pathologic metrics, and after correcting for multiple comparisons, there were no differences. Furthermore, none of the cerebellar pathologic metrics significantly differed when ET cases were grouped by rest tremor or intention tremor. This study demonstrated there were no pathological differences in motor cerebellar cortex between ET versus ET-plus cases, and thus do not currently support the notion that ET and ET-plus represent distinct clinical-pathological entities.

6. ET as a degenerative disorder: Insight into disease mechanisms

The tissue-based research studies in ET and supportive findings from clinical features and neuroimaging studies now provide a patho-mechanistic model of ET—the cerebellar degenerative model. The confluence of abnormalities in the cerebellum suggest it may be a critical hub for generation of tremor. Although much work remains to be done to fully understand the underlying basis for this disease, two potential disease mechanisms are currently highlighted: (1) a diminution of GABAergic control and (2) reorganization of cerebellar cortex.

PCs provide the entire output from cerebellar cortex and provide dense GABAergic inhibitory inputs to cerebellar nuclei. Pathological changes observed in PCs, such as torpedoes, dendritic swellings, pruning of the PC dendritic arbor, and eventual PC loss might represent a primary and degenerative change that would decrease GABAergic inputs to cerebellar nuclei. In addition, the increased PC recurrent collateral formation as well as basket cell axonal plexus hypertrophy would act synergistically to further reduce PC inhibitory output onto remaining PCs (Louis & Faust, 2020c). These events may be additive and destructive to result in a hyperexcitable state in cerebellar nuclei that drives abnormal burst firing, as both the rate and timing of firing by cerebellar nuclear neurons under normal conditions is precisely regulated, at least in part, by the coordinated firing of PCs (Person & Raman, 2012).

Cerebellar control of movement requires precise and coordinated firing of its neuronal populations. A key feature of this control is the coordination, or synchronization, of PC oscillations, which converge in cerebellar nuclei to temporally organize cerebellar output (de Solages, Szapiro, Brunel, et al., 2008). Several pathological changes observed in ET cerebellum would lead to rewiring of cerebellar circuits and may facilitate generation of coordinated PC oscillations. Synaptic pruning deficits in CFs like that seen in human ET have been linked to tremor generation in a mouse model with ET-like tremor (Pan et al., 2020), and gap-junction coupling of ION neurons promotes synchronous PC firing. PC recurrent collateral synapses facilitate coordinated PC oscillations (Witter, Rudolph, Pressler, Lahlaf, & Regehr, 2016). Molecular layer interneurons shape PC firing patterns (Brown, Arancillo, Lin, et al., 2019), and their gap-junction mediated coupling within the molecular layer and ephaptic coupling at PC axonal initial segments by basket cells both enhance synchronous firing in cerebellar cortex (Han et al., 2018; Hoehne, McFadden, & DiGregorio, 2020). While many of these plasticity-type changes in PC inputs may initially represent compensatory responses, they are likely insufficient and possibly maladaptive or destructive, leading to progression of clinical signs in ET.

Our current understanding of ET pathology is largely based on extensive characterization within the motor region of cerebellar cortex, which is the region of greatest interest given the primary motoric nature of ET. Whether the morphological disease pattern seen in this region applies to other anatomic and functional compartments of cerebellum remains to be determined. The current disease model has revealed the greatest number of changes in the cerebellum, although it is theoretically possible that more subtle changes could extend more widely within the tremor circuit. Indeed, neuroimaging studies have identified a wide array of changes outside the cerebellum in ET, but their role in motor and non-motor features in this disease remain to be defined. Additional studies examining synaptic, axonal and other changes across regions in the physiologic tremor network is a challenge for future studies.

7. Conclusions

ET is a progressive, aging-associated condition with selective neuronal loss (affecting PCs) and other underlying structural changes in the cerebellum that disrupt normal function, all of which are features of a degenerative disorder. Both clinically and pathologically, there are numerous similarities and intersection points between ET and other disorders of cerebellar degeneration. Furthermore, considerable clinical associations suggest that ET may share disease mechanisms with other degenerative disorders, including PD and AD. The constellation of clinical non-motor findings in ET raises the possibility there is dysfunction in other brain regions, as seen in other degenerative disorders. A wide array of neuroimaging studies has identified functional and structural abnormalities in the cerebellum, although it remains to be established to what degree these contribute to clinical symptoms of ET. In the end, there are likely considerable overlaps between the degenerative and electric hypotheses for ET, including a current focus on central involvement of the cerebellum for tremor generation. There may be future merging of aspects of these hypotheses as we begin to explore whether cellular reactions occur elsewhere within the tremor network. A major challenge is to apply uniform and stringent inclusion criteria in ET disease studies and to organize the considerable clinical, pathological, and molecular heterogeneity within ET.

References

Axelrad, J. E., Louis, E. D., Honig, L. S., Flores, I., Ross, G. W., Pahwa, R., et al. (2008). Reduced Purkinje cell number in essential tremor: A postmortem study. *Archives of Neurology*, *65*(1), 101–107.

Babij, R., Lee, M., Cortés, E., Vonsattel, J. P., Faust, P. L., & Louis, E. D. (2013). Purkinje cell axonal anatomy: Quantifying morphometric changes in essential tremor versus control brains. *Brain*, *136*(10), 3051–3061.

Bagepally, B. S., Bhatt, M. D., Chandran, V., Saini, J., Bharath, R. D., Vasudev, M. K., et al. (2012). Decrease in cerebral and cerebellar gray matter in essential tremor: A voxel-based morphometric analysis under 3T MRI. *Journal of Neuroimaging*, *22*, 275–278.

Béliveau, E., Tremblay, C., Aubry-Lafontaine, É., Paris-Robidas, S., Delay, C., Robinson, C., et al. (2015). Accumulation of amyloid-β in the cerebellar cortex of essential tremor patients. *Neurobiology of Disease*, *82*, 397–408.

Benito-León, J. (2014). Essential tremor: A neurodegenerative disease? *Tremor and Other Hyperkinetic Movements (N. Y.)*, *4*, 252.

Benito-León, J., Alvarez-Linera, J., Hernández-Tamames, J. A., Alonso-Navarro, H., Jiménez-Jiménez, F. J., & Louis, E. D. (2009). Brain structural changes in essential tremor: Voxel-based morphometry at 3-Tesla. *Journal of the Neurological Sciences*, *287*(1–2), 138–142.

Benito-León, J., & Labiano-Fontcuberta, A. (2016). Linking essential tremor to the cerebellum: Clinical evidence. *Cerebellum*, *15*(3), 253–262.

Bhalsing, K. S., Upadhyay, N., Kumar, K. J., Saini, J., Yadav, R., Gupta, A. K., et al. (2014). Association between cortical volume loss and cognitive impairments in essential tremor. *European Journal of Neurology*, *21*, 874–883.

Bhatia, K. P., Bain, P., Bajaj, N., Elble, R. J., Hallett, M., Louis, E. D., et al. (2018). Consensus statement on the classification of tremors from the task force on tremor of the International Parkinson and Movement Disorder Society. *Movement Disorders*, *33*, 75–87.

Boecker, H., Weindl, A., Brooks, D. J., Ceballos-Baumann, A. O., Liedtke, C., et al. (2010). GABAergic dysfunction in essential tremor: An 11C-flumazenil PET study. *Journal of Nuclear Medicine*, *51*, 1030–1035.

Brown, A. M., Arancillo, M., Lin, T., et al. (2019). Molecular layer interneurons shape the spike activity of cerebellar Purkinje cells. *Scientific Reports*, *9*, 1742.

Carulli, D., Buffo, A., & Strata, P. (2004). Reparative mechanisms in the cerebellar cortex. *Progress in Neurobiology*, *72*, 373–398.

Cerasa, A., Messina, D., Nicoletti, G., Novellino, F., Lanza, P., Condino, F., et al. (2009). Cerebellar atrophy in essential tremor using an automated segmentation method. *American Journal of Neuroradiology*, *30*, 1240–1243.

Cerasa, A., & Quattrone, A. (2016). Linking essential tremor to the cerebellum-neuroimaging evidence. *Cerebellum*, *15*(3), 263–275.

Choe, M., Cortés, E., Vonsattel, J. P., Kuo, S. H., Faust, P. L., & Louis, E. D. (2016). Purkinje cell loss in essential tremor: Random sampling quantification and nearest neighbor analysis. *Movement Disorders*, *31*(3), 393–401.

Choi, S.-M., Kim, B. C., Chang, J., Choi, K.-H., Nam, T.-S., Kim, J.-T., et al. (2015). Comparison of the brain volume in essential tremor and Parkinson's disease tremor using an automated segmentation method. *European Neurology*, *73*, 303–309.

Clark, L. N., & Louis, E. D. (2018). Essential tremor. *Handbook of Clinical Neurology*, *147*, 229–239.

Cohen, O., Pullman, S., Jurewicz, E., Watner, D., & Louis, E. D. (2003). Rest tremor in patients with essential tremor: Prevalence, clinical correlates, and electrophysiologic characteristics. *Archives of Neurology*, *60*, 405–410.

Critchley, M. (1949). Observations on essential (heredofamilal) tremor. *Brain*, *72*, 113–139.

Critchley, M., & Greenfield, J. G. (1948). Olivoponto-cerebellar atrophy. *Brain*, *71*(4), 344–364.

de Solages, C., Szapiro, G., Brunel, N., et al. (2008). High-frequency organization and synchrony of activity in the Purkinje cell layer of the cerebellum. *Neuron*, *58*(5), 775–788.

Delay, C., Tremblay, C., Brochu, E., Paris-Robidas, S., Emond, V., Rajput, A. H., et al. (2014). Increased LINGO1 in the cerebellum of essential tremor patients. *Movement Disorders*, *29*(13), 1637–1647.

Duval, C., Daneault, J. F., Hutchison, W. D., & Sadikot, A. F. (2016). A brain network model explaining tremor in Parkinson's disease. *Neurobiology of Disease*, *85*, 49–59.

Dyke, J. P., Cameron, E., Hernandez, N., Dydak, U., & Louis, E. D. (2017). Gray matter density loss in essential tremor: A lobule by lobule analysis of the cerebellum. *Cerebellum & Ataxias, 4*, 10.

Elkouzi, A., Kattah, J. C., & Elble, R. J. (2016). Hypertrophic olivary degeneration does not reduce essential tremor. *Movement Disorders Clinical Practice, 3*, 209–211.

Erickson-Davis, C. R., Faust, P. L., Vonsattel, J. P., Gupta, S., Honig, L. S., & Louis, E. D. (2010). "Hairy baskets" associated with degenerative Purkinje cell changes in essential tremor. *Journal of Neuropathology and Experimental Neurology, 69*(3), 262–271.

Farrell, K., Cosentino, S., Iida, M. A., Chapman, S., Bennett, D. A., Faust, P. L., et al. (2019). Quantitative assessment of pathological tau burden in essential tremor: A postmortem study. *Journal of Neuropathology and Experimental Neurology, 78*(1), 31–37.

Ferrer, I., Fabregues, I., Pineda, M., Gracia, I., & Ribalta, T. (1984). A Golgi study of cerebellar atrophy in human chronic alcoholism. *Neuropathology and Applied Neurobiology, 10*, 245–253.

Gionco, J. T., Hartstone, W. G., Martuscello, R. T., Kuo, S. H., Faust, P. L., & Louis, E. D. (2021). Essential tremor versus "ET-plus": A detailed postmortem study of cerebellar pathology. *Cerebellum, 20*(6), 904–912.

Gironell, A. (2014). The GABA hypothesis in essential tremor: Lights and shadows. *Tremor And Other Hyperkinetic Movements (N. Y.), 4*, 254.

Gironell, A., Figueiras, F. P., Pagonabarraga, J., Herance, J. R., Pascual-Sedano, B., Trampal, C., et al. (2012). Gaba and serotonin molecular neuroimaging in essential tremor: a clinical correlation study. *Parkinsonism and Related Disorders, 18*, 876–880.

Grimaldi, G., & Manto, M. (2013). Is essential tremor a Purkinjopathy? The role of the cerebellar cortex in its pathogenesis. *Movement Disorders, 28*(13), 1759–1761.

Han, K. S., Guo, C., Chen, C. H., Witter, L., Osorno, T., & Regehr, W. G. (2018). Ephaptic coupling promotes synchronous firing of cerebellar Purkinje cells. *Neuron, 100*, 564–578.

Hartstone, W. G., Brown, M. H., Kelly, G. C., Tate, W. J., Kuo, S. H., Dwork, A. J., et al. (2021). Dentate nucleus neuronal density: A postmortem study of essential tremor versus control brains. *Movement Disorders, 36*(4), 995–999.

Helmchen, C., Hagenow, A., Miesner, J., Sprenger, A., Rambold, H., Wenzelburger, R., et al. (2003). Eye movement abnormalities in essential tremor may indicate cerebellar dysfunction. *Brain, 126*(Pt. 6), 1319–1332.

Helmich, R. C., Toni, I., Deuschl, G., & Bloem, B. R. (2013). The pathophysiology of essential tremor and Parkinson's tremor. *Current Neurology and Neuroscience Reports, 13*(9), 378.

Hoehne, A., McFadden, M. H., & DiGregorio, D. A. (2020). Feed-forward recruitment of electrical synapses enhances synchronous spiking in the mouse cerebellar cortex. *eLife, 9*, e57344.

Janicki, S. C., Cosentino, S., & Louis, E. D. (2013). The cognitive side of essential tremor: What are the therapeutic implications? *Therapeutic Advances in Neurological Disorders, 6*(6), 353–368.

Jia, L., Jia-Lin, S., Qin, D., Qing, L., & Yan, Z. (2011). A diffusion tensor imaging study in essential tremor. *Journal of Neuroimaging, 21*, 370–374.

Kano, M., Watanabe, T., Uesaka, N., & Watanabe, M. (2018). Multiple phases of climbing fiber synapse elimination in the developing cerebellum. *Cerebellum, 17*, 722–734.

Kim, S. H., Farrell, K., Cosentino, S., Vonsattel, J. G., Faust, P. L., Cortes, E. P., et al. (2021). Tau isoform profile in essential tremor diverges from other tauopathies. *Journal of Neuropathology and Experimental Neurology, 80*(9), 835–843.

Klein, J. C., Lorenz, B., Kang, J. S., Baudrexel, S., Seifried, C., van de Loo, S., et al. (2011). Diffusion tensor imaging of whitematter involvement in essential tremor. *Human Brain Mapping, 32*, 896–904.

Kuo, S. H., Erickson-Davis, C., Gillman, A., Faust, P. L., Vonsattel, J. P., & Louis, E. D. (2011). Increased number of heterotopic Purkinje cells in essential tremor. *Journal of Neurology Neurosurgery and Psychiatry, 82*(9), 1038–1040.

Kuo, S. H., Faust, P. L., Vonsattel, J. P., Ma, K., & Louis, E. D. (2011). Parallel fiber counts and parallel fiber integrated density are similar in essential tremor cases and controls. *Acta Neuropathologica, 121*(2), 287–289.

Kuo, S. H., Lin, C. Y., Wang, J., Liou, J. Y., Pan, M. K., Louis, R. J., et al. (2016). Deep brain stimulation and climbing fiber synaptic pathology in essential tremor. *Annals of Neurology, 80*(3), 461–465.

Kuo, S. H., Lin, C. Y., Wang, J., Sims, P. A., Pan, M. K., Liou, J. Y., et al. (2017). Climbing fiber-Purkinje cell synaptic pathology in tremor and cerebellar degenerative diseases. *Acta Neuropathologica, 133*(1), 121–138.

Kuo, S. H., Tang, G., Louis, E. D., Ma, K., Babji, R., Balatbat, M., et al. (2013). Lingo-1 expression is increased in essential tremor cerebellum and is present in the basket cell pinceau. *Acta Neuropathologica, 125*(6), 879–889.

Kuo, S. H., Wang, J., Tate, W. J., Pan, M. K., Kelly, G. C., Gutierrez, J., et al. (2017). Cerebellar pathology in early onset and late onset essential tremor. *Cerebellum, 16*(2), 473–482.

Larsson, T., & Sjogren, T. (1960). Essential tremor: A clinical and genetic population study. *Acta Psychiatrica Scandinavica. Supplementum, 36*, 1–176.

Lee, M., Cheng, M. M., Lin, C. Y., Louis, E. D., Faust, P. L., & Kuo, S. H. (2014). Decreased EAAT2 protein expression in the essential tremor cerebellar cortex. *Acta Neuropathologica Commununications, 2*, 157.

Lee, D., Gan, S. R., Faust, P. L., Louis, E. D., & Kuo, S. H. (2018). Climbing fiber-Purkinje cell synaptic pathology across essential tremor subtypes. *Parkinsonism and Related Disorders, 51*, 24–29.

Lee, P. J., Kerridge, C. A., Chatterjee, D., Koeppen, A. H., Faust, P. L., & Louis, E. D. (2019). A quantitative study of empty baskets in essential tremor and other motor neurodegenerative diseases. *Journal of Neuropathology and Experimental Neurology, 78*(2), 113–122.

Lin, C. Y., Louis, E. D., Faust, P. L., Koeppen, A. H., Vonsattel, J. P., & Kuo, S. H. (2014). Abnormal climbing fibre-Purkinje cell synaptic connections in the essential tremor cerebellum. *Brain, 137*(1), 3149–3159.

Louis, E. D. (2016). Non-motor symptoms in essential tremor: A review of the current data and state of the field. *Parkinsonism and Related Disorders, 22*(Suppl. 1), S115–S118.

Louis, E. D. (2019). The roles of age and aging in essential tremor: An epidemiological perspective. *Neuroepidemiology, 52*(1–2), 111–118.

Louis, E. D. (2021). The essential tremors: Evolving concepts of a family of diseases. *Frontiers in Neurology, 12*, 650601.

Louis, E. D., Agnew, A., Gillman, A., Gerbin, M., & Viner, A. S. (2011). Estimating annual rate of decline: Prospective, longitudinal data on arm tremor severity in two groups of essential tremor cases. *Journal of Neurology Neurosurgery and Psychiatry, 82*, 761–765.

Louis, E. D., Asabere, N., Agnew, A., Moskowitz, C. B., Lawton, A., Cortes, E., et al. (2011). Rest tremor in advanced essential tremor: A post-mortem study of nine cases. *Journal of Neurology Neurosurgery and Psychiatry, 82*(3), 261–265.

Louis, E. D., Babij, R., Cortés, E., Vonsattel, J. P., & Faust, P. L. (2013). The inferior olivary nucleus: A postmortem study of essential tremor cases versus controls. *Movement Disorders, 28*(6), 779–786.

Louis, E. D., Babij, R., Lee, M., Cortés, E., & Vonsattel, J. P. (2013). Quantification of cerebellar hemispheric Purkinje cell linear density: 32 ET cases versus 16 controls. *Movement Disorders, 28*(13), 1854–1859.

Louis, E. D., Diaz, D. T., Kuo, S. H., Gan, S. R., Cortes, E. P., Vonsattel, J. P. G., et al. (2018). Inferior Olivary nucleus degeneration does not lessen tremor in essential tremor. *Cerebellum and Ataxias*, *5*, 1.

Louis, E. D., & Faust, P. L. (2020a). Essential tremor: The most common form of cerebellar degeneration? *Cerebellum & Ataxias*, 7, 12.

Louis, E. D., & Faust, P. L. (2020b). Essential tremor within the broader context of other forms of cerebellar degeneration. *Cerebellum*, *19*(6), 879–896.

Louis, E. D., & Faust, P. L. (2020c). Essential tremor pathology: Neurodegeneration and reorganization of neuronal connections. *Nature Reviews Neurology*, *16*(2), 69–83.

Louis, E. D., Faust, P. L., Ma, K. J., Yu, M., Cortes, E., & Vonsattel, J. P. (2011). Torpedoes in the cerebellar vermis in essential tremor cases vs. controls. *Cerebellum*, *10*(4), 812–819.

Louis, E. D., Faust, P. L., Vonsattel, J. P., & Erickson-Davis, C. (2009). Purkinje cell axonal torpedoes are unrelated to advanced aging and likely reflect cerebellar injury. *Acta Neuropathologica*, *117*(6), 719–721.

Louis, E. D., Faust, P. L., Vonsattel, J. P., Honig, L. S., Henchcliffe, C., Pahwa, R., et al. (2009a). Older onset essential tremor: More rapid progression and more degenerative pathology. *Movement Disorders*, *24*(11), 1606–1612.

Louis, E. D., Faust, P. L., Vonsattel, J. P., Honig, L. S., Rajput, A., Rajput, A., et al. (2009b). Torpedoes in Parkinson's disease, Alzheimer's disease, essential tremor, and control brains. *Movement Disorders*, *24*(11), 1600–1605.

Louis, E. D., Faust, P. L., Vonsattel, J. P., Honig, L. S., Rajput, A., Robinson, C. A., et al. (2007). Neuropathological changes in essential tremor: 33 cases compared with 21 controls. *Brain*, *130*(Pt. 12), 3297–3307.

Louis, E. D., Ford, B., & Barnes, L. F. (2000). Clinical subtypes of essential tremor. *Archives of Neurology*, *57*(8), 1194–1198.

Louis, E. D., Huang, C. C., Dyke, J. P., Long, Z., & Dydak, U. (2014). Neuroimaging studies of essential tremor: How well do these studies support/refute the neurodegenerative hypothesis? *Tremor and Other Hyperkinetic Movements (N. Y.)*, *4*, 235.

Louis, E. D., Jurewicz, E. C., & Watner, D. (2003). Community-based data on associations of disease duration and age with severity of essential tremor: Implications for disease pathophysiology. *Movement Disorders*, *18*, 90–93.

Louis, E. D., Kerridge, C. A., Chatterjee, D., Martuscello, R. T., Diaz, D. T., Koeppen, A. H., et al. (2019). Contextualizing the pathology in the essential tremor cerebellar cortex: A patholog-omics approach. *Acta Neuropathologica*, *138*(5), 859–876.

Louis, E. D., Kuo, S. H., Tate, W. J., Kelly, G. C., & Faust, P. L. (2017). Cerebellar pathology in childhood-onset vs. adult-onset essential tremor. *Neuroscience Letters*, *659*, 69–74.

Louis, E. D., Kuo, S. H., Tate, W. J., Kelly, G. C., Gutierrez, J., Cortes, E. P., et al. (2018). Heterotopic Purkinje cells: A comparative postmortem study of essential tremor and spinocerebellar ataxias 1, 2, 3, and 6. *Cerebellum*, *17*(2), 104–110.

Louis, E. D., Kuo, S. H., Vonsattel, J. P., & Faust, P. L. (2014). Torpedo formation and Purkinje cell loss: Modeling their relationship in cerebellar disease. *Cerebellum*, *13*(4), 433–439.

Louis, E. D., Kuo, S. H., Wang, J., Tate, W. J., Pan, M. K., Kelly, G. C., et al. (2017). Cerebellar pathology in familial vs. sporadic essential tremor. *Cerebellum*, *16*(4), 786–791.

Louis, E. D., Lee, M., Babij, R., Ma, K., Cortés, E., Vonsattel, J. P., et al. (2014). Reduced Purkinje cell dendritic arborization and loss of dendritic spines in essential tremor. *Brain*, *137*(12), 3142–3148.

Louis, E. D., Lee, M., Cortés, E., Vonsattel, J. P., & Faust, P. L. (2014). Matching asymmetry of tremor with asymmetry of postmortem cerebellar hemispheric changes in essential tremor. *Cerebellum*, *13*(4), 462–470.

Louis, E. D., & Lenka, A. (2017). The olivary hypothesis of essential tremor: Time to lay this model to rest? *Tremor and Other Hyperkinetic Movements (N. Y.)*, 7, 473.

Louis, R. J., Lin, C. Y., Faust, P. L., Koeppen, A. H., & Kuo, S. H. (2015). Climbing fiber synaptic changes correlate with clinical features in essential tremor. *Neurology*, *84*(22), 2284–2286.

Louis, E. D., Ma, K., Babij, R., Cortés, E., Liem, R. K., Vonsattel, J. P., et al. (2012). Neurofilament protein levels: Quantitative analysis in essential tremor cerebellar cortex. *Neuroscience Letters*, *518*(1), 49–54.

Louis, E. D., & McCreary, M. (2021). How common is essential tremor? Update on the worldwide prevalence of essential tremor. *Tremor and Other Hyperkinetic Movements (N. Y.)*, *11*, 28.

Louis, E. D., Shungu, D. C., Chan, S., Mao, X., Jurewicz, E. C., & Watner, D. (2002). Metabolic abnormality in the cerebellum in patients with essential tremor: A proton magnetic resonance spectroscopic imaging study. *Neuroscience Letters*, *333*, 17–20.

Louis, E. D., & Vonsattel, J. P. (2008). The emerging neuropathology of essential tremor. *Movement Disorders*, *23*(2), 174–182.

Louis, E. D., Vonsattel, J. P., Honig, L. S., Lawton, A., Moskowitz, C., Ford, B., et al. (2006). Essential tremor associated with pathologic changes in the cerebellum. *Archives of Neurology*, *63*(8), 1189–1193.

Louis, E. D., Vonsattel, J. P., Honig, L. S., Ross, G. W., Lyons, K. E., & Pahwa, R. (2006). Neuropathologic findings in essential tremor. *Neurology*, *66*(11), 1756–1759.

Louis, E. D., Yi, H., Erickson-Davis, C., Vonsattel, J. P., & Faust, P. L. (2009). Structural study of Purkinje cell axonal torpedoes in essential tremor. *Neuroscience Letters*, *450*(3), 287–291.

Nicoletti, G., Manners, D., Novellino, F., Condino, F., Malucelli, E., Barbiroli, B., et al. (2010). Diffusion tensor MRI changes in cerebellar structures of patients with familial essential tremor. *Neurology*, *74*, 988–994.

Novellino, F., Cherubini, A., Chiriaco, C., Morelli, M., Salsone, M., Arabia, G., et al. (2013). Brain iron deposition in essential tremor: A quantitative 3-Tesla magnetic resonance imaging study. *Movement Disorders*, *28*, 196–200.

Novellino, F., Nicoletti, G., Cherubini, A., Caligiuri, M. E., Nisticò, R., Salsone, M., et al. (2016). Cerebellar involvement in essential tremor with and without resting tremor: A diffusion tensor imaging study. *Parkinsonism and Related Disorders*, *27*, 61–66.

Pagan, F. L., Butman, J. A., Dambrosia, J. M., & Hallett, M. (2003). Evaluation of essential tremor with multi-voxel magnetic resonance spectroscopy. *Neurology*, *60*, 1344–1347.

Pan, J. J., Lee, M., Honig, L. S., Vonsattel, J. P., Faust, P. L., & Louis, E. D. (2014). Alzheimer's-related changes in non-demented essential tremor patients vs. controls: Links between tau and tremor? *Parkinsonism and Related Disorders*, *20*(6), 655–658.

Pan, M. K., Li, Y. S., Wong, S. B., Ni, C. L., Wang, Y. M., Liu, W. C., et al. (2020). Cerebellar oscillations driven by synaptic pruning deficits of cerebellar climbing fibers contribute to tremor pathophysiology. *Science Translational Medicine*, *12*(526), eaay1769.

Paris-Robidas, S., Brochu, E., Sintes, M., et al. (2012). Defective dentate nucleus GABA receptors in essential tremor. *Brain*, *135*, 105–116.

Person, A. L., & Raman, I. M. (2012). Synchrony and neural coding in cerebellar circuits. *Frontiers in Neural Circuits*, *6*, 97.

Petzold, A. (2005). Neurofilament phosphoforms: Surrogate markers for axonal injury, degeneration and loss. *Journal of the Neurological Sciences*, *233*, 183–198.

Pietracupa, S., Bologna, M., Bharti, K., Pasqua, G., Tommasin, S., Elifani, F., et al. (2019). White matter rather than gray matter damage characterizes essential tremor. *European Radiology*, *29*, 6634–6642.

Pietracupa, S., Bologna, M., Tommasin, S., Berardelli, A., & Pantano, P. (2021). The contribution of neuroimaging to the understanding of essential tremor pathophysiology: A systematic review. *Cerebellum*.

Putzke, J. D., Whaley, N. R., Baba, Y., Wszolek, Z. K., & Uitti, R. J. (2006). Essential tremor: Predictors of disease progression in a clinical cohort. *Journal of Neurology Neurosurgery and Psychiatry*, *77*, 1235–1237.

Quattrone, A., Cerasa, A., Messina, D., Nicoletti, G., Hagberg, G. E., Lemieux, L., et al. (2008). Essential head tremor is associated with cerebellar vermis atrophy: A volumetric and voxel-based morphometry MR imaging study. *American Journal of Neuroradiology*, *29*, 1692–1697.

Radler, K. H., Zdrodowska, M. A., Dowd, H., Cersonsky, T. E. K., Huey, E. D., Cosentino, S., et al. (2020). Rate of progression from mild cognitive impairment to dementia in an essential tremor cohort: A prospective, longitudinal study. *Parkinsonism and Related Disorders*, *74*, 38–42.

Raethjen, J., & Deuschl, G. (2012). The oscillating central network of essential tremor. *Clinical Neurophysiology*, *123*(1), 61–64.

Rajput, A., Robinson, C. A., & Rajput, A. H. (2004). Essential tremor course and disability: A clinicopathologic study of 20 cases. *Neurology*, *62*, 932–936.

Rajput, A. H., Robinson, C. A., Rajput, M. L., & Rajput, A. (2011). Cerebellar Purkinje cell loss is not pathognomonic of essential tremor. *Parkinsonism and Related Disorders*, *17*(1), 16–21.

Ross, G. W., Dickson, D. W., Cersosimo, M., Aires, B., Tsuboi, Y., Katsuse, O., et al. (2004). Pathological Investigation of Essential Tremor. *Neurology*, *62*(Suppl 5), A537–A538.

Rossi, F., Gianola, S., & Corvetti, L. (2006). The strange case of Purkinje axon regeneration and plasticity. *Cerebellum*, *5*, 174–182.

Saini, J., Bagepally, B. S., Bhatt, M. D., Chandran, V., Bharath, R. D., Prasad, C., et al. (2012). Diffusion tensor imaging: Tract based spatial statistics study in essential tremor. *Parkinsonism and Related Disorders*, *18*, 477–482.

Shill, H. A., Adler, C. H., Sabbagh, M. N., Connor, D. J., Caviness, J. N., Hentz, J. G., et al. (2008). Pathologic findings in prospectively ascertained essential tremor subjects. *Neurology*, *70*(Pt. 2), 1452–1455.

Shin, H., Lee, D.-K., Lee, J.-M., Huh, Y.-E., Youn, J., Louis, E. D., et al. (2016). Atrophy of the cerebellar vermis in essential tremor: Segmental volumetric MRI analysis. *Cerebellum*, *15*, 174–181.

Shintaku, M., & Kaneda, D. (2009). Chromosome 16q22.1-linked autosomal dominant cerebellar ataxia: an autopsy case report with some new observations on cerebellar pathology. *Neuropathology*, *29*, 285–292.

Song, P., Zhang, Y., Zha, M., Yang, Q., Ye, X., Yi, Q., et al. (2021). The global prevalence of essential tremor, with emphasis on age and sex: A meta-analysis. *Journal of Global Health*, *11*, 04028.

Symanski, C., Shill, H. A., Dugger, B., Hentz, J. G., Adler, C. H., Jacobson, S. A., et al. (2014). Essential tremor is not associated with cerebellar Purkinje cell loss. *Movement Disorders*, *29*(4), 496–500.

Thenganatt, M. A., & Jankovic, J. (2016). The relationship between essential tremor and Parkinson's disease. *Parkinsonism and Related Disorders*, *22*(Suppl. 1), S162–S165.

Wang, J., Kelly, G. C., Tate, W. J., Li, Y. S., Lee, M., Gutierrez, J., et al. (2016). Excitatory amino acid transporter expression in the essential tremor dentate nucleus and cerebellar cortex: A postmortem study. *Parkinsonism and Related Disorders*, *32*, 87–93.

Watanabe, M. (2008). Molecular mechanisms governing competitive synaptic wiring in cerebellar Purkinje cells. *Tohoku Journal of Experimental Medicine*, *214*, 175–190.

Wills, A. J., Jenkins, I. H., Thompson, P. D., Findley, L. J., & Brooks, D. J. (1994). Red nuclear and cerebellar but no olivary activation associated with essential tremor: A positron emission tomographic study. *Annals of Neurology*, *36*, 636–642.

Witter, L., Rudolph, S., Pressler, R. T., Lahlaf, S. I., & Regehr, W. G. (2016). Purkinje cell collaterals enable output signals from the cerebellar cortex to feed back to Purkinje cells and interneurons. *Neuron, 91*, 312–319.
Wu, Y. C., Louis, E. D., Gionco, J., Pan, M. K., Faust, P. L., & Kuo, S. H. (2021). Increased climbing fiber lateral crossings on Purkinje cell dendrites in the cerebellar hemisphere in essential tremor. *Movement Disorders, 36*(6), 1440–1445.
Yang, Q., et al. (2000). Morphological Purkinje cell changes in spinocerebellar ataxia type 6. *Acta Neuropathologica, 100*, 371–376.
Yu, M., Ma, K., Faust, P. L., Honig, L. S., Cortés, E., Vonsattel, J. P., et al. (2012). Increased number of Purkinje cell dendritic swellings in essential tremor. *European Journal of Neurology, 19*(4), 625–630.

CHAPTER FOUR

Is essential tremor a degenerative or an electrical disorder? Electrical disorder

Hyun Joo Cho*
Human Motor Control Section, National Institute of Neurological Disorders and Stroke, National Institutes of Health, Bethesda, MD, United States
*Corresponding author: e-mail address: hyunjoo.cho@nih.gov

Contents

Abstract

Essential tremor (ET) is one of the most common movement disorders, yet we do not have a complete understanding of its pathophysiology. From a phenomenology standpoint, ET is an isolated tremor syndrome of bilateral upper limb action tremor with or without tremor in other body locations. ET is a pathological tremor that arises from excessive oscillation in the central motor network. The tremor network comprises of multiple brain regions including the inferior olive, cerebellum, thalamus, and motor cortex, and there is evidence that a dynamic oscillatory disturbance within this network leads to tremor. ET is a chronic disorder, and the natural history shows a slow progression of tremor intensity with age. There are reported data suggesting that ET follows the disease model of a neurodegenerative disorder, however whether ET is a degenerative or electrical disorder has been a subject of debate. In this chapter, we will review cumulative evidence that ET as a syndrome is a fundamentally electric disorder. The etiology is likely heterogenous and may not be primarily neurodegenerative.

International Review of Neurobiology, Volume 163
ISSN 0074-7742
https://doi.org/10.1016/bs.irn.2022.02.004

1. Introduction

Essential tremor (ET) is defined as an isolated tremor syndrome of bilateral upper limb action tremor with or without tremor in other body locations (Bhatia et al., 2018). Neurophysiological properties of ET are consistent with a central oscillator with some influence of mechanical-reflex pathways (Hallett, 1998). ET has a tremor frequency of 4–10 Hz and its frequency decreases with aging (Elble, 1995). In younger people, tremor frequency is often 8–12 Hz, which coincides with the frequency of synchronous entrainment of motor units in normal adults (Elble, Higgins, & Hughes, 1994). The inferior olivary nucleus (ION) was considered as a main oscillator for ET given that its neurons have intrinsic oscillatory properties around the frequency of ET (Llinás, 1988). However, with advances in the understanding of the oscillatory nature of the human nervous system, especially in human motor control, it is believed that ET does not emerge from one single oscillator but rather an oscillatory network comprising most parts of the physiological central motor network (Raethjen & Deuschl, 2012).

ET is the most prevalent movement disorder, effecting 1.33% of the general population at all ages and 5.79% for age ≥ 65 (Louis & McCreary, 2021), yet there is no curative treatment for this condition in part due to a lack of clear understanding of its pathophysiology. It is often considered a "benign tremor syndrome," however it causes significant disability for patients due to both motor and psychosocial limitations (Busenbark, Nash, Nash, Hubble, & Koller, 1991; Thangavelu, Talk, Clark, & Dissanayaka, 2020). It is imperative to better understand the pathophysiology of ET to develop biomarkers and ultimately effective treatments. ET is clearly a chronic movement disorder, but whether ET is a degenerative disorder or not has been a subject of controversy in the field (Benito-León, 2014; Bermejo-Pareja, 2011; Deuschl & Elble, 2009; Louis, 2009; Rajput, Adler, Shill, & Rajput, 2012).

In this manuscript, we will review the existing evidence to show that ET is an electric disorder. We will review neurophysiology, neuroimaging, and neuropathology studies to understand the nature of electric disorder in ET. Finally, we will argue that ET may not be a primary neurodegenerative disorder.

2. ET is an electric disorder

2.1 Genesis of tremor

Tremor is a common movement disorder that is defined as rhythmical, involuntary, oscillatory movement of a body part (Puschmann & Wszolek, 2011). There are two basic principles when we address tremor genesis—*where* and *how* the tremor is generated. As for the tremor localization, tremor may come from the regional pathology in the brain that generates the oscillation (Louis & Faust, 2020c), or excessive oscillation of the broad neural network, either by hyperexcitability of the neuronal loop or decoupling/instability of the network (Raethjen & Deuschl, 2012). There are also peripheral pathways that are involved in tremor genesis (McAuley & Marsden, 2000) which will not be discussed in detail in this paper. The second question is what underlies excessive oscillation either regionally or network-wise that leads to tremors. Some tremors could be from degenerative pathology (Lai et al., 2019), inflammatory changes (Koch, Mostert, Heersema, & De Keyser, 2007), or a pure electrical perturbation (Köster et al., 1998). In this review, we will argue that ET is a network disorder (localization) and is intrinsically electric disorder (pathophysiology).

Two sets of neural pathways in motor control are of particular interest (Fig. 1). One is the cortico-striato-thalamo-cortical pathway, which is a neuronal circuit that controls movement, selection, initiation, reinforcement, and reward (Graybiel, Aosaki, Flaherty, & Kimura, 1994). The other is the cerebello-thalamo-cortical pathway, which is responsible for fine-tuning voluntary precision movements (Horne & Butler, 1995). The cerebello-thalamo-cortical pathway is generally linked to the generation of ET and the cortico-striato-thalamo-cortical pathway is implicated in Parkinson's tremor (Sharifi, Nederveen, Booij, & van Rootselaar, 2014). Where the inferior olivary–cerebellar network interacts with the cerebello-thalamo-cortical pathway is a key constituent of the ET network.

Oscillations of neuronal activity are fundamental for different parts of the Central Nervous System (CNS) to communicate effectively (Engel & Fries, 2010). CNS oscillation has an important role in human motor control as well (Farmer, Bremner, Halliday, Rosenberg, & Stephens, 1993). Early animal studies showed the presence of synchronized corticomotor neuron activities at certain frequency to modulate descending pathways controlling

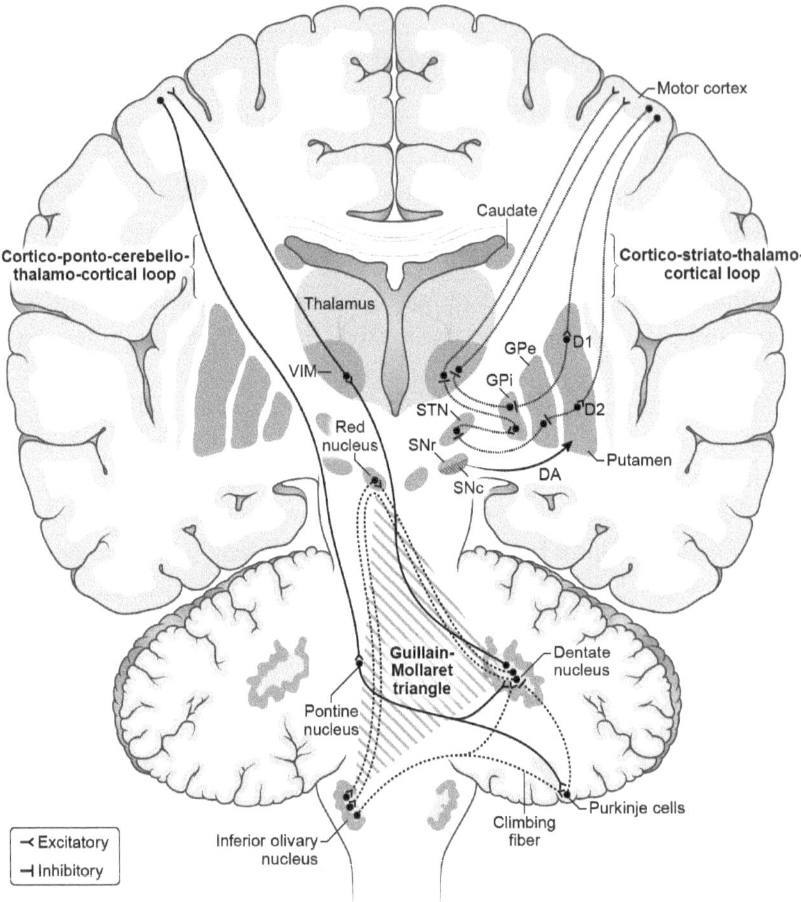

Fig. 1 Schematic illustration of the brain regions and pathways involved in the pathophysiology of tremors. See text for details. (VIM, ventralis intermedius; GPe, globus pallidus externa; GPi, globus pallidus interna; STN, subthalamic nucleus; SNr, substantia nigra pars reticulata; SNc, substantia nigra pars compacta; DA, dopamine; D1, dopamine receptor type 1; D2, dopamine receptor type 2).

motor output (Murthy & Fetz, 1992; Nicolelis, Baccala, Lin, & Chapin, 1995). In humans, 8–12 Hz peak frequency in EMG reflects a centrally-tuned oscillatory operation (McAuley & Marsden, 2000).

Multiple oscillators in the frequency range of tremor are found throughout the nervous system. The sensorimotor cortex, thalamus, inferior olive, cerebellar nuclei, brainstem raphe, reticular nuclei, globus pallidus, and the spinal cord are just a few of the many that possess the intrinsic ability to generate rhythmic activity, which if pathologically altered could produce an abnormal tremor (Elble, 1996). These central oscillators have network properties that promote neuronal entrainment through recurrent inhibition and excitation which can perpetuate itself in a rhythmic fashion. This oscillating central motor network takes part in normal physiological motor control. Pathologic tremors may arise from distortion or amplification of these central oscillations, either due to an unstable loop circuit or a nucleus with excessive spontaneous rhythmicity.

2.2 Neurophysiological features of essential tremor

Tremors can be analyzed by electrophysiological measures such as accelerometer and electromyography (Deuschl, Krack, Lauk, & Timmer, 1996). Tremor recordings can help to differentiate between types of tremor syndromes and the source of tremors, whether it is mechanical, mechanical-reflex, or central oscillator (For review, please refer to "How to do an electrophysiological study of tremor" (Vial, Kassavetis, Merchant, Haubenberger, & Hallett, 2019)). A prospective study by Gironell posed six neurophysiological criteria for ET including EMG burst frequency, latency of postural tremor, effect of weight load and mental concentration on tremor (Table 1). This diagnostic criteria was tested in 300 consecutive outpatients who presented with postural tremor of the hands as the predominant symptoms, and it showed a sensitivity of 97.7%, a specificity of 82.3%, a positive predictive value of 95.1%, and a negative predictive value of 91.1% (Gironell, Kulisevsky, Pascual-Sedano, & Barbanoj, 2004). ET frequency does not change more than 1 Hz by adding weight to hands, which confirms the presence of the central oscillator. ET can have a mechanical component, which often has a similar

Table 1 Neurophysiological features of essential tremor.

Neurophysiological features of Essential Tremor
1. Rhythmic bursts of postural tremor on EMG
2. Tremor frequency greater than or equal to 4 Hz
3. Rest tremor absent or, if present, frequency 1.5 Hz lower with respect to the postural tremor
4. Absence of latency from rest to postural position
5. Changes of the dominant frequency peak less or equal to 1 Hz after the weight load test
6. No changes in tremor amplitude after mental concentration

frequency as the central component and will merge, but the mechanical component can be separated with weight (Vial et al., 2019). ET can be differentiated with Parkinson's tremor based on the absence of dominant resting tremor and lack of lag between resting and posturing. In other words, ET seems to be turned on by voluntary movement. EMG bursts in agonist and antagonist muscles that produce ET can be either synchronous or alternating (Milanov, 2001), or the pattern may slip over a period of a few minutes. Therefore, EMG pattern is not helpful as a diagnostic feature.

2.3 Oscillators of essential tremor

2.3.1 Inferior olive

Since the 1970s, ION has received the greatest attention as a potential oscillator of ET. The neurons in the ION have a set of calcium channels that can produce rhythmic discharges (Park et al., 2010). The olivary hypothesis is based on the similarities between ET and the tremor induced by harmaline in laboratory primates (Llinás, 1988). According to the olivary hypothesis, patients with essential tremor may have enhanced synchronization and 4–12 Hz neuronal rhythmicity in their olives (Deuschl & Elble, 2000). This could result from altered olivary network properties (e.g., increased dendrodendritic gap junction), altered neuromodulation of the olivary network (e.g., serotonin and GABA), abnormal enhancement of the membrane conductance that underlie the neuronal oscillations, or a combination of these mechanisms (Llinas, 2014). Olivary oscillation is transmitted and possibly amplified through the cerebellum, resulting in an entrainment of the thalamus, motor cortex, and brainstem nuclei in mammalian brain although this is still a hypothesis in human (Llinás, 2014).

However, there is an argument that ION is not the primary source of tremor in ET based on negative neuroimaging and pathology study results as well as clinical discrepancies between harmaline induced tremor and ET (Louis & Lenka, 2017). Whether ION is central to the pathophysiology of ET or not will be addressed in more detail in another paper of this book.

2.3.2 Cerebellum

The cerebellum has gained considerable attention in many movement disorders, but especially in ET. Deep cerebellar nuclei and Purkinje cells interact with ION closely via the olivocerebellar pathway, which comprises of the Guillain Mollaret triangle along with rubro-olivary fibers and the dentato-rubro tract (Fig. 1). From the ION, olivocerebellar fibers cross the contralateral inferior cerebellar peduncle to reach the cerebellar cortex, then pass from the cerebellar cortex to the contralateral dentate nucleus. The terminal portion of the axon of an ION, climbing fiber, makes powerful synapses with Purkinje cells, triggering a burst of spikes that is referred to as the complex spike (Llinas, 2014).

Recording electrophysiological activity of the cerebellum has been challenging due to its anatomical configuration. However, in recent years there have been considerable advances in studying the electrophysiology of the cerebellum using magnetoencephalogram (MEG) and electroencephalogram (EEG) (Andersen, Jerbi, & Dalal, 2020; Samuelsson, Sundaram, Khan, Sereno, & Hamalainen, 2020). The MEG study revealed a tremor network of areas consisting of the contralateral primary motor cortex, premotor cortex, thalamus, brainstem, and ipsilateral cerebellum (Schnitzler, Munks, Butz, Timmermann, & Gross, 2009). In a more recent study, cerebellar EEG were successfully recorded in ET patients that showed 4–12 Hz oscillatory activity. This was distinct from the occipital alpha rhythm and was correlated with tremor severity (Pan et al., 2020).

The role of the cerebellum in ET as an electrical disorder can also be studied by non-invasive modulation of cerebellar activity with transcranial magnetic stimulation (TMS), transcranial direct current stimulation (tDCS), or transcranial alternating current stimulation (tACS) (Maas, Helmich, & van de Warrenburg, 2020). The functional influence of the cerebellum on the cerebello-thalamo-cortical pathway can be assessed by cerebellar inhibition (CBI), which quantifies the inhibitory projection from the Purkinje cells to the dentato-thalamo-cortical pathway (Ugawa, Uesaka, Terao, Hanajima, & Kanazawa, 1995). CBI studies in ET have reported conflicting outcomes. One study found CBI to be significantly reduced compared to healthy controls (Hanajima et al., 2016), whereas another found it to be normal (Pinto, Lang, & Chen, 2003). Various interventional studies have been conducted in ET targeting the cerebellum. An open-label trial of 1-Hz repetitive TMS of the cerebellum showed transient but significant improvement in tremor clinical rating scale and accelerometric values after one active session of rTMS compared to sham stimulation (Gironell et al., 2002), but a single-blinded, randomized, sham-controlled study did not show any significant difference in tremor score or TMS measures between active and sham stimulation groups (Shin, Hallett, & Sohn, 2019). Cathodal tDCS over bilateral cerebellar hemisphere for 10 sessions did not result in clinical benefit (Gironell et al., 2014). More recently, tACS phase-locked to hand tremor showed that ET amplitude can be efficiently suppressed within a few seconds in a double-blinded, sham-controlled experiment, although there were some non-responders which reflects the heterogeneity of ET (Schreglmann et al., 2021). We may soon anticipate seeing more effective non-invasive brain stimulation targeting the cerebellum in more precise mechanisms as we gain better understandings of the electrophysiologic role of the cerebellum in ET.

2.3.3 Thalamus

The thalamus is a key structure for oscillatory entrainment of subcortical and cortical nodes. The ventral intermediate nucleus (Vim) receives primarily excitatory input from the cerebellum and projects to the primary motor cortex (Fig. 1). The thalamus is considered to play a role in mechanisms for switching from a non-oscillatory pattern during rest to oscillatory patterns during tremor. Functional imaging studies in ET patients have demonstrated increased bilateral cerebellar activity not only during tremors but also at rest without tremors (Passamonti, Cerasa, & Quattrone, 2012; Sharifi et al., 2014), yet tremors in ET mostly occur with movement only. Thalamus may act as a gate for tremors in ET between the cerebellum and motor cortex (Milosevic et al., 2018). Vim is known to exhibit tonic irregular firing at rest and kinesthetic movement-related responses with both passive and active movements (Lenz et al., 1990). Vim also shows periodic activity at tremor frequency (Lenz et al., 1994) but do not exhibit oscillatory activity at this frequency at rest (Hua & Lenz, 2005). Periodic activity of Vim neurons is coherent with EMG during posture but not rest (Marsden, Ashby, Limousin-Dowsey, Rothwell, & Brown, 2000). Among different functional classes of neurons in the thalamus, neurons that fire during voluntary movement have the highest proportion of neurons that show tremor-related activity in Vim (Hua & Lenz, 2005). Taken together, the evidence shows that postural tremor is mediated and modulated by voluntary motor neurons in Vim.

Deep-brain stimulation (DBS) targeting the Vim is used for the treatment of medically-intractable ET (Haubenberger & Hallett, 2018). Vim DBS presumably resets the oscillatory rhythm in the cerebellar-thalamo-cortical circuit and prevents propagation of the tremor oscillations to the limbs (Anderson, Hu, Pittman, & Kiss, 2004). Microelectrode recording during DBS surgery shows that thalamic neural inhibition is necessary for tremor suppression (Milosevic et al., 2018). Vim seems to function as a gate within the cerebello-thalamo-cortical network and Vim DBS may work as a filter to uncouple the thalamo-cortical loop from the cortico-spinal reflex loops (Milosevic et al., 2018). This gating property of Vim can be employed to read out cerebellar outflow in ET patients. One study recorded Vim activity during DBS surgery in two groups of ET patients—postural tremor and intention tremor—as well as cerebellar tremor (Zakaria et al., 2013). Thalamic neurons in patients with either intention ET or cerebellar tremor had lower firing rates and lower spike × EMG coherence than those in patients with postural ET. This result suggests that postural tremor in ET may be due to excitatory oscillatory input from the cerebellum, but intention tremor in ET may be associated with cerebellar dysfunction resulting in deafferentation. This difference in mechanism seems to be consistent with clinical observation of cases in which postural ET progresses to intention tremor over time (Deuschl, Wenzelburger, Loffler, Raethjen, & Stolze, 2000).

2.3.4 Motor cortex

The primary motor cortex (M1) receives projections from the thalamus as well as other cortical regions including the premotor cortex, the supplementary motor area, and multiple parietal regions. Several EEG/MEG studies clearly demonstrated rhythmic cortical activity in the motor cortex that is coherent with ET (Hellwig et al., 2001; Raethjen, Govindan, Kopper, Muthuraman, & Deuschl, 2007; Schnitzler et al., 2009). M1 is the area showing the strongest cerebro-muscular coupling in ET in an MEG study (Schnitzler et al., 2009). However, some studies that investigated corticomuscular coherence over time showed only intermittent corticomuscular coherence despite sufficient tremor power (Raethjen et al., 2007; Sharifi et al., 2017). These results suggest that motor cortex activity may not directly drive the tremor in ET, at least in sustaining tremors. A recent EEG-EMG study used a partial directed coherence method, similar to Granger causality analysis; this showed that corticomuscular coherence was much higher in the afferent direction (EMG to EEG) than the efferent flow (EEG to EMG) (Sharifi et al., 2021). This result suggests that corticomuscular coherence seen in ET could be mostly due to an afferent signal.

Even with some controversy regarding the importance of cortical contribution in ET, M1 has been tested multiple times as a target of non-invasive neurostimulation given its easy accessibility and read-out. Supra-threshold single-pulse TMS has been shown to reset the tremor (Britton et al., 1993), implying that brain stimulation over the motor cortex could potentially modulate the oscillatory mechanisms responsible for tremors. Inhibitory subthreshold continuous theta burst stimulation (inhibitory rTMS) over M1 in ET patients reduced tremor amplitude but did not alter the tremor frequency (Chuang, Huang, Lu, & Chen, 2014; Hellriegel, Schulz, Siebner, Deuschl, & Raethjen, 2012). Also, continuous theta burst stimulation applied to the M1 reduced short interval intracortical inhibition and prolonged cortical silent period only in the healthy control group and not in the ET group (Chuang et al., 2014). This suggests that inhibitory circuits within M1 are aberrant and less modifiable in patients with ET.

2.4 ET is a network disorder

Another potential mechanism of tremor is that oscillations may be generated within loops secondary to network instability, rather than the lesion or dysfunction of the individual component of the motor network. It appears that all the above mentioned centers become active simultaneously as the central tremor network in ET (Schnitzler et al., 2009). From EEG–EMG coherence studies, we know that the tremor in ET can be maintained without cortical participation (Raethjen et al., 2007). It implies that there may be a dynamic shifting within the motor network and that all network components may act as oscillators and resonators on their own, entraining each other.

The same topographic motor network is involved for different tremors including ET, Parkinson's tremor, and voluntary rhythmic movements. However, the dynamics and interaction within the network differ between voluntary tremor and pathologic tremor. Using a partial directed coherence method, an EEG-EMG study revealed a mainly unidirectional flow of information from the diencephalon to cortex in voluntary tremor, as opposed to a bidirectional subcortico-cortical flow in Parkinson's tremor and ET promoting uncontrollable loop oscillations (Muthuraman et al., 2012). Thus, a bidirectional oscillatory loop-type of interaction appears to be an important mechanism in pathological tremors. Although the main landscape of the tremor network has been confirmed repeatedly in many studies, there is evidence that different tremors may be related with cerebellar activity coming from slightly different area of the cerebellum (Muthuraman et al., 2018). A study using high resolution EEG showed that the topography of cerebellar involvement differed among Parkinson's, essential, and mimicked tremors, suggesting different cerebellar mechanisms in tremor genesis. The same study showed that the cerebellar sources mapped to motor cortex in ET and Parkinson's tremor and to premotor cortex in controls mimicking tremors based on the Buckner atlas (Buckner, Krienen, Castellanos, Diaz, & Yeo, 2011). Additionally, the flow of oscillatory activity was primarily from the cerebellum to the cerebral cortex and EMG in ET and Parkinson's tremor, but activity flow was in the opposite direction in controls mimicking tremors.

Neuroimaging studies give insight into the pathophysiology of ET as a network disorder. Volumetric imaging so far shows no consistent pattern of atrophy in a particular region of the brain (Holtbernd & Shah, 2021). However, diffusion tensor imaging (DTI) shows more consistent microstructural alterations of the cerebellothalamocortical tract. Mean diffusivity of fiber tracts connecting the Vim with motor and supplementary motor cortical regions was found decreased (Revuelta, McGill, Jensen, & Bonilha, 2019). Fraction anisotropy was decreased and mean diffusivity was increased in cerebellar peduncles in ET compared to healthy controls (Tikoo et al., 2020). Functional Magnetic Resonance Imaging (fMRI) can be helpful to detect circuit-level tremor-related activity. The first task-based fMRI study in ET patients identified increased activity in the contralateral sensorimotor cortices, thalamus, and globus pallidus and bilateral overactivation of the cerebellar hemispheres and dentate nuclei during arm posturing (Bucher, Seelos, Dodel, Reiser, & Oertel, 1997). fMRI with simultaneous EMG recording can detect brain activity associated with changes in tremor power during scanning (Madelein van der Stouwe, Nieuwhof, & Helmich, 2020). With this technique, studies showed increased cerebellar activity associated with tremors, supporting the pathophysiological role of the cerebellum as a potential driving oscillator (Broersma et al., 2016; Bucher et al., 1997; Buijink, Broersma, et al., 2015; Buijink, van der Stouwe, et al., 2015). On the other hand, studies that investigated functional connectivity at rest showed more widespread changes in functional connectivity between the cerebellum and motor network, implicating that ET might be due to cerebellar dysfunction and subsequent decoupling with the motor network (Buijink, Broersma, et al., 2015; Buijink, van der Stouwe, et al., 2015; Fang et al., 2016; Tikoo et al., 2020). Vim seed-based functional connectivity analysis of resting-state fMRI showed decreased functional connectivity in the cerebellum and increased connectivity in the motor cortex (Holtbernd & Shah, 2021). A dentate nucleus seed-based approach revealed a decrease in functional connectivity with cortical, subcortical, and cerebellar areas in ET patients (Hellriegel et al., 2012). In the view of widespread functional alterations reported above, it is likely that there is no sole driver of tremor genesis but rather constitutes a major hub within a multiple oscillator tremor network, in line with neurophysiological data.

3. Essential tremor is not a primary neurodegenerative disorder

3.1 Definition and characteristics of neurodegenerative disorders

We should clarify the definition of neurodegeneration before we argue whether ET is a neurogenerative disease or not. The term neurodegeneration is a combination of two words—"neuro," referring to nerve cells (i.e., neurons) and "degeneration," referring to progressive damage. There are many different neurodegenerative diseases, estimated to be a few hundreds, and each neurodegenerative disease has defining clinical features that are explained by a characteristic distribution of neuronal loss (Przedborski, Vila, & Jackson-Lewis, 2003).

The risk of being affected by a neurodegenerative disease increases with age. However, there can be young-onset neurodegenerative diseases, especially with genetic conditions such as Huntington's disease. The exact cause for many neurodegenerative diseases remains unknown, but it essentially comes down to relative roles of genetic and environmental factors.

Clinically, the onset of neurodegenerative disorders is usually insidious. Disease onset always precedes the symptom onset, often by many years, because of significant cellular redundancy in neuronal pathways. All neurodegenerative diseases have a preclinical phase. A symptomatic phase begins when the number of residual neurons in a given pathway decreases below the threshold needed to maintain normal functioning of the affected pathway (Bloem, Okun, & Klein, 2021; Scheltens et al., 2021). The symptom onset is followed by relentless progression of the disease, and the progression and severity of symptoms are expected to be greater than ones that come with normal aging. Some degree of degeneration can be seen with normal aging. For example, the dopaminergic neurons are lost with aging (Kaasinen & Rinne, 2002) and beta amyloid is seen in up to 20% of healthy older people (Rodrigue, Kennedy, & Park, 2009) who do not have Parkinson's disease (PD) or Alzheimer's disease (AD). Therefore, progressive neuronal death throughout the natural course of the disease constitutes the core definition of neurodegenerative disorders.

Pathologically, a focal loss of neurons with reactive gliosis is commonly seen in neurodegenerative disorders. However, in many neurodegenerative diseases, the different lesions appear to be functionally and anatomically interconnected, perhaps via transneuronal transmission of the disease (Cauda et al., 2018; Franzmeier et al., 2020; Saper, Wainer, & German, 1987).

3.2 Is ET a neurodegenerative disease?

To answer this question, we will examine the disease onset, progression, association with other neurodegenerative diseases, and neuropathology of ET.

3.2.1 Disease onset and progression

The prevalence of ET dramatically increases with aging, just like any other neurodegenerative disease. However, ET is different from other neurological disorders because the age of onset has a bimodal distribution (Louis & Dogu, 2007). Positive family history of ET and alcohol responsiveness are more common among young-onset ET (Hopfner et al., 2016; Louis et al., 2001). Older age of onset is independently associated with a faster rate of tremor progression (Hopfner et al., 2016; Louis et al., 2009; Louis, Ford, & Barnes, 2000). More recent postmortem study showed that both early onset ET and late onset ET had similar cerebellar pathological changes, suggesting that late onset ET has more rapid progression thereby reaching the same endpoint at a similar age of death (Kuo et al., 2017). At first glance, it seems that late-onset ET follows typical features of other neurodegenerative disorders such as Parkinson's disease (Kempster et al., 2007). But late-onset ET may represent more heterogenous subgroups of patients than young-onset ET. Action tremors in the elderly can be commonly labeled as ET. However, there is some evidence that there are non-ET tremors in the elderly associated with age-related decline in fine motor skills (Deuschl, Petersen, Lorenz, & Christensen, 2015). In a large epidemiological study, tremors without lack of definite ET features among elderly (>70 years) was associated with higher mortality and cognitive impairment, whereas definite ET was not (Deuschl et al., 2015). This non-ET tremors in the elderly was referred to as aging-related tremor by Deuschl et al. (2015) or as senile tremor by others (Friedlander, 1956; Louis, Wendt, & Ford, 2000; Marshall, 1961). There is evidence that aging-related tremors and

ET may come from different tremor networks (Muthuraman et al., 2015). An EEG study using a source analysis revealed the typical cortico-ponto-cerebello-thalamo-cortical network for classical ET, but patients with aging-related tremor showed a cortico-thalamic network only (Muthuraman et al., 2015). However, the term of "age-related tremor" or 'senile tremor' have not been widely accepted as its nosologic entity is unclear.

Clinical symptoms of ET evolve with time. Tremor severity increases with aging and disease duration, albeit slowly (Louis, Jurewicz, & Watner, 2003). Existing data suggest that the progression rate of tremor severity is approximately 3%–5% per year (Elble, Higgins, & Hughes, 1992; Louis, Agnew, Gillman, Gerbin, & Viner, 2011). ET frequency is also strongly correlated with the patient's age in a linear negative relationship (−0.06 to 0.08 Hz/year), but there is no correlation with duration of illness (Elble, 2000; Elble, Higgins, Leffler, & Hughes, 1994). Postural tremor occurs in the beginning and intention tremor develops later during the course of the disease, which causes more significant disability. The onset of intention tremor was associated with age but not with disease duration in one cohort (Deuschl et al., 2000), but this was not the case in another cohort where intention tremor was associated with disease duration and tremor severity but not with age (Louis, Frucht, & Rios, 2009; Louis, Huey, & Cosentino, 2021). There are other features of ET that correlate with disease duration. Resting tremor in ET is associated with both age and disease duration (in some studies) as well as tremor severity (Louis, Hernandez, & Michalec, 2015). Head tremor seems to start as an infrequent tremor in one direction and becomes more frequent and complex as the disease progress with longer duration and higher severity (Robakis & Louis, 2016).

Some of the characteristics of disease progression in ET showed that aging seems to play as a common denominator in clinical pictures of ET: increase in tremor amplitude, decrease in tremor frequency, and onset of intention and rest tremor. It may be hard to tease out whether the primary drive of clinical progression is aging or disease progression or combination of both. Normal age-related neural loss can be superimposed on a non-progressive lesion, which could result in an increase in clinical severity and an apparent worsening with age. However, there seems to be enough clinical evidence for progressive cerebellar dysregulation associated with disease (Deuschl et al., 2000; Hanajima et al., 2016; Helmchen et al., 2003; Louis & Faust, 2020c; Stolze, Petersen, Raethjen, Wenzelburger, & Deuschl, 2001).

3.2.2 Association with other neurodegenerative diseases

The relationship between ET and other neurodegenerative diseases, mainly PD and dementia, has been a subject of debate for some time. Prevalence of ET increases with age; therefore one could potentially develop two separate common disorders. Alternatively, and more convincingly, there could be a biological link that explains the association.

Clinical observations seem clear that some ET patients go on to develop PD later in life (Koller, Busenbark, & Miner, 1994; Minen & Louis, 2008). Epidemiologic studies provide a model that ET is a risk factor for PD (Benito-Leon, Louis, Bermejo-Pareja, & Neurological Disorders in Central Spain Study, 2009; Tan, Lee, Fook-Chong, & Lum, 2008). Family studies also suggest that family members of those with PD have an increased risk of ET (Costello, Bordelon, Bronstein, & Ritz, 2010; Spanaki & Plaitakis, 2009). Some ET patients showed Lewy body pathology in the locus coeruleus, suggesting a common pathology with PD (Louis et al., 2007), but some argue that this is largely an incidental finding with a similar frequency of Lewy body seen in controls (Shill, Adler, & Beach, 2012). Existing studies report long latency between onset of ET and parkinsonism—mean duration 14–21 years (Bellows & Jankovic, 2021; Minen & Louis, 2008), although one of these study showed that it was less than 5 years in 38.5% of ET-PD patients (Minen & Louis, 2008). It is unclear whether underlying pathogenic mechanism for ET-PD is different for short and long latencies.

Association between ET and dementia has also been noted in the Neurological Disorders in Central Spain [NEDICES] Study. A population-based case-control study in the NEDICES cohort showed that patients with ET performed slightly but significantly more poorly on formal neuropsychological testing (Benito-León, Louis, & Bermejo-Pareja, 2006b). A prospective study in this cohort demonstrated that ET patients had a faster rate of cognitive decline in a 3-year span (Louis, Benito-León, Vega-Quiroga, & Bermejo-Pareja, 2010). The same group also reported significant cognitive decline even in a premotor phase of ET compared to controls (Benito-León, Louis, Sánchez-Ferro, & Bermejo-Pareja, 2013). Similar results were reported from a separate cohort (Radler et al., 2020; Thawani, Schupf, & Louis, 2009). But another NEDICES study showed that ET cases with tremor onset after age 65 were 70% more likely to develop dementia than were controls, whereas ET cases with tremor onset at age ≤ 65 and controls were equally likely to develop dementia (Benito-León, Louis, & Bermejo-Pareja, 2006a).

The abovementioned evidence certainly points toward an association between ET and other neurodegenerative diseases. Late-onset ET seems to be more associated with dementia but this relationship is not clear for the association with PD.

3.2.3 Pathology of ET

Although ION received the most attention as a potential ET generator in early years, no histopathological abnormalities have been found in ION to date (Louis, Babij, Cortés, Vonsattel, & Faust, 2013). As more evidence emerged pointing toward cerebellar disturbance in ET based on clinical observation and electrophysiological studies (Deuschl et al., 2000; Helmchen et al., 2003; Louis & Faust, 2020b; Stolze et al., 2001), the cerebellum became a promising target of pathology studies in recent years. Indeed, many studies have shown pathological neuronal changes in the cerebellar cortex encompassing swelling and thickening of PC axons; swellings, regressive changes and loss of spines in PC dendrites; remodeling of PC axons; PC death; remodeling of basket cell axons; heterotopic PCs; reorganization of climbing fibers; and alterations in astroglia (for review, see "Essential tremor pathology: neurodegeneration and reorganization of neuronal connections" (Louis & Faust, 2020a)). Although there are clinical differences in early vs late onset ET, pathology reports in two groups showed similar findings with a notion that they seem to reach the same pathological endpoints at a similar age of death (Kuo, Wang, et al., 2017).

The underlying pathophysiology of these observations could be a neurodegenerative process. Alternative view would be that cerebellar pathology may be secondary findings of ET, in a similar manner how seizures can cause neuronal damage (Duncan, 2002; Holmes, 2002; Meldrum, 1997). However, this hypothesis was never tested in tremor. It seems clear, though, that cerebellar pathology in ET is different from other degenerative cerebellar diseases such as spinocerebellar ataxia (SCA) or multiple system atrophy (MSA) (Kuo et al., 2017; Louis et al., 2019; Louis, Kuo, Vonsattel, & Faust, 2014). There is difference in terms of degree of pathological changes in PC morphology with ET generally at the low end and SCA1/2/6 or MSA at the high end of severity (Louis et al., 2019). But there are also distinctive features seen in ET but not in SCA or MSA including redistribution of climbing fiber synapses extending into the outer 20% of molecular layer, increased length of climbing fiber and more complex branching (Kuo, Lin, et al., 2017; Louis et al., 2019).

Finally, pathologic staging of ET as a neurodegenerative disease model is yet to be verified. If we believe that ET is a neurodegenerative disorder, one would expect to see progressive neuronal changes from early to late stage as in Braak staging in PD and AD. Louis et al. provided a provisional model for the degenerative cascade in ET pathology (Louis & Faust, 2020a), but more work is needed to consolidate this model. Current limitations are caused by the small number of postmortem ET brains and the fact that pathological changes associated with ET have only been observed in cross-sectional studies conducted at the time of death rather than in longitudinal studies. Methodological limitation comes from the fact that ET brains with short disease duration are extremely rare, which limits the ability to perform staging. Therefore, whether the variety of histopathological abnormalities observed in ET brains represents different stages of the disease course or heterogeneity of the disease needs further verification.

4. Conclusion and future direction

This author concludes that ET is conceptually an electric disorder with excessive oscillatory activity within the motor network that includes the inferior olive, cerebellum, thalamus, and motor cortex as major hubs. The evidence indicates that ET may not arise from a single center oscillator and that there is a dynamic oscillatory disturbance of the cerebello-thalamo-cortical network resulting in ET. However, we also reviewed evidence of ET progression over time, evolution from postural tremor to intention tremor with signs of cerebellar dysfunction, association with other neurodegenerative disorders, and pathological changes in the brain, especially in the cerebellum.

ET can be considered as a multi-layer disorder. In the new consensus statement, tremor is classified by two axes: clinical features (Axis 1) and etiology (Axis 2) (Bhatia et al., 2018). ET is defined as a tremor syndrome in axis 1 with largely homogenous phenotype, but there can be heterogenous etiologies in axis 2. We could consider three different theoretical disease models of ET from electrical and neurodegenerative disorder perspectives. The first model is that ET itself is fundamentally an electrical disorder from a pathophysiology point of view and that the effects of aging on top of the non-neurodegenerative ET process that ultimately leads to neurodegeneration. This aging effect on ET may result in changes in oscillatory motor network function, presented as decreased tremor frequency and increased tremor severity with aging. Another possibility is that underlying neurodegeneration leads to functional and electrical changes that manifests as tremor in ET. Tremor in PD could be explained by this second model (Helmich, Hallett, Deuschl, Toni, & Bloem, 2012). Lastly, ET as an electric disorder may accelerate the neurodegenerative process, which then involves the cerebellum and leads to cerebellar dysfunction. As an analogy, metabolic syndrome itself is not considered a degenerative disorder, but it causes chronic multi-system inflammation in the final pathway and increases the risk of neurodegenerative disorders such as PD and dementia (Craft, 2009; Zhang & Tian, 2014). Chronic motor network dysfunction and excessive oscillation may somehow lead to accelerated aging of the involved system and secondary pathological changes.

Although our understanding of the biology of ET has advanced in recent years, further work is clearly needed as there are still many unanswered questions. To determine whether ET is a neurodevelopmental electrical circuit disorder or a neurodegenerative disorder, we need more longitudinal neurophysiology, neuroimaging, and pathology studies accompanied with careful phenotyping.

References

Andersen, L. M., Jerbi, K., & Dalal, S. S. (2020). Can EEG and MEG detect signals from the human cerebellum? *NeuroImage*, *215*, 116817. https://doi.org/10.1016/j.neuroimage.2020.116817.

Anderson, T., Hu, B., Pittman, Q., & Kiss, Z. H. (2004). Mechanisms of deep brain stimulation: An intracellular study in rat thalamus. *The Journal of Physiology*, *559*(Pt 1), 301–313. https://doi.org/10.1113/jphysiol.2004.064998.

Bellows, S. T., & Jankovic, J. (2021). Phenotypic features of isolated essential tremor, essential tremor plus, and essential tremor-Parkinson's disease in a movement disorders clinic. *Tremor and Other Hyperkinetic Movements (N. Y.)*, *11*, 12. https://doi.org/10.5334/tohm.581.

Benito-León, J. (2014). Essential tremor: A neurodegenerative disease? *Tremor and Other Hyperkinetic Movements (N. Y.)*, *4*, 252. https://doi.org/10.7916/d8765cg0.

Benito-León, J., Louis, E. D., & Bermejo-Pareja, F. (2006a). Elderly-onset essential tremor is associated with dementia. *Neurology*, *66*(10), 1500–1505. https://doi.org/10.1212/01.wnl.0000216134.88617.de.

Benito-León, J., Louis, E. D., & Bermejo-Pareja, F. (2006b). Population-based case-control study of cognitive function in essential tremor. *Neurology*, *66*(1), 69–74. https://doi.org/10.1212/01.wnl.0000192393.05850.ec.

Benito-Leon, J., Louis, E. D., Bermejo-Pareja, F., & Neurological Disorders in Central Spain Study Group. (2009). Risk of incident Parkinson's disease and parkinsonism in essential tremor: A population based study. *Journal of Neurology, Neurosurgery, and Psychiatry*, *80*(4), 423–425. https://doi.org/10.1136/jnnp.2008.147223.

Benito-León, J., Louis, E. D., Sánchez-Ferro, Á., & Bermejo-Pareja, F. (2013). Rate of cognitive decline during the premotor phase of essential tremor: A prospective study. *Neurology*, *81*(1), 60–66. https://doi.org/10.1212/WNL.0b013e318297ef2b.

Bermejo-Pareja, F. (2011). Essential tremor- -A neurodegenerative disorder associated with cognitive defects? *Nature Reviews. Neurology*, 7(5), 273–282. https://doi.org/10.1038/nrneurol.2011.44.

Bhatia, K. P., Bain, P., Bajaj, N., Elble, R. J., Hallett, M., Louis, E. D., et al. (2018). Consensus statement on the classification of tremors. From the task force on tremor of the International Parkinson and Movement Disorder Society. *Movement Disorders*, *33*(1), 75–87. https://doi.org/10.1002/mds.27121.

Bloem, B. R., Okun, M. S., & Klein, C. (2021). Parkinson's disease. *Lancet*, *397*(10291), 2284–2303. https://doi.org/10.1016/s0140-6736(21)00218-x.

Britton, T. C., Thompson, P. D., Day, B. L., Rothwell, J. C., Findley, L. J., & Marsden, C. D. (1993). Modulation of postural wrist tremors by magnetic stimulation of the motor cortex in patients with Parkinson's disease or essential tremor and in normal subjects mimicking tremor. *Annals of Neurology*, *33*(5), 473–479. https://doi.org/10.1002/ana.410330510.

Broersma, M., van der Stouwe, A. M. M., Buijink, A. W. G., de Jong, B. M., Groot, P. F. C., Speelman, J. D., et al. (2016). Bilateral cerebellar activation in unilaterally challenged essential tremor. *NeuroImage: Clinical*, *11*, 1–9. https://doi.org/10.1016/j.nicl.2015.12.011.

Bucher, S. F., Seelos, K. C., Dodel, R. C., Reiser, M., & Oertel, W. H. (1997). Activation mapping in essential tremor with functional magnetic resonance imaging. *Annals of Neurology*, *41*(1), 32–40. https://doi.org/10.1002/ana.410410108.

Buckner, R. L., Krienen, F. M., Castellanos, A., Diaz, J. C., & Yeo, B. T. (2011). The organization of the human cerebellum estimated by intrinsic functional connectivity. *Journal of Neurophysiology*, *106*(5), 2322–2345. https://doi.org/10.1152/jn.00339.2011.

Buijink, A. W., Broersma, M., van der Stouwe, A. M., van Wingen, G. A., Groot, P. F., Speelman, J. D., et al. (2015). Rhythmic finger tapping reveals cerebellar dysfunction in essential tremor. *Parkinsonism & Related Disorders*, *21*(4), 383–388. https://doi.org/10.1016/j.parkreldis.2015.02.003.

Buijink, A. W., van der Stouwe, A. M., Broersma, M., Sharifi, S., Groot, P. F., Speelman, J. D., et al. (2015). Motor network disruption in essential tremor: A functional and effective connectivity study. *Brain*, *138*(Pt. 10), 2934–2947. https://doi.org/10.1093/brain/awv225.

Busenbark, K. L., Nash, J., Nash, S., Hubble, J. P., & Koller, W. C. (1991). Is essential tremor benign? *Neurology*, *41*(12), 1982–1983. https://doi.org/10.1212/wnl.41.12.1982.

Cauda, F., Nani, A., Manuello, J., Premi, E., Palermo, S., Tatu, K., et al. (2018). Brain structural alterations are distributed following functional, anatomic and genetic connectivity. *Brain*, *141*(11), 3211–3232. https://doi.org/10.1093/brain/awy252.

Chuang, W. L., Huang, Y. Z., Lu, C. S., & Chen, R. S. (2014). Reduced cortical plasticity and GABAergic modulation in essential tremor. *Movement Disorders*, *29*(4), 501–507. https://doi.org/10.1002/mds.25809.

Costello, S., Bordelon, Y., Bronstein, J., & Ritz, B. (2010). Familial associations of Alzheimer disease and essential tremor with Parkinson disease. *European Journal of Neurology*, *17*(6), 871–878. https://doi.org/10.1111/j.1468-1331.2010.02974.x.

Craft, S. (2009). The role of metabolic disorders in Alzheimer disease and vascular dementia: Two roads converged. *Archives of Neurology*, *66*(3), 300–305. https://doi.org/10.1001/archneurol.2009.27.

Deuschl, G., & Elble, R. J. (2000). The pathophysiology of essential tremor. *Neurology*, *54*(11 Suppl. 4), S14–S20.

Deuschl, G., & Elble, R. (2009). Essential tremor- -neurodegenerative or nondegenerative disease towards a working definition of ET. *Movement Disorders*, *24*(14), 2033–2041. https://doi.org/10.1002/mds.22755.

Deuschl, G., Krack, P., Lauk, M., & Timmer, J. (1996). Clinical neurophysiology of tremor. *Journal of Clinical Neurophysiology*, *13*(2), 110–121. https://doi.org/10.1097/00004691-199603000-00002.

Deuschl, G., Petersen, I., Lorenz, D., & Christensen, K. (2015). Tremor in the elderly: Essential and aging-related tremor. *Movement Disorders*, *30*(10), 1327–1334. https://doi.org/10.1002/mds.26265.

Deuschl, G., Wenzelburger, R., Loffler, K., Raethjen, J., & Stolze, H. (2000). Essential tremor and cerebellar dysfunction clinical and kinematic analysis of intention tremor. *Brain*, *123*(Pt. 8), 1568–1580. https://doi.org/10.1093/brain/123.8.1568.

Duncan, J. S. (2002). Seizure-induced neuronal injury: Human data. *Neurology*, *59*(9 Suppl. 5), S15–S20. https://doi.org/10.1212/wnl.59.9_suppl_5.s15.

Elble, R. J. (1995). The role of aging in the clinical expression of essential tremor. *Experimental Gerontology*, *30*(3–4), 337–347. https://doi.org/10.1016/0531-5565(94)00055-8.

Elble, R. J. (1996). Central mechanisms of tremor. *Journal of Clinical Neurophysiology*, *13*(2), 133–144. https://doi.org/10.1097/00004691-199603000-00004.

Elble, R. J. (2000). Essential tremor frequency decreases with time. *Neurology*, *55*(10), 1547–1551. https://doi.org/10.1212/wnl.55.10.1547.

Elble, R. J., Higgins, C., & Hughes, L. (1992). Longitudinal study of essential tremor. *Neurology*, *42*(2), 441–443. https://doi.org/10.1212/wnl.42.2.441.

Elble, R. J., Higgins, C., & Hughes, L. (1994). Essential tremor entrains rapid voluntary movements. *Experimental Neurology*, *126*(1), 138–143. https://doi.org/10.1006/exnr.1994.1050.

Elble, R. J., Higgins, C., Leffler, K., & Hughes, L. (1994). Factors influencing the amplitude and frequency of essential tremor. *Movement Disorders*, *9*(6), 589–596. https://doi.org/10.1002/mds.870090602.

Engel, A. K., & Fries, P. (2010). Beta-band oscillations- -signalling the status quo? *Current Opinion in Neurobiology*, *20*(2), 156–165. https://doi.org/10.1016/j.conb.2010.02.015.

Fang, W., Chen, H., Wang, H., Zhang, H., Puneet, M., Liu, M., et al. (2016). Essential tremor is associated with disruption of functional connectivity in the ventral intermediate nucleus- -motor cortex- -cerebellum circuit. *Human Brain Mapping*, *37*(1), 165–178. https://doi.org/10.1002/hbm.23024.

Farmer, S. F., Bremner, F. D., Halliday, D. M., Rosenberg, J. R., & Stephens, J. A. (1993). The frequency content of common synaptic inputs to motoneurones studied during voluntary isometric contraction in man. *The Journal of Physiology*, *470*, 127–155. https://doi.org/10.1113/jphysiol.1993.sp019851.

Franzmeier, N., Neitzel, J., Rubinski, A., Smith, R., Strandberg, O., Ossenkoppele, R., et al. (2020). Functional brain architecture is associated with the rate of tau accumulation in Alzheimer's disease. *Nature Communications*, *11*(1), 347. https://doi.org/10.1038/s41467-019-14159-1.

Friedlander, W. J. (1956). Characteristics of postural tremor in normal and in various abnormal states. *Neurology*, *6*(10), 716–724. https://doi.org/10.1212/wnl.6.10.716.

Gironell, A., Kulisevsky, J., Lorenzo, J., Barbanoj, M., Pascual-Sedano, B., & Otermin, P. (2002). Transcranial magnetic stimulation of the cerebellum in essential tremor: A controlled study. *Archives of Neurology*, *59*(3), 413–417. https://doi.org/10.1001/archneur.59.3.413.

Gironell, A., Kulisevsky, J., Pascual-Sedano, B., & Barbanoj, M. (2004). Routine neurophysiologic tremor analysis as a diagnostic tool for essential tremor: A prospective study. *Journal of Clinical Neurophysiology*, *21*(6), 446–450. Retrieved from https://journals.lww.com/clinicalneurophys/Fulltext/2004/11000/Routine_Neurophysiologic_Tremor_Analysis_as_a.9.aspx.

Gironell, A., Martínez-Horta, S., Aguilar, S., Torres, V., Pagonabarraga, J., Pascual-Sedano, B., et al. (2014). Transcranial direct current stimulation of the cerebellum in essential tremor: A controlled study. *Brain Stimulation*, 7(3), 491–492. https://doi.org/10.1016/j.brs.2014.02.001.

Graybiel, A. M., Aosaki, T., Flaherty, A. W., & Kimura, M. (1994). The basal ganglia and adaptive motor control. *Science*, *265*(5180), 1826–1831. https://doi.org/10.1126/science.8091209.

Hallett, M. (1998). Overview of human tremor physiology. *Movement Disorders*, *13*(Suppl. 3), 43–48. https://doi.org/10.1002/mds.870131308.

Hanajima, R., Tsutsumi, R., Shirota, Y., Shimizu, T., Tanaka, N., & Ugawa, Y. (2016). Cerebellar dysfunction in essential tremor. *Movement Disorders*, *31*(8), 1230–1234. https://doi.org/10.1002/mds.26629.

Haubenberger, D., & Hallett, M. (2018). Essential tremor. *The New England Journal of Medicine*, *378*(19), 1802–1810. https://doi.org/10.1056/NEJMcp1707928.

Hellriegel, H., Schulz, E. M., Siebner, H. R., Deuschl, G., & Raethjen, J. H. (2012). Continuous theta-burst stimulation of the primary motor cortex in essential tremor. *Clinical Neurophysiology*, *123*(5), 1010–1015. https://doi.org/10.1016/j.clinph.2011.08.033.

Hellwig, B., Haussler, S., Schelter, B., Lauk, M., Guschlbauer, B., Timmer, J., et al. (2001). Tremor-correlated cortical activity in essential tremor. *Lancet*, *357*(9255), 519–523. https://doi.org/10.1016/s0140-6736(00)04044-7.

Helmchen, C., Hagenow, A., Miesner, J., Sprenger, A., Rambold, H., Wenzelburger, R., et al. (2003). Eye movement abnormalities in essential tremor may indicate cerebellar dysfunction. *Brain*, *126*(Pt. 6), 1319–1332. https://doi.org/10.1093/brain/awg132.

Helmich, R. C., Hallett, M., Deuschl, G., Toni, I., & Bloem, B. R. (2012). Cerebral causes and consequences of parkinsonian resting tremor: A tale of two circuits? *Brain*, *135*(Pt. 11), 3206–3226. https://doi.org/10.1093/brain/aws023.

Holmes, G. L. (2002). Seizure-induced neuronal injury: Animal data. *Neurology*, *59*(9 Suppl. 5), S3–S6. https://doi.org/10.1212/wnl.59.9_suppl_5.s3.

Holtbernd, F., & Shah, N. J. (2021). Imaging the pathophysiology of essential tremor-a systematic review. *Frontiers in Neurology*, *12*, 680254. https://doi.org/10.3389/fneur.2021.680254.

Hopfner, F., Ahlf, A., Lorenz, D., Klebe, S., Zeuner, K. E., Kuhlenbäumer, G., et al. (2016). Early- and late-onset essential tremor patients represent clinically distinct subgroups. *Movement Disorders*, *31*(10), 1560–1566. https://doi.org/10.1002/mds.26708.

Horne, M. K., & Butler, E. G. (1995). The role of the cerebello-thalamo-cortical pathway in skilled movement. *Progress in Neurobiology*, *46*(2–3), 199–213. Retrieved from https://www.ncbi.nlm.nih.gov/pubmed/7568913.

Hua, S. E., & Lenz, F. A. (2005). Posture-related oscillations in human cerebellar thalamus in essential tremor are enabled by voluntary motor circuits. *Journal of Neurophysiology*, *93*(1), 117–127. https://doi.org/10.1152/jn.00527.2004.

Kaasinen, V., & Rinne, J. O. (2002). Functional imaging studies of dopamine system and cognition in normal aging and Parkinson's disease. *Neuroscience and Biobehavioral Reviews*, *26*(7), 785–793. https://doi.org/10.1016/s0149-7634(02)00065-9.

Kempster, P. A., Williams, D. R., Selikhova, M., Holton, J., Revesz, T., & Lees, A. J. (2007). Patterns of levodopa response in Parkinson's disease: A clinico-pathological study. *Brain*, *130*(Pt. 8), 2123–2128. https://doi.org/10.1093/brain/awm142.

Koch, M., Mostert, J., Heersema, D., & De Keyser, J. (2007). Tremor in multiple sclerosis. *Journal of Neurology*, *254*(2), 133–145. https://doi.org/10.1007/s00415-006-0296-7.

Koller, W. C., Busenbark, K., & Miner, K. (1994). The relationship of essential tremor to other movement disorders: Report on 678 patients. Essential Tremor Study Group. *Annals of Neurology*, *35*(6), 717–723. https://doi.org/10.1002/ana.410350613.

Köster, B., Lauk, M., Timmer, J., Winter, T., Guschlbauer, B., Glocker, F. X., et al. (1998). Central mechanisms in human enhanced physiological tremor. *Neuroscience Letters*, *241*(2–3), 135–138. https://doi.org/10.1016/s0304-3940(98)00015-9.

Kuo, S. H., Lin, C. Y., Wang, J., Sims, P. A., Pan, M. K., Liou, J. Y., et al. (2017). Climbing fiber-Purkinje cell synaptic pathology in tremor and cerebellar degenerative diseases. *Acta Neuropathologica*, *133*(1), 121–138. https://doi.org/10.1007/s00401-016-1626-1.

Kuo, S. H., Wang, J., Tate, W. J., Pan, M. K., Kelly, G. C., Gutierrez, J., et al. (2017). Cerebellar pathology in early onset and late onset essential tremor. *Cerebellum*, *16*(2), 473–482. https://doi.org/10.1007/s12311-016-0826-5.

Lai, R. Y., Tomishon, D., Figueroa, K. P., Pulst, S. M., Perlman, S., Wilmot, G., et al. (2019). Tremor in the degenerative cerebellum: Towards the understanding of brain circuitry for tremor. *Cerebellum*, *18*(3), 519–526. https://doi.org/10.1007/s12311-019-01016-6.

Lenz, F. A., Kwan, H. C., Dostrovsky, J. O., Tasker, R. R., Murphy, J. T., & Lenz, Y. E. (1990). Single unit analysis of the human ventral thalamic nuclear group. Activity correlated with movement. *Brain*, *113*(Pt. 6), 1795–1821. https://doi.org/10.1093/brain/113.6.1795.

Lenz, F. A., Kwan, H. C., Martin, R. L., Tasker, R. R., Dostrovsky, J. O., & Lenz, Y. E. (1994). Single unit analysis of the human ventral thalamic nuclear group. Tremor-related activity in functionally identified cells. *Brain*, *117*(Pt. 3), 531–543. https://doi.org/10.1093/brain/117.3.531.

Llinás, R. R. (1988). The intrinsic electrophysiological properties of mammalian neurons: Insights into central nervous system function. *Science*, *242*(4886), 1654–1664. https://doi.org/10.1126/science.3059497.

Llinas, R. (2014). The olivo-cerebellar system: A key to understanding the functional significance of intrinsic oscillatory brain properties. *Frontiers in Neural Circuits*, 7(96). https://doi.org/10.3389/fncir.2013.00096.

Llinás, R. R. (2014). Intrinsic electrical properties of mammalian neurons and CNS function: A historical perspective. *Frontiers in Cellular Neuroscience*, *8*, 320. https://doi.org/10.3389/fncel.2014.00320.

Louis, E. D. (2009). Essential tremors: A family of neurodegenerative disorders? *Archives of Neurology*, *66*(10), 1202–1208. https://doi.org/10.1001/archneurol.2009.217.

Louis, E. D., Agnew, A., Gillman, A., Gerbin, M., & Viner, A. S. (2011). Estimating annual rate of decline: Prospective, longitudinal data on arm tremor severity in two groups of essential tremor cases. *Journal of Neurology, Neurosurgery, and Psychiatry*, *82*(7), 761–765. https://doi.org/10.1136/jnnp.2010.229740.

Louis, E. D., Babij, R., Cortés, E., Vonsattel, J. P., & Faust, P. L. (2013). The inferior olivary nucleus: A postmortem study of essential tremor cases versus controls. *Movement Disorders*, *28*(6), 779–786. https://doi.org/10.1002/mds.25400.

Louis, E. D., Benito-León, J., Vega-Quiroga, S., & Bermejo-Pareja, F. (2010). Faster rate of cognitive decline in essential tremor cases than controls: A prospective study. *European Journal of Neurology*, *17*(10), 1291–1297. https://doi.org/10.1111/j.1468-1331.2010.03122.x.

Louis, E. D., & Dogu, O. (2007). Does age of onset in essential tremor have a bimodal distribution? Data from a tertiary referral setting and a population-based study. *Neuroepidemiology*, *29*(3–4), 208–212. https://doi.org/10.1159/000111584.

Louis, E. D., & Faust, P. L. (2020a). Essential tremor pathology: Neurodegeneration and reorganization of neuronal connections. *Nature Reviews. Neurology*, *16*(2), 69–83. https://doi.org/10.1038/s41582-019-0302-1.

Louis, E. D., & Faust, P. L. (2020b). Essential tremor within the broader context of other forms of cerebellar degeneration. *Cerebellum*, *19*(6), 879–896. https://doi.org/10.1007/s12311-020-01160-4.

Louis, E. D., & Faust, P. L. (2020c). Essential tremor: The most common form of cerebellar degeneration? *Cerebellum Ataxias*, 7, 12. https://doi.org/10.1186/s40673-020-00121-1.

Louis, E. D., Faust, P. L., Vonsattel, J. P., Honig, L. S., Henchcliffe, C., Pahwa, R., et al. (2009). Older onset essential tremor: More rapid progression and more degenerative pathology. *Movement Disorders*, *24*(11), 1606–1612. https://doi.org/10.1002/mds.22570.

Louis, E. D., Faust, P. L., Vonsattel, J. P., Honig, L. S., Rajput, A., Robinson, C. A., et al. (2007). Neuropathological changes in essential tremor: 33 cases compared with 21 controls. *Brain*, *130*(Pt. 12), 3297–3307. https://doi.org/10.1093/brain/awm266.

Louis, E. D., Ford, B., & Barnes, L. F. (2000). Clinical subtypes of essential tremor. *Archives of Neurology*, *57*(8), 1194–1198. https://doi.org/10.1001/archneur.57.8.1194.

Louis, E. D., Ford, B., Frucht, S., Barnes, L. F., Ming, X. T., & Ottman, R. (2001). Risk of tremor and impairment from tremor in relatives of patients with essential tremor: A community-based family study. *Annals of Neurology*, *49*(6), 761–769. https://doi.org/10.1002/ana.1022.

Louis, E. D., Frucht, S. J., & Rios, E. (2009). Intention tremor in essential tremor: Prevalence and association with disease duration. *Movement Disorders*, *24*(4), 626–627. https://doi.org/10.1002/mds.22370.

Louis, E. D., Hernandez, N., & Michalec, M. (2015). Prevalence and correlates of rest tremor in essential tremor: Cross-sectional survey of 831 patients across four distinct cohorts. *European Journal of Neurology*, *22*(6), 927–932. https://doi.org/10.1111/ene.12683.

Louis, E. D., Huey, E. D., & Cosentino, S. (2021). Features of "ET plus" correlate with age and tremor duration: "ET plus" may be a disease stage rather than a subtype of essential tremor. *Parkinsonism & Related Disorders*, *91*, 42–47. https://doi.org/10.1016/j.parkreldis.2021.08.017.

Louis, E. D., Jurewicz, E. C., & Watner, D. (2003). Community-based data on associations of disease duration and age with severity of essential tremor: Implications for disease pathophysiology. *Movement Disorders*, *18*(1), 90–93. https://doi.org/10.1002/mds.10302.

Louis, E. D., Kerridge, C. A., Chatterjee, D., Martuscello, R. T., Diaz, D. T., Koeppen, A. H., et al. (2019). Contextualizing the pathology in the essential tremor cerebellar cortex: A patholog-omics approach. *Acta Neuropathologica*, *138*(5), 859–876. https://doi.org/10.1007/s00401-019-02043-7.

Louis, E. D., Kuo, S. H., Vonsattel, J. P., & Faust, P. L. (2014). Torpedo formation and Purkinje cell loss: Modeling their relationship in cerebellar disease. *Cerebellum*, *13*(4), 433–439. https://doi.org/10.1007/s12311-014-0556-5.

Louis, E. D., & Lenka, A. (2017). The olivary hypothesis of essential tremor: Time to lay this model to rest? *Tremor and Other Hyperkinetic Movements (N. Y.)*, 7, 473. https://doi.org/10.7916/D8FF40RX.

Louis, E. D., & McCreary, M. (2021). How common is essential tremor? Update on the worldwide prevalence of essential tremor. *Tremor and Other Hyperkinetic Movements (N. Y.)*, *11*, 28. https://doi.org/10.5334/tohm.632.

Louis, E. D., Wendt, K. J., & Ford, B. (2000). Senile tremor. What is the prevalence and severity of tremor in older adults? *Gerontology*, *46*(1), 12–16. https://doi.org/10.1159/000022127.

Maas, R., Helmich, R. C. G., & van de Warrenburg, B. P. C. (2020). The role of the cerebellum in degenerative ataxias and essential tremor: Insights from noninvasive modulation of cerebellar activity. *Movement Disorders*, *35*(2), 215–227. https://doi.org/10.1002/mds.27919.

Madelein van der Stouwe, A. M., Nieuwhof, F., & Helmich, R. C. (2020). Tremor pathophysiology: Lessons from neuroimaging. *Current Opinion in Neurology*, *33*(4), 474–481. https://doi.org/10.1097/wco.0000000000000829.

Marsden, J. F., Ashby, P., Limousin-Dowsey, P., Rothwell, J. C., & Brown, P. (2000). Coherence between cerebellar thalamus, cortex and muscle in man: Cerebellar thalamus interactions. *Brain, 123*(Pt. 7), 1459–1470. https://doi.org/10.1093/brain/123.7.1459.

Marshall, J. (1961). The effect of ageing upon physiological tremor. *Journal of Neurology, Neurosurgery, and Psychiatry, 24*(1), 14–17. https://doi.org/10.1136/jnnp.24.1.14.

McAuley, J. H., & Marsden, C. D. (2000). Physiological and pathological tremors and rhythmic central motor control. *Brain, 123*(8), 1545–1567. https://doi.org/10.1093/brain/123.8.1545.

Meldrum, B. S. (1997). First Alfred Meyer memorial lecture. Epileptic brain damage: A consequence and a cause of seizures. *Neuropathology and Applied Neurobiology, 23*(3), 185–201. discussion 201-182.

Milanov, I. (2001). Electromyographic differentiation of tremors. *Clinical Neurophysiology, 112*(9), 1626–1632. https://doi.org/10.1016/s1388-2457(01)00629-0.

Milosevic, L., Kalia, S. K., Hodaie, M., Lozano, A. M., Popovic, M. R., & Hutchison, W. D. (2018). Physiological mechanisms of thalamic ventral intermediate nucleus stimulation for tremor suppression. *Brain, 141*(7), 2142–2155. https://doi.org/10.1093/brain/awy139.

Minen, M. T., & Louis, E. D. (2008). Emergence of Parkinson's disease in essential tremor: A study of the clinical correlates in 53 patients. *Movement Disorders, 23*(11), 1602–1605. https://doi.org/10.1002/mds.22161.

Murthy, V. N., & Fetz, E. E. (1992). Coherent 25- to 35-Hz oscillations in the sensorimotor cortex of awake behaving monkeys. *Proceedings of the National Academy of Sciences of the United States of America, 89*(12), 5670–5674. https://doi.org/10.1073/pnas.89.12.5670.

Muthuraman, M., Deuschl, G., Anwar, A. R., Mideksa, K. G., von Helmolt, F., & Schneider, S. A. (2015). Essential and aging-related tremor: Differences of central control. *Movement Disorders, 30*(12), 1673–1680. https://doi.org/10.1002/mds.26410.

Muthuraman, M., Heute, U., Arning, K., Anwar, A. R., Elble, R., Deuschl, G., et al. (2012). Oscillating central motor networks in pathological tremors and voluntary movements. What makes the difference? *NeuroImage, 60*(2), 1331–1339. https://doi.org/10.1016/j.neuroimage.2012.01.088.

Muthuraman, M., Raethjen, J., Koirala, N., Anwar, A. R., Mideksa, K. G., Elble, R., et al. (2018). Cerebello-cortical network fingerprints differ between essential, Parkinson's and mimicked tremors. *Brain, 141*(6), 1770–1781. https://doi.org/10.1093/brain/awy098.

Nicolelis, M. A., Baccala, L. A., Lin, R. C., & Chapin, J. K. (1995). Sensorimotor encoding by synchronous neural ensemble activity at multiple levels of the somatosensory system. *Science, 268*(5215), 1353–1358. https://doi.org/10.1126/science.7761855.

Pan, M. K., Li, Y. S., Wong, S. B., Ni, C. L., Wang, Y. M., Liu, W. C., et al. (2020). Cerebellar oscillations driven by synaptic pruning deficits of cerebellar climbing fibers contribute to tremor pathophysiology. *Science Translational Medicine, 12*(526). https://doi.org/10.1126/scitranslmed.aay1769.

Park, Y. G., Park, H. Y., Lee, C. J., Choi, S., Jo, S., Choi, H., et al. (2010). Ca(V)3.1 is a tremor rhythm pacemaker in the inferior olive. *Proceedings of the National Academy of Sciences of the United States of America, 107*(23), 10731–10736. https://doi.org/10.1073/pnas.1002995107.

Passamonti, L., Cerasa, A., & Quattrone, A. (2012). Neuroimaging of essential tremor: What is the evidence for cerebellar involvement? *Tremor and Other Hyperkinetic Movements (N. Y.), 2*. https://doi.org/10.7916/d8f76b8g.

Pinto, A. D., Lang, A. E., & Chen, R. (2003). The cerebellothalamocortical pathway in essential tremor. *Neurology, 60*(12), 1985–1987. https://doi.org/10.1212/01.wnl.0000065890.75790.29.

Przedborski, S., Vila, M., & Jackson-Lewis, V. (2003). Neurodegeneration: What is it and where are we? *The Journal of Clinical Investigation, 111*(1), 3–10. https://doi.org/10.1172/JCI17522.

Puschmann, A., & Wszolek, Z. K. (2011). Diagnosis and treatment of common forms of tremor. *Seminars in Neurology, 31*(1), 65–77. https://doi.org/10.1055/s-0031-1271312.

Radler, K. H., Zdrodowska, M. A., Dowd, H., Cersonsky, T. E. K., Huey, E. D., Cosentino, S., et al. (2020). Rate of progression from mild cognitive impairment to dementia in an essential tremor cohort: A prospective, longitudinal study. *Parkinsonism & Related Disorders, 74*, 38–42. https://doi.org/10.1016/j.parkreldis.2020.04.008.

Raethjen, J., & Deuschl, G. (2012). The oscillating central network of essential tremor. *Clinical Neurophysiology, 123*(1), 61–64. https://doi.org/10.1016/j.clinph.2011.09.024.

Raethjen, J., Govindan, R. B., Kopper, F., Muthuraman, M., & Deuschl, G. (2007). Cortical involvement in the generation of essential tremor. *Journal of Neurophysiology, 97*(5), 3219–3228. https://doi.org/10.1152/jn.00477.2006.

Rajput, A. H., Adler, C. H., Shill, H. A., & Rajput, A. (2012). Essential tremor is not a neurodegenerative disease. *Neurodegenerative Disease Management, 2*(3), 259–268. https://doi.org/10.2217/nmt.12.23.

Revuelta, G., McGill, C., Jensen, J. H., & Bonilha, L. (2019). Characterizing thalamo-cortical structural connectivity in essential tremor with diffusional kurtosis imaging tractography. *Tremor and Other Hyperkinetic Movements (N. Y.), 9*. https://doi.org/10.7916/tohm.v0.690.

Robakis, D., & Louis, E. D. (2016). Head tremor in essential tremor: "Yes-yes", "no-no", or "round and round"? *Parkinsonism & Related Disorders, 22*, 98–101. https://doi.org/10.1016/j.parkreldis.2015.11.002.

Rodrigue, K. M., Kennedy, K. M., & Park, D. C. (2009). Beta-amyloid deposition and the aging brain. *Neuropsychology Review, 19*(4), 436–450. https://doi.org/10.1007/s11065-009-9118-x.

Samuelsson, J. G., Sundaram, P., Khan, S., Sereno, M. I., & Hamalainen, M. S. (2020). Detectability of cerebellar activity with magnetoencephalography and electroencephalography. *Human Brain Mapping, 41*(9), 2357–2372. https://doi.org/10.1002/hbm.24951.

Saper, C. B., Wainer, B. H., & German, D. C. (1987). Axonal and transneuronal transport in the transmission of neurological disease: Potential role in system degenerations, including Alzheimer's disease. *Neuroscience, 23*(2), 389–398. https://doi.org/10.1016/0306-4522(87)90063-7.

Scheltens, P., De Strooper, B., Kivipelto, M., Holstege, H., Chételat, G., Teunissen, C. E., et al. (2021). Alzheimer's disease. *Lancet, 397*(10284), 1577–1590. https://doi.org/10.1016/s0140-6736(20)32205-4.

Schnitzler, A., Munks, C., Butz, M., Timmermann, L., & Gross, J. (2009). Synchronized brain network associated with essential tremor as revealed by magnetoencephalography. *Movement Disorders, 24*(11), 1629–1635. https://doi.org/10.1002/mds.22633.

Schreglmann, S. R., Wang, D., Peach, R. L., Li, J., Zhang, X., Latorre, A., et al. (2021). Non-invasive suppression of essential tremor via phase-locked disruption of its temporal coherence. *Nature Communications, 12*(1), 363. https://doi.org/10.1038/s41467-020-20581-7.

Sharifi, S., Luft, F., Potgieter, S., Heida, T., Mugge, W., Schouten, A. C., et al. (2021). Directionality of corticomuscular coupling in essential tremor and cortical myoclonic tremor. *Clinical Neurophysiology, 132*(8), 1878–1886. https://doi.org/10.1016/j.clinph.2021.04.011.

Sharifi, S., Luft, F., Verhagen, R., Heida, T., Speelman, J. D., Bour, L. J., et al. (2017). Intermittent cortical involvement in the preservation of tremor in essential tremor. *Journal of Neurophysiology, 118*(5), 2628–2635. https://doi.org/10.1152/jn.00848.2016.

Sharifi, S., Nederveen, A. J., Booij, J., & van Rootselaar, A. F. (2014). Neuroimaging essentials in essential tremor: A systematic review. *NeuroImage: Clinical, 5*, 217–231. https://doi.org/10.1016/j.nicl.2014.05.003.

Shill, H. A., Adler, C. H., & Beach, T. G. (2012). Pathology in essential tremor. *Parkinsonism & Related Disorders*, *18*(Suppl. 1), S135–S137. https://doi.org/10.1016/s1353-8020(11)70042-6.

Shin, H. W., Hallett, M., & Sohn, Y. H. (2019). Cerebellar repetitive transcranial magnetic stimulation for patients with essential tremor. *Parkinsonism & Related Disorders*, *64*, 304–307. https://doi.org/10.1016/j.parkreldis.2019.03.019.

Spanaki, C., & Plaitakis, A. (2009). Essential tremor in Parkinson's disease kindreds from a population of similar genetic background. *Movement Disorders*, *24*(11), 1662–1668. https://doi.org/10.1002/mds.22655.

Stolze, H., Petersen, G., Raethjen, J., Wenzelburger, R., & Deuschl, G. (2001). The gait disorder of advanced essential tremor. *Brain*, *124*(Pt. 11), 2278–2286. https://doi.org/10.1093/brain/124.11.2278.

Tan, E. K., Lee, S. S., Fook-Chong, S., & Lum, S. Y. (2008). Evidence of increased odds of essential tremor in Parkinson's disease. *Movement Disorders*, *23*(7), 993–997. https://doi.org/10.1002/mds.22005.

Thangavelu, K., Talk, A. C., Clark, G. I., & Dissanayaka, N. N. W. (2020). Psychosocial factors and perceived tremor disability in essential tremor. *Neuroscience and Biobehavioral Reviews*, *108*, 246–253. https://doi.org/10.1016/j.neubiorev.2019.10.021.

Thawani, S. P., Schupf, N., & Louis, E. D. (2009). Essential tremor is associated with dementia: Prospective population-based study in New York. *Neurology*, *73*(8), 621–625. https://doi.org/10.1212/WNL.0b013e3181b389f1.

Tikoo, S., Pietracupa, S., Tommasin, S., Bologna, M., Petsas, N., Bharti, K., et al. (2020). Functional disconnection of the dentate nucleus in essential tremor. *Journal of Neurology*, *267*(5), 1358–1367. https://doi.org/10.1007/s00415-020-09711-9.

Ugawa, Y., Uesaka, Y., Terao, Y., Hanajima, R., & Kanazawa, I. (1995). Magnetic stimulation over the cerebellum in humans. *Annals of Neurology*, *37*(6), 703–713. https://doi.org/10.1002/ana.410370603.

Vial, F., Kassavetis, P., Merchant, S., Haubenberger, D., & Hallett, M. (2019). How to do an electrophysiological study of tremor. *Clinical Neurophysiology Practice*, *4*, 134–142. https://doi.org/10.1016/j.cnp.2019.06.002.

Zakaria, R., Lenz, F. A., Hua, S., Avin, B. H., Liu, C. C., & Mari, Z. (2013). Thalamic physiology of intentional essential tremor is more like cerebellar tremor than postural essential tremor. *Brain Research*, *1529*, 188–199. https://doi.org/10.1016/j.brainres.2013.07.011.

Zhang, P., & Tian, B. (2014). Metabolic syndrome: An important risk factor for Parkinson's disease. *Oxidative Medicine and Cellular Longevity*, *2014*, 729194. https://doi.org/10.1155/2014/729194.

SECTION III

How important is the inferior olive in essential tremor? An evolving story

Sheng-Han Kuo[a,b] and Elan D. Louis[c]

[a]Department of Neurology, College of Physicians and Surgeons, Columbia University, New York, NY, United States
[b]Initiative for Columbia Ataxia and Tremor, Columbia University, New York, NY, United States
[c]Department of Neurology, University of Texas Southwestern Medical Center, Dallas, TX, United States

As part of the current series, which highlights current controversies in the essential tremor (ET) field, we commissioned authors to submit papers with opposing viewpoints on the question of how important the inferior olive (IO) is in the pathophysiology of ET, one using available data to put forth the argument that the IO is a central brain region involved in the pathophysiology of ET (Lang & Handforth, 2022) and the other, using available data to put forth the argument that the IO is not a central region involved in ET (Manto, 2022). We are grateful to both sets of authors for thoughtfully making their cases and here we provide additional commentary.

IOs, located in the brainstem, have a physiological feature subthreshold oscillatory activity. Upon the exposure to harmaline, a chemical that enhances gap junction coupling, IO neuronal firing patterns become highly synchronous, which triggers downstream Purkinje cells in the cerebellar cortex to fire rhythmically and create tremor in animals (Cheng, Tang, & Kuo, 2013; Llinas & Volkind, 1973; Llinas & Yarom, 1986). This tremor has been used as an animal model for ET because it shares several features with ET. IO neurons also receive GABAergic input from the deep cerebellar nuclear neurons, which can regulate the coupling of IO neuronal firing (Lang, 2002). These physiological studies help us to understand that the loop involving IO neurons-Purkinje cells-deep cerebellar nuclear neurons is capable of generating tremor in animals (Handforth & Lang, 2021). For decades, the olivary model of ET was the reigning model, but with additional data, the situation has evolved, and the model has come under increasing scrutiny and criticism. This sets up the question, from where does ET arise?

Indeed, there is a lack of solid evidence that the pathophysiology of harmaline-induced tremor is in any way relevant to what is happening in ET patients. First, a study using positron emission tomography to determine the brain circuitry of ET did not find the involvement of IOs (Wills, Jenkins, Thompson, Findley, & Brooks, 1995). Second, the postmortem analysis in ET brain did not show any neuronal changes in IOs, although no specific analysis on the number of gap junctions between IO neurons or GABAergic inputs from the deep cerebellar nuclei was performed (Elkouzi, Kattah, & Elble, 2015). Third, case reports showed that two ET patients have hypertrophic IO degeneration (Louis et al., 2018; Louis, Babij, Cortés, Vonsattel, & Faust, 2013), suggesting IOs in these ET patients are not important for tremor generation. These studies have raised the question as to whether the IOs are involved in ET at all (Louis & Lenka, 2017). Aside from all of this is that the harmaline model for ET differs from ET in a number of important ways including its clinical features, pharmacological responsive phenotype and postmortem features.

Whether the IO is the central brain region in ET has critical implications for how we understand the pathophysiology of ET and how we develop therapeutic approaches. The two sides of the argument each present evidence and, in the meantime, also raise unresolved issues, which are outlined in the current series (Lang & Handforth, 2022; Manto, 2022). Specifically, more human evidence is needed to support the claim that the IO is a critical brain region in ET pathophysiology. One piece of human evidence is that harmane, a naturally occurring chemical that is similar to harmaline, can also induce similar tremor in animals, and concentrations have been found to be elevated in the blood and brain of ET patients (Louis et al., 2012; Louis et al., 2013).

If the IO is not central to the pathophysiology of ET, which brain regions does tremor in ET patients come from? In other words, investigators will need to develop new theoretical frameworks, coupled with animal studies and postmortem studies to prove ET-like tremor can be induced outside of the IO. Purkinje cell loss and Purkinje cell axonal sprouting have been identified in the ET brain (Babij, Lee, Cortés, Vonsattel, et al., 2013; Louis, Faust, Vonsattel, et al., 2007). Whether these pathological changes can create ET-like tremor in animal models has yet to be demonstrated and these studies are of some value. These questions still remain open for future investigations.

Perhaps the situation in ET is more complex than initially conceptualized and the pathophysiology involves both the IO and the Purkinje cells. Recent postmortem work has identified, among other changes in ET, an extension of climbing fibers onto more distal regions of the Purkinje cell dendrites in the ET cerebellum (Lin et al., 2014). Whether this is a primary or secondary change is not clear. Interestingly, mice with this unique pathological feature, *hotfoot17J*, develop cerebellar oscillatory activity and tremor that shares several features with ET (Pan et al., 2020), demonstrating that abnormal wiring of climbing fibers, axons of IO neurons, on Purkinje cells can produce a tremor that shares some features of ET. Since climbing fibers come from IO neurons, the climbing fiber-Purkinje cell synaptic pathology could be regarded as IO involvement. On the other hand, the morphological alteration occurs at the climbing fiber-Purkinje cell interface, located in the cerebellar cortex, and this could also be considered outside IOs. Although studies of tremor in animals must be interpreted in their proper context, these data are of interest.

As ET is a heterogeneous disorder and possibly even a family of diseases (Riboldi & Frucht, 2022; Pandey, 2022), it may be important to consider that no single pathophysiology may explain tremor in *all* ET patients. It is possible that the cerebellum can generate oscillatory activity in different ways, such as via (1) IO gap junction coupling (2) abnormal climbing fiber-Purkinje cell synaptic connections, (3) Purkinje cell degeneration with axonal sprouting. Future research directions for ET may be to develop pathological, physiological or neuroimaging tools to separate different ET subtypes, which may have therapeutic implications. What is clear, though, is that the field has evolved beyond the simple olivary model.

References

Babij, R., Lee, M., Cortés, E., Vonsattel, J. P., et al. (2013). Purkinje cell axonal anatomy: Quantifying morphometric changes in essential tremor versus control brains. *Brain, 136*, 3051–3061.

Cheng, M. M., Tang, G., & Kuo, S. H. (2013). Harmaline-induced tremor in mice: Videotape documentation and open questions about the model. *Tremor and Other Hyperkinetic Movements*. 3:tre-03-205-4668-1.

Elkouzi, A., Kattah, J. C., & Elble, R. J. (2015). Hypertrophic olivary degeneration does not reduce essential tremor. *Movement Disorders Clinical Practice, 3*, 209–211.

Handforth, A., & Lang, E. J. (2021). Increased Purkinje cell complex spike and deep cerebellar nucleus synchrony as a potential basis for syndromic essential tremor. A review and synthesis of the literature. *Cerebellum, 20*, 266–281.

Lang, E. J. (2002). GABAergic and glutamatergic modulation of spontaneous and motor-cortex-evoked complex spike activity. *Journal of Neurophysiology, 87*, 1993–2008.

Lang, E. J. & Handforth A. (2022). Is the inferior olive central to essential tremor? Yes. *International Review of Neurobiology, 163*, 133–165.

Lin, C. Y., Louis, E. D., Faust, P. L., Koeppen, A. H., Vonsattel, J. P., & Kuo, S. H. (2014). Abnormal climbing fibre-Purkinje cell synaptic connections in the essential tremor cerebellum. *Brain, 137*, 3149–3159.

Llinas, R., & Volkind, R. A. (1973). The olivo-cerebellar system: Functional properties as revealed by harmaline-induced tremor. *Experimental Brain Research, 18*, 69–87.

Llinas, R., & Yarom, Y. (1986). Oscillatory properties of guinea-pig inferior olivary neurones and their pharmacological modulation: An in vitro study. *The Journal of Physiology, 376*, 163–182.

Louis, E. D., Babij, R., Cortés, E., Vonsattel, J. P., & Faust, P. L. (2013). The inferior olivary nucleus: A postmortem study of essential tremor cases versus controls. *Movement Disorders, 28*, 779–786.

Louis, E. D., Diaz, D. T., Kuo, S. H., Gan, S. R., Cortes, E. P., Vonsattel, J. P. G., et al. (2018). Inferior Olivary nucleus degeneration does not lessen tremor in essential tremor. *Cerebellum Ataxias., 5*, 1.

Louis, E. D., Factor-Litvak, P., Liu, X., Vonsattel, J. P., Galecki, M., Jiang, W., et al. (2013). Elevated brain harmane (1-methyl-9H-pyrido[3,4-b]indole) in essential tremorcases vs. controls. *Neurotoxicology, 38*, 131–135.

Louis, E. D., Faust, P. L., Vonsattel, J. P., et al. (2007). Neuropathological changes in essential tremor: 33 Cases compared with 21 controls. *Brain, 130*, 3297–3307.

Louis, E. D., Jiang, W., Gerbin, M., Viner, A. S., Factor-Litvak, P., & Zheng, W. (2012). Blood harmane (1-methyl-9H-pyrido[3,4-b]indole) concentrations in essentialtremor: Repeat observation in cases and controls in New York. *Journal of Toxicology and Environmental Health. Part A, 75*, 673–683.

Louis, E. D., & Lenka, A. (2017). The olivary hypothesis of essential tremor: Time to lay this model to rest? *Tremor and Other Hyperkinetic Movements*, 7, 473.

Manto, M. (2022). Essential tremor is a genuine cerebellar disorder and the cerebellar cortex is the culprit. *International Review of Neurobiology, 163*, 167–187.

Pan, M. K., Li, Y. S., Wong, S. B., Ni, C. L., Wang, Y. M., Liu, W. C., et al. (2020). Cerebellar oscillations driven by synaptic pruning deficits of cerebellar climbing fibers contribute to tremor pathophysiology. *Science Translational Medicine, 12*. eaay1769.

Pandey, S. (2022). Is essential tremor a family of diseases or a syndrome? A syndrome. *International Review of Neurobiology, 163*, 31–59.

Riboldi, G. M. & Frucht, S. J. (2022). Is essential tremor a family of diseases or a syndrome? A family of diseases. *International Review of Neurobiology, 163*, 7–29.

Wills, A. J., Jenkins, I. H., Thompson, P. D., Findley, L. J., & Brooks, D. J. (1995). A positron emission tomography study of cerebral activation associated with essential and writing tremor. *Archives of Neurology, 52*, 299–305.

CHAPTER FIVE

Is the inferior olive central to essential tremor? Yes

Eric J. Lang[a,*] and Adrian Handforth[b]

[a]Department of Neuroscience and Physiology, Neuroscience Institute, New York University, New York, NY, United States

[b]Neurology Service (W127), Veterans Affairs Greater Los Angeles Healthcare System, Los Angeles, CA, United States

*Corresponding author: e-mail address: eric.lang@nyumc.org

Contents

Abstract

We consider the question whether the inferior olive (IO) is required for essential tremor (ET). Much evidence shows that the olivocerebellar system is the main system capable of generating the widespread synchronous oscillatory Purkinje cell (PC) complex spike (CS) activity across the cerebellar cortex that would be capable of generating the type of bursting cerebellar output from the deep cerebellar nuclei (DCN) that could underlie tremor. Normally, synchronous CS activity primarily reflects the effective electrical coupling of IO neurons by gap junctions, and traditionally, ET research has focused on the hypothesis of increased coupling of IO neurons as the cause of hypersynchronous CS

International Review of Neurobiology, Volume 163
ISSN 0074-7742
https://doi.org/10.1016/bs.irn.2022.02.009

activity underlying tremor. However, recent pathology studies of brains from humans with ET and evidence from mutant mice, particularly the *hotfoot17* mouse, that largely replicate the pathology of ET, suggest that the abnormal innervation of multiple Purkinje cells (PCs) by climbing fibers (Cfs) is related to tremor. In addition, ET brains show partial PC loss and axon terminal sprouting by surviving PCs. This may provide another mechanism for tremor. It is proposed that in ET, these three mechanisms may promote tremor. They all involve hypersynchronous DCN activity and an intact IO, but the level at which excessive synchronization occurs may be at the IO level (from abnormal afferent activity to this nucleus), the PC level (via aberrant Cfs), or the DCN level (via terminal PC collateral innervation).

1. Clues to circuit dysfunction in essential tremor

1.1 Where does the tremor come from?

What drives the tremor of essential tremor (ET)? Multiple observations indicate that the tremor is due to a central oscillator whose frequency is not affected by peripheral factors. An early observation was that alcohol given orally reduces tremor amplitude, whereas infusion of alcohol directly into the brachial artery to achieve the same blood alcohol level in the upper limb does not reduce tremor (Growdon, Shahani, & Young, 1975). This contrast implies that alcohol acts within the CNS to reduce a tremorigenic output. Further evidence for the central origin of the tremor in ET is that tremor frequency is not affected by cooling the limb or by topical anesthesia (Lakie, Walsh, Arblaster, Villagra, & Roberts, 1994; Pozos & Iaizzo, 1992) or by mechanically loading the limb (Homberg, Hefter, Reiners, & Freund, 1987). Thus, a long-standing goal in ET research has been to identify the underlying central oscillator responsible for the tremor.

One hypothesis is that the inferior olive (IO) is that oscillator. However, the lack of pathology found in the IO of brains of ET patients, and the recent findings of multiple pathological changes in the cerebellar cortex in such brains, have questioned the validity of this hypothesis (Louis & Lenka, 2017). Below we review evidence about the properties of the IO, and more generally the olivocerebellar system, that have supported the IO hypothesis, and then consider whether the new evidence for pathological changes in the cerebellum obviates the need for abnormal IO activity in ET.

1.2 What brain circuits are involved in ET tremor?

Mapping of central cerebral oscillatory activity with magnetoencephalography (MEG) in correlation with electromyographically (EMG) recorded tremor revealed an activated network that includes the cerebellum, thalamus, motor and premotor cortices, and brainstem (Schnitzler, Munks, Butz, Timmermann, & Gross, 2009). While these findings by themselves identify the major brain regions involved in ET, they do not provide information on the roles they play. In addition, some small structures may have been missed because of limitations of resolution. Complementary information is provided by clinical observations that lesions at various sites in this network can abolish tremor, including infarcts involving the motor cortex or pre-motor areas, the posterior limb of the internal capsule, the thalamus, and infarct, hemorrhage or tumor of the cerebellum, including the deep cerebellar nuclei (DCN) (Dupuis, Evrard, Jacquerye, Picard, & Lermen, 2010). Such observations have led to the suggestion that there is an efferent tremor pathway from cerebellum to motor areas of the cerebrum (Dupuis et al., 2010), presumably with the central oscillator in ET located in the cerebellum or upstream of it in a precerebellar nucleus, such as the IO. However, the circuits connecting the cerebellum and cerebrum form closed loops (Kelly & Strick, 2003), making it hard to define any region as upstream in an absolute sense. Indeed, clinical studies also suggest that damage to cerebellar afferent pathways from the cerebrum may also be necessary for tremor (Dupuis et al., 2010). In sum, given the anatomical connectivity, a reasonable interpretation of both the MEG and clinical observations is that all the components of this network may be needed for tremor expression. However, the identity of the underlying oscillatory mechanism is not provided by these studies.

Multiple lines of evidence point to a dysfunction of the cerebellum. ET patients exhibit anomalies characteristic of cerebellar dysfunction, such as impaired balance, irregular finger tapping, and abnormal eye movements (reviewed by (Filip, Lungu, Manto, & Bares, 2016)). Using high density electroencephalography it has been shown that alcohol, which is highly effective against tremor in many ET patients, reduces tremor by acting on the cerebellum (Pedrosa et al., 2017).

More revealing has been pathological examination of ET brains. If we start by considering the IO, no consistent morphological or imaging abnormalities of this nucleus have been found (Louis & Lenka, 2017). However, these studies only investigated the overall size of the IO, and some changes cell density and morphology, and so do not rule out more subtle anatomical abnormalities, including changes in synaptic inputs or in the gap junction coupling of IO neurons, or changes in the electrophysiology of IO neurons.

In contrast to the usual lack of apparent IO pathology in ET brains, the cerebellar cortex is a different story, revealing many anatomical abnormalities in ET. These include reduced Purkinje cell (PC) counts, PC axon torpedoes, PC dendrite swelling, PC heterotopias, thickened PC axon profiles, empty baskets, basket cell axon hypertrophy, and other changes. This pathology is similar for cases of ET or ET plus (e.g., cases with mild parkinsonism, ataxia, cognitive changes, or dystonia) (Gionco et al., 2021). Of particular interest is the finding in ET brains of aberrant climbing fiber (Cf) innervation of PCs.

Normally, each IO neuron's axon branches within the cerebellum to form about 10–15 Cfs, where a Cf is defined as the terminal branches of an olivocerebellar axon that end on a single PC. Olivary axons also branch to innervate the DCN, and can influence other cerebellar cortical neurons (e.g., basket and stellate cells, likely by spillover transmission from Cf-PC synapses (Coddington, Rudolph, Vande Lune, Overstreet-Wadiche, & Wadiche, 2013; Szapiro & Barbour, 2007); however, here we will focus on the Cf-PC synaptic connection. Each PC is typically innervated by only one Cf, which normally forms synapses only on the large main dendrites of the PC, which are generally limited to the inner 80% of the molecular layer. In contrast, parallel fibers (PFs) from cerebellar granule cells (CGCs) synapse on the smaller, distal dendrites (spiny branchlets) of the PCs, present throughout the whole molecular layer. In contrast, in ET brains, Cfs aberrantly extend into the normal PF territory (i.e., synapse on the spiny branchlets), and thus extend into the outer 20% of the molecular layer (Lin et al., 2014). Such changes in Cf termination patterns are found in both sporadic and familial ET cases, with both younger and older age of onset, and with or without head tremor (Lee, Gan, Faust, Louis, & Kuo, 2018).

Moreover, in contrast to the normal 1:1 relationship of Cf to PC, in ET brains, increased transverse spread of Cf branches suggests that Cfs may aberrantly innervate more than one PC, although definitive evidence of synapses remains to be obtained (Wu et al., 2021). Interestingly, the number of PC crossings by Cfs correlates positively with the severity of tremor, leading to the suggestion that this anomaly may result in increases in synchronous brain activity and tremor (Wu et al., 2021).

Would the abnormal patterns of Cf innervation of PC dendrites seen in brains from ET patients lead to more synchrony, and in particular, would that increase be large enough to drive tremor? To answer these questions, we need to understand the normal patterns and control of PC synchrony and how these patterns would be altered by pathological changes in the Cf-PC synaptic relationship found in ET brains.

2. IO-induced PC synchrony as related to tremor

2.1 How the IO controls PC complex spike synchrony

The normal Cf-PC synapse consists of hundreds of boutons spread over the proximal dendrites of the PC (Llinas, Bloedel, & Hillman, 1969). As a result, each IO discharge, itself having one to several spikes that propagate down the IO cell's axon, generates such a large EPSP (or series of EPSPs) that the PC fires a burst of spikes termed the complex spike (CS) (Eccles, Llinas, & Sasaki, 1966).

In rodents, CSs occur at an average firing rate of ~1 Hz, but have brief periods of 10-Hz activity. These rates are far lower than those of simple spikes (SSs: 50–100 Hz), leading many to question whether CS activity has a significant direct impact on cerebellar output (DCN activity). However, CSs are typically synchronized among PCs located within ~250-μm-wide parasagittal cortical bands that can extend across lobules (Lang, Sugihara, Welsh, & Llinas, 1999; Sasaki, Bower, & Llinas, 1989; Sugihara, Lang, & Llinas, 1993). This synchronization provides a mechanism to allow CS activity to affect DCN activity under normal conditions, and potentially to underlie tremor. (There are other potential ways CSs might distinguish themselves from the ongoing SS activity, but here we will focus on synchrony.)

The parasagittal (or more accurately, rostrocaudal) banding pattern reflects both the topography of the olivocerebellar pathway, as local regions of the IO project to rostrocaudally-running cortical strips, and the pattern of coupling among IO neurons. In particular, even though the topography of the olivocerebellar pathway is fixed, and the average synchrony distribution reflects this topography, the moment-to-moment synchrony distribution varies markedly. This variation reflects changes in the effective electrical coupling between IO neurons. This electrical coupling is mediated by gap junctions, which are predominantly formed by the protein connexin36 (Belluardo et al., 2000; Condorelli et al., 1998). The gap junctions are mostly located between the dendritic spines of neighboring neurons within encapsulated structures called glomeruli, which also contain presynaptic terminals that are believed to control the efficacy of this coupling (Llinas, Baker, & Sotelo, 1974; Llinas & Yarom, 1981a; Sotelo, Llinas, & Baker, 1974). In this way, the IO is organized into changeable populations of synchronously active cells, with the degree of coupling, and hence population membership, controlled by afferents to the IO.

In brainstem slices, IO neurons display ~6–10 Hz subthreshold oscillations of their membrane potential due to the interaction of various membrane conductances (Llinas & Yarom, 1986), including potassium and calcium conductances and the hyperpolarization-activated cation conductance (I_h) (Bal & McCormick, 1997; Llinas & Yarom, 1981a, 1981b). The oscillations of clusters of neighboring cells, are normally in phase (Leznik, Makarenko, & Llinas, 2002). The synchronization of the oscillation across cells, but not the oscillation itself, is eliminated by blocking gap junctions (Leznik & Llinas, 2005), indicating the critical role of electrical coupling in mediating coherent IO activity. Consistent with this, in brainstem slices from connexin36 knockout (KO) mice, IO neurons are not electrically coupled, and the subthreshold oscillations are not synchronized (Long, Deans, Paul, & Connors, 2002).

To what extent does gap junction-mediated electrical coupling of IO neurons underlie CS synchrony? When glutamatergic and GABAergic afferents to the IO are blocked, spontaneous CS activity is still synchronized (albeit with a somewhat modified distribution), showing that the IO can be an autonomous pacemaker and that synchronized IO output is not necessarily driven by synchronous input (Lang, 2002). Conversely, intra-IO injection of the gap junction blocker carbenoxolone reduces CS synchrony (Blenkinsop & Lang, 2006), and CS synchrony is absent in connexin36 KO mice, which lack IO coupling (Marshall, van der Giessen, de Zeeuw, & Lang, 2007). Together, these lines of evidence indicate that spontaneous CS synchrony is largely due to the electrical coupling of clusters of IO neurons whose axons project to the same parasagittal cortical strip. Of course, synchronous input to the IO can also cause CS synchrony; however, even in this case, coupling of IO neurons appears to exert a major influence on the pattern, as shown by the similar spatial distributions for spontaneous activity and for CSs evoked by electrical stimulation of the motor cortex (Lang, 2002).

It should also be noted that the Cf branching mentioned earlier contributes to CS synchrony. However, this normally represents a small contribution. Synchrony due to the olivocerebellar axon branching to provide Cfs to multiple (~10–15) PCs would result in near perfect synchronization and an unvarying distribution pattern. Yet, in multiple electrode recordings of PCs in the same cortical region, such high synchrony is rarely if ever observed, because the pairs of PCs receiving input from a common olivocerebellar axon are a tiny fraction of all possible PC pairs in a specific region, particularly as the Cfs of an olivocerebellar axon are typically distributed across multiple lobules.

The dynamic nature of CS synchrony raises the question of how the effective coupling of IO activity is modulated. GABAergic afferents to the IO play a major role, as the addition of the $GABA_A$ receptor antagonist picrotoxin enlarges synchronized IO clusters, suggesting that GABAergic afferents reduce the extent of IO synchrony (Leznik et al., 2002). Intra-IO injection of picrotoxin in rats greatly increases CS synchrony, so that synchronized CSs may be found across a much larger extent of cerebellar cortex (Lang, 2002; Lang, Sugihara, & Llinas, 1996). The same occurs when lesions of the DCN are made (Lang et al., 1996), which removes the major source of GABAergic afferents to the IO (De Zeeuw, Holstege, Ruigrok, & Voogd, 1989; Fredette & Mugnaini, 1991; Nelson & Mugnaini, 1988). Thus, the normal patterns of CS synchrony are the result of GABAergic input from the DCN limiting the effective coupling to local clusters of IO cells whose axons then project to the same rostrocaudal strip of cortex.

Other afferent systems also play important roles. Intra-IO injection of the AMPA receptor antagonist NBQX (2,3-dioxo-6-nitro-1,2,3,4-tetrahydrobenzo[*f*]quinoxaline-7-sulfonamide) enhances CS synchrony within 250-μm-wide bands, but reduces interband synchrony (i.e., causing high intracluster synchrony among IO cells and at the same time a decoupling of activity between IO clusters) (Lang, 2001, 2002). In addition, glutamate affects coupling of IO neurons via NMDA receptor activation (Turecek et al., 2014). Serotonergic afferents also affect IO coupling and CS synchrony, as discussed below. There may be yet other neurotransmitters that affect IO coupling and CS synchrony.

In summary, the extent of CS synchrony within cerebellar cortex reflects the size of IO clusters that are electrically coupled with gap junctions. The effective strength and extent of IO coupling is dynamic, and is a function of the activity of IO afferents, including GABAergic inhibitory feedback from the DCN and excitatory input from many other regions.

2.2 Is CS synchrony related to movement?

The PC is the sole output neuron of the cerebellar cortex, is GABAergic, and projects mainly to DCN, where it provides the large majority of the synaptic input to DCN cells (De Zeeuw & Berrebi, 1995; Palkovits, Mezey, Hamori, & Szentagothai, 1977). The DCN, in turn, are the major output structure of the cerebellum and contain maps of the body's muscles

(Cicirata, Angaut, Serapide, Panto, & Nicotra, 1992). Thus, for CSs to directly influence ongoing movements, they must affect DCN activity. As mentioned earlier, CSs are vastly outnumbered by SSs, and so their presence adds little to the overall PC firing rate. However, most DCN neurons, and in particular, the large non-GABAergic cells that likely include those projecting to the thalamus, have extremely short-duration PC-mediated IPSCs and fast membrane time constants (Najac & Raman, 2015b; Uusisaari, Obata, & Knopfel, 2007), which should make them especially responsive to highly synchronized input, such as that characteristic of CSs, not SSs.

The banding distribution of CS synchrony is important to consider in order to understand the effect of CS activity on the DCN and its potential to cause tremor. The synchrony bands align with the zebrin compartments of the cerebellar cortex (Sugihara, Marshall, & Lang, 2007), and PCs within a zebrin compartment project to the same DCN region (Chung, Marzban, & Hawkes, 2009; Sugihara, 2011). (Zebrin is an antibody to the metabolic enzyme aldolase C that is expressed in some PCs such that it produces a striped staining pattern across much of the cerebellar cortex.) DCN neurons are thus likely to receive convergent input from synchronized PCs. Critically, such synchronous CS activity strongly inhibits DCN activity, whereas asynchronous CSs are much less effective (Blenkinsop & Lang, 2011; Lang & Blenkinsop, 2011; Tang, Blenkinsop, & Lang, 2019). Thus, highly synchronous CS activity impacts cerebellar output, and could thereby affect movement. Indeed, rhythmic hypersynchronized CS activity, such as occurs with harmaline injections, causes tremor (Lamarre & Mercier, 1971; Llinas & Volkind, 1973).

Spontaneous CS synchrony patterns in awake animals have a similar rostrocaudal distribution as described above (Lang et al., 1999), but during movements CS synchrony increases (De Gruijl, Hoogland, & De Zeeuw, 2014; Mukamel, Nimmerjahn, & Schnitzer, 2009; Welsh, Lang, Suglhara, & Llinas, 1995). Thus, synchronous CS activity is likely to shape the outgoing motor signals from the DCN, because the effect of CS activity correlates with the extent of its synchronization (Tang et al., 2019; Tang, Suh, Blenkinsop, & Lang, 2016). Consistent with a role of synchrony for motor control, connexin36 KO mice fail to display normal movement-associated CS synchrony and have abnormal execution of spinocerebellar reflexes (De Gruijl et al., 2014).

A study in which whisker movements were evoked by motor cortex stimulation provides further evidence for CS synchrony affecting motor commands (Lang, Sugihara, & Llinas, 2006). Here, stimulus trains at various frequencies were delivered to motor cortex in anesthetized rats, and the amplitude of the induced whisker movement was measured. The IO was found to act like a central oscillator that gated the effectiveness of the stimuli in evoking movements. Specifically, the evoked whisker movements showed resonance and beat patterns consistent with motor cortex output being modulated by a central ~10 Hz oscillator. Moreover, pharmacological manipulation of IO activity altered this modulation, providing evidence that the IO was the oscillator in question. For example, intra-IO injection of picrotoxin, which increases CS synchrony, increased the amplitude of the evoked whisker movements and altered their beat patterns (which are variations in amplitude that reflect the difference in frequency of two interacting oscillators, in this case one being the motor cortex stimuli and the other being the IO).

These observations indicate that while parts of the brain such as the cerebral cortex may make executive decisions to initiate movement, the IO can act as a central oscillator that modulates movement amplitude. Although this study focused on whisker movement, a similar ~10 Hz gating of motor cortex stimulation-evoked limb movements has been shown to occur in other species, including monkeys (Horsley & Schafer, 1886). Likewise, the olivocerebellar system is likely the central oscillator responsible for the ~8–10 Hz submovements that comprise voluntary slow finger movements in humans (Vallbo & Wessberg, 1993). (Submovements refers to the fact that the acceleration profile of slow finger movements consists of a series of transient accelerations at 100 ms intervals, implying that such movements are actually generated as a rhythmic series of motor commands that produce a succession of small, short duration movements.) MEG mapping during such slow finger movements has revealed activation of an oscillatory circuit very similar to the oscillatory circuit identified for ET (Schnitzler et al., 2009), with activation of the cerebellum, thalamus, and premotor and motor cortex (Schnitzler, Timmermann, & Gross, 2006).

Given that the olivocerebellar system, in which dynamically varying clusters of IO neurons govern the extent of CS synchrony, plays a role in modulating the efficacy of motor cortex output in inducing movement, the question arises whether changes in IO coupling, if sufficient, can induce tremor. That is, tremor may be the result of a normal movement generation mechanism that is functioning in an abnormal, exaggerated manner. We address this topic below.

2.3 Does increased IO coupling and CS synchrony induce tremor?

Here, we examine three lines of evidence that increased CS synchrony causes tremor. It may be noted that with the IO as the central oscillator, the frequency of the tremor would be expected to be insensitive to a mechanical load.

2.3.1 Tremor from reduced IO GABAergic activity

Intra-IO injection of picrotoxin was observed to induce whisker tremor in some anesthetized rats. The occurrence of the tremor was associated with increased CS synchrony (Lang et al., 1996). As this GABA antagonist is known to increase IO coupling on local administration and CS synchrony, this observation suggests that increased IO coupling can be a mechanism for tremor.

2.3.2 Tremor and the IO serotonin 2a receptor: a link between IO/CS synchrony and tremor

Intra-IO serotonin (5-HT) microinjection induces CS activity to be both more rhythmic and synchronous, especially in the vermis (Sugihara, Lang, & Llinas, 1995). 5-HT$_{2a}$ receptors on IO neurons appear to mediate these effects, as intra-IO co-administration with 5-HT$_{2a}$ antagonists prevents 5-HT's effect on CS synchrony (Sugihara et al., 1995), and intra-IO injection of selective 5-HT$_{2a}$ receptor agonists increases CS synchrony. Systemic administration of the 5-HT$_{2A}$ receptor agonist quipazine in rabbits induces tremor, an effect blocked by the 5-HT receptor antagonist methysergide (Barragan, Delhaye-Bouchaud, & Laget, 1985). Rats given systemic injections of tryptophan plus a monoamine oxidase inhibitor display tremor that is associated with highly rhythmic synchronous CSs; lesions of the IO abolish this tremor (Wiklund, Sjolund, & Bjorklund, 1981).

These observations indicate that 5-HT activates IO 5-HT$_{2a}$ receptors to increase coupling and rhythmicity, leading to increased CS synchrony and tremor. It may be noted that clinically, 5-HT-enhancing agents enhance physiological tremor. Whether this is due to the enhancement of the mechanical/reflex component or the central oscillatory component is not known. Patients with ET who take such agents may report their tremor is worse.

2.4 Tremor from harmaline, a powerful synchronizer of CSs

Harmaline tremor is an intensively studied animal model of ET. Harmaline affects IO membrane conductances in such a way (Llinas & Yarom, 1986) as to make IO neurons burst-fire synchronously at 6–12 Hz (De Montigny & Lamarre, 1973; Llinas & Volkind, 1973) with the caudal medial and caudal dorsal accessory olive (MAO, DAO) subnuclei especially activated (De Montigny & Lamarre, 1975). There is corresponding synchronous 6–12 Hz CS firing, especially in parasagittal bands of vermis (Beitz & Saxon, 2004), but also in hemispheric lobules (Sasaki et al., 1989). The convergent projection of these synchronized PCs to DCN neurons results in hyperpolarization and rebound bursting of DCN neurons (Llinas & Muhlethaler, 1988). The prevention of GABA release from PC terminals within DCN prevents harmaline-induced DCN bursting (Brown et al., 2020), and sectioning of the cerebellar peduncles abolishes harmaline-induced rhythmicity everywhere except in the IO (De Montigny & Lamarre, 1973; Llinas & Volkind, 1973).

In association with this harmaline-induced rhythmic IO/CS/DCN activity is tremor at the same frequency, as indicated by spinal motor neuron recordings (Llinas & Volkind, 1973). Not surprisingly, the same brain structures required for rhythmic activity in this axis are required for harmaline tremor. The tremor is abolished by lesions of IO (Sharabi et al., 2019; Simantov, Snyder, & Oster-Granite, 1976), or when the IO ceases to respond to repeated harmaline with burst-firing (Lorden, Stratton, Mays, & Oltmans, 1988), or when gap junction blockers are administered (Martin & Handforth, 2006). Genetic strains that lack normal CSs, such as dystonic rats (LeDoux & Lorden, 2002) or Kv3.3 KO mice (Hurlock, McMahon, & Joho, 2008; Zagha, Lang, & Rudy, 2008) fail to show tremor (Lorden, Oltmans, McKeon, Lutes, & Beales, 1985; McMahon et al., 2004; Stratton & Lorden, 1991). Harmaline tremor is disrupted or attenuated by cooling of the cerebellar cortex (Llinas & Volkind, 1973) or massive reduction of PCs in PC degeneration (*pcd*) mice (Milner, Cadoret, Lessard, & Smith, 1995). Mutant mice in which PC terminals are unable to release GABA within the DCN display greatly attenuated harmaline tremor, indicating that the rhythmic DCN bursting that is lost in these mice is required for tremor (Brown et al., 2020).

In addition to gap junction blockers affecting tremor by a potential effect on IO coupling, as cited above, three other pharmacologic interventions may affect tremor by acting on the IO.

Lesions that reduce 5-HT afferents to the IO lessen harmaline or harmine tremor (Costall, Kelly, & Naylor, 1976; Wiklund et al., 1981). Similarly, harmaline tremor is reduced by administration of the broad receptor antagonist methysergide or the 5-HT synthesis inhibitor para-chlorophenylalanine

(Arshaduddin, Kadasah, Al Deeb, Al Moutaery, & Tariq, 2005; Mehta, Saravanan, & Mohanakumar, 2003). Conversely, harmine tremor is exacerbated by the 5-HT precursor 5-hydroxytryptophan (Costall et al., 1976) and by the serotonin uptake inhibitors citalopram and imipramine (Arshaduddin et al., 2004, 2005). As discussed in the previous section, 5-HT appears to engender tremor by activating IO $5HT_{2a}$ receptors, with effects similar to those of harmaline and intra-IO picrotoxin: increased IO coupling, CS synchrony, and tremor. This not only provides an understanding why 5-HT enhances harmaline tremor or, when sufficiently intense, itself induces tremor, but provides a potential explanation why 5-HT uptake inhibitors promote tremor in humans. We should note that although 5-HT's actions overlap with those of harmaline, and are additive with it, they act on different binding sites, as the $5\text{-}HT_{2a}$ antagonist LY53857 does not block harmaline's effects on CSs (Sugihara et al., 1995).

Multiple T-type calcium channel antagonists administered systemically suppress harmaline tremor, consistent with the role of calcium channels in rhythmic IO oscillations (Handforth et al., 2010). While it is possible that these drugs acted on calcium channels elsewhere in the tremor circuit, such as the DCN or thalamus, intra-IO injection of the T-type calcium channel antagonist mibefradil also suppressed harmaline tremor (Park et al., 2010). Some early evidence from clinical trials for ET suggests these agents hold promise (Ondo, 2020; Papapetropoulos et al., 2021).

Another parallel between the olivocerebellar system and ET, as indicated by the harmaline model, involves the AMPA receptor. As mentioned above, intra-IO block of AMPA receptors narrows the spatial extent of CS synchrony (Lang, 2001), an action that would lessen the synchronization of input between DCN neurons receiving PC input from different zebrin stripes. To the extent tremor requires large scale synchronization of DCN neurons, as opposed to synchronization among local clusters of DCN neurons, this would be predicted to reduce tremor. Indeed, systemic administration of AMPA receptor antagonists has been shown to suppress harmaline tremor (Mignani et al., 2002; Paterson, Malekiani, Foreman, Olivier, & Hanania, 2009). Conversely, the allosteric AMPA receptor activator PF-4778574, which, like harmaline, increases cerebellar cortical cyclic GMP, induces tremor in multiple species (Shaffer et al., 2013). Clinical trials with the AMPA antagonist perampanel have shown efficacy for ET (Gironell, Pascual-Sedano, & Marin-Lahoz, 2019; Handforth, Tse, & Elble, 2020). However, it is possible that AMPA antagonists oppose tremor by acting elsewhere in the tremor circuit than in the IO, such as at the Cf-PC synapse, or on cerebellar granule cells (Masugi-Tokita et al., 2007; Tomiyama, Palacios, Cortes, & Mengod, 1999).

Although we have discussed the role of rhythmic IO/CS/DCN activity as underlying tremor in the harmaline model, it should be recognized that the broader circuit engaged by harmaline overlaps substantially with the oscillatory circuit in ET revealed by MEG, involving brainstem, cerebellum, thalamus, premotor and motor cortex (Schnitzler et al., 2009). Metabolic mapping reveals activation of the medial cerebellar cortex and DCN in both ET (Boecker et al., 1996) and after harmaline in cats (Batini, Buisseret-Delmas, & Conrath-Verrier, 1981). In addition, inducible early gene mapping reveals activation of PCs, CGCs, IO neurons, motor thalamus, and motor cortex by harmaline (Beitz & Saxon, 2004; Kosmowska et al., 2020; Kosmowska, Ossowska, Glowacka, & Wardas, 2017; Miwa, Nishi, Fuwa, & Mizuno, 2000; Tian & Bishop, 2002). Evidently, cerebral circuitry is important for the expression of harmaline tremor, as high-frequency thalamic stimulation suppresses the tremor (Bekar et al., 2008; Lee & Chang, 2019) comparable to the therapeutic effect of deep brain stimulation (DBS) of the thalamus in ET. In yet another parallel, intra-thalamic injection of a $GABA_A$ receptor agonist, muscimol, suppresses both harmaline tremor in mice (Bekar et al., 2008) and tremor in ET patients (Pahapill et al., 1999).

If harmaline tremor and ET share common circuitry, they should share common pharmacology. In fact, drugs that suppress tremor in ET also suppress harmaline tremor, including primidone, propranolol, benzodiazepines, gabapentin, gamma-hydroxybutyrate, alcohol, and zonisamide (Handforth, 2012; Ondo, 2020). Conversely, drugs that ET patients may report worsen tremor (tricyclics, caffeine) enhance harmaline tremor (Handforth, 2012). Moreover, the harmaline model predicted efficacy by 1-octanol (Nahab et al., 2011; Sinton, Krosser, Walton, & Llinas, 1989), T-type calcium antagonism (Ondo, 2020; Papapetropoulos et al., 2021), and by AMPA receptor antagonism (Handforth et al., 2020; Paterson et al., 2009).

In conclusion, harmaline causes an action tremor by inducing rhythmic, coupled IO firing and markedly synchronized CSs that engender DCN neuronal burst-firing. Each of these components is required for the tremor. The overall circuit activated by harmaline substantially overlaps that of ET, consistent with a similar pharmacologic response profile. The circuitry and pharmacologic similarity of ET to harmaline renders it plausible that tremor in ET, like that of harmaline, is driven by increased CS synchrony. It may be noted that in several aspects, harmaline tremor differs from that of ET in being a whole-body tremor that is drug-induced and transient, and for some drugs that have been reported to suppress harmaline tremor that do affect ET, such as dopamine agonists and antiepileptic drugs, it is possible that they were administered in doses high enough to exert nonspecific tremor reduction.

2.5 How is the enhanced CS synchrony conveyed to the rest of the brain?

So far, we have discussed how increased CS synchrony, as a consequence of increased IO coupling, may be associated with increased movement amplitude and tremor. In this scheme, there remains a gap in our description of how CS hypersynchrony translates into altered circuit dynamics that underlie tremor, namely, the likely outflow pathway. We now consider this issue.

As the DCN is both the major target of PCs and the major outflow nucleus from cerebellum, it is logical to start with the role of the DCN. We have already discussed above how PCs converge on DCN neurons. However, it may initially seem implausible that enhanced CS synchrony among such PCs could promote large amplitude movement and tremor, given the inhibitory GABAergic nature of PCs. This apparent paradox may be resolved by considering that rhythmic hypersynchronous CS activity causes a profound inhibition of DCN neurons due to the convergence of many PCs onto each DCN neuron, and that this inhibition can be followed by a rebound depolarization that triggers a burst of action potentials, as described earlier. Thus, in situations of hypersynchronous CS activity there will be an entrainment of DCN neurons into a synchronized burst firing mode that is propagated to, and entrains, thalamic and brainstem regions that feed into descending motor pathways (Lamarre & Mercier, 1971; Llinas & Volkind, 1973).

In support of this notion, rebound DCN cell firing has been observed in physiological conditions associated with high levels of synchrony (Bengtsson, Ekerot, & Jorntell, 2011; Dykstra, Engbers, Bartoletti, & Turner, 2016; Hoebeek, Witter, Ruigrok, & De Zeeuw, 2010). We noted above that harmaline induces hyperpolarization and burst firing of DCN neurons (Llinas & Muhlethaler, 1988), and interference with PC terminal GABA release that impairs harmaline-induced DCN bursting also attenuates harmaline tremor (Brown et al., 2020). It has also been found that rhythmic optogenetic stimulation of GABA release from PC terminals produces DCN neuronal bursting and tremor in mice (Pan et al., 2020). Bilateral deep brain stimulation of the interpositus nucleus also suppresses harmaline tremor in mice (Brown et al., 2020).

Insofar as (1) the DCN project to the thalamus, (2) the thalamus is part of the oscillatory circuit revealed by MEG in ET (Schnitzler et al., 2009), and (3) thalamic DBS reduces tremor in ET and the harmaline model, the thalamus is the most likely projection site for burst-firing DCN neurons that cause tremor in ET. In support of this inference, micro-dialysis in the motor thalamus of rats has revealed increased glutamate release during harmaline tremor (Kosmowska et al., 2020). In parallel, magnetic resonance spectroscopy has shown elevated glutamate plus glutamine levels in the thalamus of ET patients, with the degree of change correlated with tremor severity (Barbagallo et al., 2018). These findings are consistent with CS hypersynchrony-induced inhibition-rebound DCN burst-spiking resulting in pulsatile glutamate release in the thalamus.

Burst-firing of DCN neurons projecting to the thalamus would propagate synchrony to forebrain motor areas. Indeed, "Tremor cells" have been recorded in the thalamic ventralis intermedius (Vim) and ventralis oralis posterior (Vop) that discharge coherently with EMG-recorded tremor in ET patients (Kobayashi et al., 2003). Moreover, Vim/Vop cells expressing tremor-associated oscillations are highly synchronized within ensembles (Hanson, Fuller, Lebedev, Turner, & Nicolelis, 2012). Ultimately, the hypersynchrony in the olivocerebellar system would be propagated via descending motor pathways to spinal motor neurons, as shown by findings in both the harmaline and *hotfoot17* models (see below), to produce a tremor that is coherent with hypersynchronous IO oscillatory activity.

3. The olivocerebellar-DCN circuit and tremor: Potential players

3.1 Potential role of aberrant Cfs in tremor

In the previous section we discussed the considerable evidence that increased CS synchrony that is propagated to the DCN and then onto the thalamus can cause tremor. In the standard harmaline model of ET, increased CS synchrony is thought to be solely due to the hypersynchronization of IO activity by harmaline. However, the findings that in ET brains Cfs have increased numbers of transverse branches that potentially innervate multiple PCs (though actual synaptic connections have yet to be demonstrated) and that the increase in such branches correlates with tremor severity, has led to the suggestion that increased synchrony caused by excessive Cf branching underlies the tremor (Wu et al., 2021). Of course, these mechanisms are not mutually exclusive, and could act synergistically if both are present.

Animal studies provide a number of results related to the possibility that aberrant branching underlies tremor. Cfs and PFs compete to innervate PC dendrites, a competition that is regulated by various molecules (Watanabe, 2008), including the GluRδ2 receptor, which is normally located on PC spines (Mishina, Uemura, Yasumura, & Yoshida, 2012). When this receptor is absent in mice, Cfs abnormally innervate the distal dendrites of the PCs, and aberrant transverse Cf branches extend mediolaterally and thus may innervate nearby PCs so that multiple PCs may share Cfs (Miyazaki et al., 2010). This pathology is like that seen in the ET cerebellum, which has a 75% depletion of GluRδ2 receptors (Pan et al., 2020). The homozygous *hotfoot17* mouse, has an approximately 90% reduction of cerebellar GluRδ2, due to genetically induced instability of the GluRδ2 molecule, and displays 20-Hz action tremor that is life-long after adult onset and progressive. The tremor is pharmacologically like that of ET in being reduced by propranolol, primidone, and alcohol (Pan et al., 2020). If a viral vector is introduced to transiently express mRNA for normal GluRδ2 in the cerebellar cortex of *hotfoot17* mice, both the abnormal Cf distribution on PCs and tremor are temporarily abolished (Pan et al., 2020). Other interventions targeting the olivocerebellar system also abolish tremor in *hotfoot17* mice, including intra-IO lidocaine injection, optogenetic silencing of the Cf to PC synapse, and optogenetic suppression of PC terminal firing within the DCN (Pan et al., 2020). Thus, in this model, the same structures are involved that were discussed in previous sections as being involved in inducing tremor through hypersynchronous IO/CS/DCN activity.

Is tremor in *hotfoot17* mice due to increased CS synchrony? Recordings in these mice show strong coherence between IO bursting and tremor (Pan et al., 2020). In addition, these mice display oscillatory cerebellar cortical local field potentials (LFPs) that are not seen in WT mice. Like the tremor, these LFP oscillations are coupled with IO bursting. Intra-IO lidocaine injections abolish the cerebellar oscillations as well as the tremor (Pan et al., 2020). Overall, these results suggest that in the *hotfoot17* mouse excessive CS synchrony occurs that results in LFP-detectable cerebellar synchrony and drives tremor. However, it is important to point out that while the LFPs require olivocerebellar activity, they may reflect changes in other cortical activity as well. Interestingly, many ET patients were found to display cerebellar oscillations that were coherent with tremor, and the intensity of the oscillations correlated with tremor severity (Pan et al., 2020).

In sum, the *hotfoot17* animal model shares with ET aberrant Cf branching and subtotal reduction of GluRδ2 receptors, and exhibits tremor that requires the aberrant Cf-PC branching and intact PC to DCN projections. As a consequence of the anomalous Cf branching, excessive CS synchrony is inferred to occur, resulting in LFP-detectable cerebellar oscillations and tremor. Importantly, for *hotfoot17* model, multi-PC innervation needs to be established, and its frequency, spatial extent, and physiological strength need to be characterized.

3.2 Does increased IO coupling play a role in ET?

In the above section we discussed evidence from animal models that aberrant Cf innervation of multiple PCs may form the basis of CS hypersynchrony, leading to tremor. In this scheme, it may be supposed that this anomaly is sufficient to explain the tremor; that is, there is no need to invoke increased IO coupling in ET. However, it is possible that such branching may play an important role yet not be sufficient by itself, particularly because the transverse Cf branches in ET are relatively limited in extent, and thus would connect, and potentially synchronize, only closely spaced PCs. In contrast, DCN regions, and likely individual DCN cells, receive convergent input from PCs located within an extended rostrocaudally oriented cortical strips. The CS activity of such PCs can be synchronized via the coupling of IO neurons as discussed earlier, raising the possibility that increased IO coupling might be required, or at least play a contributory role. In deciding between these alternatives, a key unknown parameter is the degree (and spatial extent) of synchrony needed to drive DCN neurons into a synchronized bursting pattern. However, simultaneous recordings of CS and DCN activity suggest that the strong inhibition of the DCN, as would likely be needed to drive rebound bursting, requires synchronization of at least 50% of the PC input to a DCN cell (Tang et al., 2019), suggesting that the increase in synchrony directly due to aberrant branching may not be sufficient by itself. Thus, we consider how increased IO coupling may be caused by changes in cerebellar cortical output and/or changes in other parts of the oscillatory circuit in ET.

The IO receives afferent activity from parts of the oscillatory circuit, such as the motor cortex and DCN, and changes in this input could increase IO coupling. Of these, the best understood is the massive cerebellar GABAergic DCN-olivary projection that is the main inhibitory input to the IO

(De Zeeuw et al., 1989; Nelson & Mugnaini, 1989). The DCN neurons that project to the IO in turn receive a GABAergic projection from PCs as their main input. The IO neurons then project Cfs to the region of cerebellar cortex containing PCs that project via the DCN to the same IO region from which it receives Cfs. These tri-synaptic circuits thus connect the cerebellar cortex, DCN and IO in a series of parallel closed loops. Whereas most DCN neurons are preferentially sensitive to synchronous activity, those DCN neurons that project to the IO are designed to integrate ongoing activity (Najac & Raman, 2015a; Uusisaari et al., 2007), such as that provided by PC simple spikes (SSs). These features of the tri-synaptic circuit provide a mechanism by which PF-induced SSs in each region of cerebellar cortex can influence, via the IO, its own CS activity.

This idea of the cerebellum and IO being connected in a closed loop fashion was tested by recording SSs and CSs in anesthetized rats and applying drugs to the cerebellar cortex to increase or decrease local SS activity (Marshall & Lang, 2009). When picrotoxin was applied to block GABA-A mediated inhibition due to local interneurons in a small region of the cerebellar cortex, SS frequency increased, as did CS synchrony and firing rate. This can be explained as follows. Picrotoxin disinhibits PCs, increasing SS firing. This increased SS activity leads to more GABA release within DCN, inhibiting the GABAergic DCN neurons that project to the IO. The decreased release of GABA release in IO glomeruli permits increased IO coupling, so that enhanced CS synchrony ensues. Conversely, application of the $GABA_A$ receptor agonist muscimol decreases SSs and reduces CS synchrony. These results indicate that CS synchrony, and hence tremor, could be influenced by alterations of cerebellar cortical function that affect SS firing, such as changes in CGC activity or changes in the local inhibitory interneuron activity, with the effects exerted through this double inhibition pathway.

Positron emission tomography (PET) studies have shown increased regional blood flow in cerebellar cortex in ET (Boecker et al., 1996; Jenkins et al., 1993; Wills, Jenkins, Thompson, Findley, & Brooks, 1994), indicating increased local synaptic activity. Given the massive CGC population and its projections, the increased metabolic activity likely includes increased PF terminal firing. This would cause increased SS activity, which would, via the double inhibitory pathway, increase CS synchrony and exacerbate tremor. In these metabolic scans, the IO would not necessarily be expected to reveal increased signal, as the metabolic demand of mildly increased IO firing rate is modest and would likely be compensated by the reduced energy requirements of decreased DCN-IO terminal firing. Thus, the absence of IO signal changes in ET subjects (Wills et al., 1994) is not surprising nor evidence against a role of the olivocerebellar system in ET.

The notion that excess SS activity can contribute to tremor is consistent with the conclusion by Dupuis et al. (2010) that lesions of the pons abolish tremor in ET cases by interrupting signals from the motor and premotor cortices that would drive mossy fiber activity. By removing mossy fiber excitatory drive to CGCs, SS firing would be reduced, decreasing CS synchrony, leading to loss of tremor. In this context, the heightened metabolic activity of the cerebellar cortex in ET raises the question whether this afferent pathway from cerebral cortex via pons is hyperactive in ET. Anomalies in the tremor circuit might contribute to increased cerebellar cortical activity. For example, in ET patients, there is evidence for a reduced GABA modulation in premotor and motor cortex (Chuang, Huang, Lu, & Chen, 2014), which could result in increased excitatory drive to mossy fiber pathways. Alternatively, it may be that there is an excessive excitatory drive to CGCs from motor cortex that is secondary to a hypersynchronous CS/DCN/thalamus outflow that occurs due to aberrant Cf innervation of PCs. Lastly, abnormalities in the basket cell-PC synapse have been reported (Kuo et al., 2013), which raises the question of whether this synapse is functioning abnormally, as defective inhibition of PCs would lead to increased SS inhibition of DCN-IO neurons.

Alcohol, well known for its efficacy in many ET patients (Growdon et al., 1975), reduces CS frequency and harmaline tremor in rodents (Martin, Thu Le, & Handforth, 2005; Rappaport, Gentry, Schneider, & Dole, 1984; Sinclair, Lo, & Harris, 1982). High-density EEG has revealed that the site of alcohol's anti-tremor action in ET is the cerebellum (Pedrosa et al., 2017). Blood flow imaging in ET patients has shown that low-dose alcohol reduces cerebellar cortical activity towards normal (Boecker et al., 1996), indicating a reduction of synaptic activity. If that reduction is associated with reduced SS firing, then reduced CS synchrony would also be expected to occur, supporting the latter's role in tremor suppression that occurs in these subjects. Supporting this idea, alcohol increased blood flow in the IO region may be attributed to the metabolic demand of increased terminal fiber firing of inhibitory DCN neurons projecting to the IO (Boecker et al., 1996). In this interpretation, alcohol has an action on the double inhibition pathway like that of the GABA receptor agonist muscimol discussed above. In fact, alcohol has been shown to selectively activate extra-synaptic $GABA_A$ receptors at low, non-intoxicating levels

(Wallner, Hanchar, & Olsen, 2003). As it happens, CGCs intensely express extra-synaptic GABA receptors, whose location enables them to control PF excitatory transmission to PCs (Hortnagl et al., 2013; Kato, 1990; Laurie, Seeburg, & Wisden, 1992; Persohn, Malherbe, & Richards, 1992; Wisden, Laurie, Monyer, & Seeburg, 1992). These likely actions of alcohol on the double inhibition pathway provide an attractive explanation how low-dose alcohol could reduce cerebellar cortical activity and reduce CS synchrony, and thereby suppress tremor.

In general support of this scheme, the drug THIP (gaboxadol; 4,5,6,7-tetrahydroisoxazolo[5,4-*c*]pyridin-3-ol), that in low doses selectively activates $GABA_A$ extra-synaptic receptors (Meera, Wallner, & Otis, 2011) was found to suppress harmaline tremor in WT mice, but not in KO mice lacking subunits of the extra-synaptic GABA receptor that are preferentially expressed on CGCs (Handforth, Kadam, Kosoyan, & Eslami, 2018). These findings suggest that THIP acts to suppress tremor by activating extra-synaptic GABA receptors that inhibit CGCs, so that there is a reduction of SS activity and via the double inhibitory pathway, CS synchrony.

In summary, in this section we have discussed evidence that enhanced CS synchrony in ET may occur from increased IO coupling, which can occur due to changes in afferent activity, particularly from CGCs via the double inhibition pathway, or possibly via altered interneuron function. The increased metabolism of the cerebellar cortex in ET strongly suggests the IO is involved. In addition, the IO is a logical site for the action of some tremor-exacerbating agents, such as 5-HT uptake inhibitors, as discussed in an earlier section.

3.3 Yet another potential cause of DCN hypersynchrony: Partial PC loss

Above, we have discussed how rhythmic synchronized DCN bursting occurs, causing tremor, as a result of hypersynchronous CSs that are convergent onto DCN neurons, either because of increased IO coupling or aberrant Cfs that yoke multiple PCs into synchrony. Yet another potential cause of convergent, synchronous CS activity is partial PC loss combined with intra-DCN collateral sprouting of the axon terminals of surviving PCs. Through the convergent effect of collateralized axon terminals, even normal levels of CS activity could result in synchronous hyperpolarization of DCN cells, whose burst firing would promote tremor at the

IO oscillation frequency. In all these scenarios the physical integrity of the IO is required, for without the IO there can be no CSs.

Rendering this notion plausible are findings that PC numbers are reduced (Louis, Babij, Lee, Cortes, & Vonsattel, 2013), and increased terminal PC axonal sprouting is present in ET brains compared to age-matched controls (Babij et al., 2013). The possibility that sporadic partial PC loss may engender action tremor in this way is relevant not only to ET, but potentially to action tremor in clinical conditions involving cerebellar damage as after head trauma and various intoxications (Biary, Cleeves, Findley, & Koller, 1989). The onset of action tremor after head trauma is not immediate, but delayed to the first few months after the injury, suggesting a role for neural plasticity.

Purkinje cell degeneration mice *(pcd)* lose most of their PCs by postnatal day 35. They show no spontaneous tremor and have a poor tremor response to harmaline (Milner et al., 1995). In the DCN, GABA is markedly reduced, consistent with loss of PC axon terminals (Roffler-Tarlov, Beart, O'Gorman, & Sidman, 1979). In addition, partial PC loss occurs in various mammalian species with aging, yet tremor does not occur (Nandy, 1981; Sturrock, 1989). In aged rats there is a loss of GABAergic fibers in DCN, indicating a lack of PC fiber sprouting (Lolova & Davidoff, 1990). In contrast, *Weaver* mice have partial PC loss and display tremor. Although they have a 50% reduction of vermal PCs and a smaller PC reduction in the cerebellar hemispheres, they likely have extensive PC axon terminal sprouting in DCN, as DCN GABA levels are normal, and in another PC projection site, the vestibular nucleus, normal PC axon terminal numbers are present (Baurle, Grover, & Grusser-Cornehls, 1992; Roffler-Tarlov & Turey, 1982).

Just as DCN GABA levels are normal in *Weaver* mice despite partial PC loss, DCN GABA levels are normal in ET patients despite partial PC loss (Louis, Hernandez, Dyke, Ma, & Dydak, 2018), consistent with the increased PC axon terminal sprouting that is known to occur in ET (Babij et al., 2013). Interestingly, in asymmetric ET, the asymmetric cerebellar pathology is positively correlated with tremor severity (Louis, Lee, Cortes, Vonsattel, & Faust, 2014). Despite this, the survival of the remaining PCs appears required for tremor expression in both species, as would be expected given their role in generating CSs. If the vermis is lesioned, tremor in *Weaver* mice is abolished (Grusser-Cornehls, Grusser, & Baurle, 1999). The requirement for PCs in ET is inferred by the therapeutic response to thalamic DBS, which blocks output from DCN, which in turn is controlled by PC activity.

These parallels render it plausible that in some ET and non-ET cases with action tremor, partial PC loss with PC terminal sprouting by remaining PCs in DCN may induce hypersynchronization and hyperpolarization of DCN neurons, so that rebound burst-firing at the IO-CS frequency occurs, leading to tremor. More research on this question is needed. For example, do DCN neurons in *Weaver* mice display hyperpolarization and burst-firing? Is the tremor in these mice abolished by interventions that interfere with the IO-PC-DCN pathway as shown in *hotfoot17* mice?

3.4 Is the inferior olive really required for tremor in ET?

Above we discussed how hypersynchronous IO/CS activity can lead to tremor, and also how normal IO oscillatory function may be required for tremor to occur even in the setting of either Cf aberrant multi-PC innervation or PC axon collateralization within DCN when partial PC loss occurs. In these three scenarios, all of which are likely relevant to ET, the integrity of the IO is a sine qua non for tremor. Of course, the most grand and elegant hypothesis can be destroyed by one ugly little fact. Here we consider two case reports that their authors judged posed great difficulty for the IO hypothesis and that we have discussed in a previous review (Handforth & Lang, 2021).

The first case involves an ET patient whose tremor was reported to have started at age 10, and progressively worsened throughout his life, including the last years of his life (ages 85–94 years) during which time he was studied by the authors of the report (Louis, Diaz, et al., 2018). The brain displayed considerable IO degeneration and it was suggested that this proved that the IO hypothesis is wrong. However, this case was atypical for ET in showing IO pathology, loss of GABA terminals in DCN, and very severe PC loss. Given the marked loss of PCs and terminals in the DCN, it is possible that ET-type tremor was replaced in his last years by ataxic tremor that is characteristic of severe cerebellar damage.

Ataxic tremor is seen, for example, in monkeys with cerebellar damage. In this setting the frequency of upper extremity tremor changes according to the position of the limb or with mechanical loading, indicating that this tremor is affected by mechanical and reflex functions (Carpenter & Stevens, 1957; Vilis & Hore, 1977), in contrast to frequency-invariant tremor of ET, owing to the latter's dependence on a central oscillator. Unfortunately, such tests that could have distinguished the two tremors were not performed. Thus, we believe IO degeneration in this case does not provide conclusive evidence against the IO being necessary for ET tremor.

A second case involves a 77-year old man with an approximately 30-year history of bilateral postural and kinetic hand tremor who developed palatal tremor and imbalance. MRI revealed olivary hypertrophy. On physical examination, tremor was observed, and EMG demonstrated a 5.3-Hz tremor whose frequency was not affected by loading the limb with a weight, consistent with centrally-generated tremor like that found in ET. The authors reasoned that this case shows that the IO is not required for ET, at least not in all cases (Elkouzi, Kattah, & Elble, 2016). However, it is conceivable that in this case parts of the IO that are required for tremor were spared.

The principal nucleus is predominantly involved in cases of olivary hypertrophy, whereas the MAO does not have to be affected, and involvement of the DAO is rare (Verhaart & Voogd, 1962). In contrast, harmaline micro-injections within the subnuclei of IO reveal that it is the caudal portions of MAO and DAO that are most activated into hypersynchronous oscillatory activity (De Montigny & Lamarre, 1975). Thus, the likely at least partial, if not total, preservation of the MAO and DAO may explain why tremor persisted in this case.

In conclusion, despite these cases, it remains plausible that portions of the IO are required for tremor, either by enabling the amplification of normal oscillatory activity into downstream hypersynchrony through the effects of Cf aberrant multi-PC innervation and/or PC terminal sprouting within DCN, or through the perpetuation of IO hypersynchrony to CS synchrony.

4. Conclusions

We first explored early evidence that tremor in ET owes its expression to an oscillator located within the brain, and noted the cerebellum as an important part of an oscillatory circuit. Recent evidence of aberrant Cf sprouting with multi-PC innervation in the brains of ET patients has suggested that this pathology as a source of synchronous activity that could underlie tremor. To understand why this might be credible, we discussed the ability of the olivocerebellar system to generate synchronous CS activity and described how this activity can modulate cerebellar output from the DCN through convergent PC projections. We discussed how ~10 Hz oscillatory activity flowing through the IO-PC-CS-DCN circuit influences the efficacy of motor cortex activity in producing movement. For example, blocking IO GABA receptors, which increases IO and CS synchrony, increases the amplitude of movements evoked by motor cortex stimulation, and can induce tremor. Similarly, harmaline or activation of IO 5-HT_{2a} receptors promote IO synchrony, CS synchrony, and tremor.

This wealth of evidence linking IO hypersynchrony to CS hypersynchrony, with downstream synchronous DCN hyperpolarization and burst-firing, and tremor, renders it plausible that multi-PC Cf innervation, which would be expected to promote CS synchrony, could promote tremor. In further support, *hotfoot17* mice, which have GluRδ2 deficiency and pathologic features like those found in ET, display tremor that depends on the integrity of the IO and its projection to PCs, and of the PC projections to DCN. Interestingly, the physiological signature in *hotfoot17* mice is cerebellar oscillations that are synchronous with tremor and IO bursting and may reflect heightened CS synchrony. Cerebellar oscillations are also present in ET patients to a degree correlated with tremor severity, supporting the importance of olivocerebellar activity in ET.

Thus, we consider it likely that CS synchrony generated at the IO level will contribute to tremor in two ways. One is hypersynchrony of IO neurons due to alterations in afferent activity to this nucleus. The increased metabolism of the cerebellar cortex that occurs in ET suggests that CGCs are more active than normal, a conjecture consistent with the finding that pontine infarcts that interrupt excitatory mossy fiber afferents to CGCs abolish tremor. Through the double inhibition pathway, increased CGC activity, triggering more SSs, would lead to reduced intra-IO GABA release, promoting or exacerbating tremor. The double inhibition pathway provides a mechanism by which afferents to the cerebellum or agents that appear to act within the cerebellum, such as alcohol, may promote or suppress tremor.

In another potential mechanism for tremor, it is suggested that in the setting of partial PC loss, terminal axon sprouting within DCN by surviving PCs engenders DCN hyperpolarization and hypersynchrony, promoting or exacerbating tremor. As the IO is required for CSs to occur, the integrity of the IO is necessary for this tremor mechanism as well.

In summary, in this chapter we have reviewed several potential mechanisms of tremor in ET, which have in common hypersynchrony of outflow from the DCN and require the integrity of the IO, but differ in the level at which hypersynchrony primarily may first occur (PC, IO or DCN). It is probable that the relevant contribution of these mechanisms may differ among ET patients, but that all three mechanisms when present could act synergistically to cause serial hypersynchronization along the IO to PC to DCN axis.

References

Arshaduddin, M., Al Kadasah, S., Biary, N., Al Deeb, S., Al Moutaery, K., & Tariq, M. (2004). Citalopram, a selective serotonin reuptake inhibitor augments harmaline-induced tremor in rats. *Behavioural Brain Research*, *153*(1), 15–20.

Arshaduddin, M., Kadasah, S., Al Deeb, S., Al Moutaery, K., & Tariq, M. (2005). Exacerbation of harmaline-induced tremor by imipramine. *Pharmacology, Biochemistry, and Behavior*, *81*(1), 9–14.

Babij, R., Lee, M., Cortes, E., Vonsattel, J. P., Faust, P. L., & Louis, E. D. (2013). Purkinje cell axonal anatomy: Quantifying morphometric changes in essential tremor versus control brains. *Brain*, *136*(Pt 10), 3051–3061.

Bal, T., & McCormick, D. A. (1997). Synchronized oscillations in the inferior olive are controlled by the hyperpolarization-activated cation current I(h). *Journal of Neurophysiology*, *77*(6), 3145–3156.

Barbagallo, G., Arabia, G., Novellino, F., Nistico, R., Salsone, M., Morelli, M., et al. (2018). Increased glutamate + glutamine levels in the thalamus of patients with essential tremor: A preliminary proton MR spectroscopic study. *Parkinsonism & Related Disorders*, *47*, 57–63.

Barragan, L. A., Delhaye-Bouchaud, N., & Laget, P. (1985). Drug-induced activation of the inferior olivary nucleus in young rabbits. Differential effects of harmaline and quipazine. *Neuropharmacology*, *24*(7), 645–654.

Batini, C., Buisseret-Delmas, C., & Conrath-Verrier, M. (1981). Harmaline-induced tremor. I. Regional metabolic activity as revealed by [14C]2-deoxyglucose in cat. *Experimental Brain Research*, *42*(3–4), 371–382.

Baurle, J., Grover, B. G., & Grusser-Cornehls, U. (1992). Plasticity of GABAergic terminals in Deiters' nucleus of weaver mutant and normal mice: A quantitative light microscopic study. *Brain Research*, *591*(2), 305–318.

Beitz, A. J., & Saxon, D. (2004). Harmaline-induced climbing fiber activation causes amino acid and peptide release in the rodent cerebellar cortex and a unique temporal pattern of Fos expression in the olivo-cerebellar pathway. *Journal of Neurocytology*, *33*(1), 49–74.

Bekar, L., Libionka, W., Tian, G. F., Xu, Q., Torres, A., Wang, X., et al. (2008). Adenosine is crucial for deep brain stimulation-mediated attenuation of tremor. *Nature Medicine*, *14*(1), 75–80.

Belluardo, N., Mudo, G., Trovato-Salinaro, A., Le Gurun, S., Charollais, A., Serre-Beinier, V., et al. (2000). Expression of connexin36 in the adult and developing rat brain. *Brain Research*, *865*(1), 121–138.

Bengtsson, F., Ekerot, C. F., & Jorntell, H. (2011). In vivo analysis of inhibitory synaptic inputs and rebounds in deep cerebellar nuclear neurons. *PLoS One*, *6*(4), e18822.

Biary, N., Cleeves, L., Findley, L., & Koller, W. (1989). Post-traumatic tremor. *Neurology*, *39*(1), 103–106.

Blenkinsop, T. A., & Lang, E. J. (2006). Block of inferior olive gap junctional coupling decreases Purkinje cell complex spike synchrony and rhythmicity. *The Journal of Neuroscience*, *26*(6), 1739–1748.

Blenkinsop, T. A., & Lang, E. J. (2011). Synaptic action of the olivocerebellar system on cerebellar nuclear spike activity. *The Journal of Neuroscience*, *31*(41), 14708–14720.

Boecker, H., Wills, A. J., Ceballos-Baumann, A., Samuel, M., Thompson, P. D., Findley, L. J., et al. (1996). The effect of ethanol on alcohol-responsive essential tremor: A positron emission tomography study. *Annals of Neurology*, *39*(5), 650–658.

Brown, A. M., White, J. J., van der Heijden, M. E., Zhou, J., Lin, T., & Sillitoe, R. V. (2020). Purkinje cell misfiring generates high-amplitude action tremors that are corrected by cerebellar deep brain stimulation. *eLife*, *9*.

Carpenter, M. B., & Stevens, G. H. (1957). Structural and functional relationships between the deep cerebellar nuclei and the brachium conjunctivum in the rhesus monkey. *The Journal of Comparative Neurology*, *107*(1), 109–163.
Chuang, W. L., Huang, Y. Z., Lu, C. S., & Chen, R. S. (2014). Reduced cortical plasticity and GABAergic modulation in essential tremor. *Movement Disorders*, *29*(4), 501–507.
Chung, S. H., Marzban, H., & Hawkes, R. (2009). Compartmentation of the cerebellar nuclei of the mouse. *Neuroscience*, *161*(1), 123–138.
Cicirata, F., Angaut, P., Serapide, M. F., Panto, M. R., & Nicotra, G. (1992). Multiple representation in the nucleus lateralis of the cerebellum: An electrophysiologic study in the rat. *Experimental Brain Research*, *89*(2), 352–362.
Coddington, L. T., Rudolph, S., Vande Lune, P., Overstreet-Wadiche, L., & Wadiche, J. I. (2013). Spillover-mediated feedforward inhibition functionally segregates interneuron activity. *Neuron*, *78*(6), 1050–1062.
Condorelli, D. F., Parenti, R., Spinella, F., Trovato Salinaro, A., Belluardo, N., Cardile, V., et al. (1998). Cloning of a new gap junction gene (Cx36) highly expressed in mammalian brain neurons. *The European Journal of Neuroscience*, *10*(3), 1202–1208.
Costall, B., Kelly, D. M., & Naylor, R. J. (1976). The importance of 5-hydroxytryptamine for the induction of harmine tremor and its antagonism by dopaminergic agonists assessed by lesions of the midbrain raphe nuclei. *European Journal of Pharmacology*, *35*(1), 109–119.
De Gruijl, J. R., Hoogland, T. M., & De Zeeuw, C. I. (2014). Behavioral correlates of complex spike synchrony in cerebellar microzones. *The Journal of Neuroscience*, *34*(27), 8937–8947.
De Montigny, C., & Lamarre, Y. (1973). Rhythmic activity induced by harmaline in the olivo-cerebello-bulbar system of the cat. *Brain Research*, *53*(1), 81–95.
De Montigny, C., & Lamarre, Y. (1975). Effects produced by local applications of harmaline in the inferior olive. *Canadian Journal of Physiology and Pharmacology*, *53*(5), 845–849.
De Zeeuw, C. I., & Berrebi, A. S. (1995). Postsynaptic targets of Purkinje cell terminals in the cerebellar and vestibular nuclei of the rat. *European Journal of Neuroscience*, 7, 2322–2333.
De Zeeuw, C. I., Holstege, J. C., Ruigrok, T. J., & Voogd, J. (1989). Ultrastructural study of the GABAergic, cerebellar, and mesodiencephalic innervation of the cat medial accessory olive: Anterograde tracing combined with immunocytochemistry. *The Journal of Comparative Neurology*, *284*(1), 12–35.
Dupuis, M. J., Evrard, F. L., Jacquerye, P. G., Picard, G. R., & Lermen, O. G. (2010). Disappearance of essential tremor after stroke. *Movement Disorders*, *25*(16), 2884–2887.
Dykstra, S., Engbers, J. D., Bartoletti, T. M., & Turner, R. W. (2016). Determinants of rebound burst responses in rat cerebellar nuclear neurons to physiological stimuli. *The Journal of Physiology*, *594*(4), 985–1003.
Eccles, J. C., Llinas, R., & Sasaki, K. (1966). The excitatory synaptic action of climbing fibres on the Purkinje cells of the cerebellum. *The Journal of Physiology*, *182*(2), 268–296.
Elkouzi, A., Kattah, J. C., & Elble, R. J. (2016). Hypertrophic olivary degeneration does not reduce essential tremor. *Movement Disorders Clinical Practice*, *3*(2), 209–211.
Filip, P., Lungu, O. V., Manto, M. U., & Bares, M. (2016). Linking essential tremor to the cerebellum: Physiological evidence. *Cerebellum*, *15*(6), 774–780.
Fredette, B. J., & Mugnaini, E. (1991). The GABAergic cerebello-olivary projection in the rat. *Anatomy and Embryology*, *184*, 225–243.
Gionco, J. T., Hartstone, W. G., Martuscello, R. T., Kuo, S. H., Faust, P. L., & Louis, E. D. (2021). Essential tremor versus "ET-plus": A detailed postmortem study of cerebellar pathology. *Cerebellum*.
Gironell, A., Pascual-Sedano, B., & Marin-Lahoz, J. (2019). Perampanel, a new hope for essential tremor: An open label trial. *Parkinsonism & Related Disorders*, *60*, 171–172.

Growdon, J. H., Shahani, B. T., & Young, R. R. (1975). The effect of alcohol on essential tremor. *Neurology, 25*(3), 259–262.

Grusser-Cornehls, U., Grusser, C., & Baurle, J. (1999). Vermectomy enhances parvalbumin expression and improves motor performance in weaver mutant mice: An animal model for cerebellar ataxia. *Neuroscience, 91*(1), 315–326.

Handforth, A. (2012). Harmaline tremor: Underlying mechanisms in a potential animal model of essential tremor. *Tremor and Other Hyperkinetic Movements, 2*.

Handforth, A., Homanics, G. E., Covey, D. F., Krishnan, K., Lee, J. Y., Sakimura, K., et al. (2010). T-type calcium channel antagonists suppress tremor in two mouse models of essential tremor. *Neuropharmacology, 59*(6), 380–387.

Handforth, A., Kadam, P. A., Kosoyan, H. P., & Eslami, P. (2018). Suppression of harmaline tremor by activation of an extrasynaptic GABAA receptor: Implications for essential tremor. *Tremor and Other Hyperkinetic Movements, 8*, 546.

Handforth, A., & Lang, E. J. (2021). Increased Purkinje cell complex spike and deep cerebellar nucleus synchrony as a potential basis for syndromic essential tremor. A review and synthesis of the literature. *Cerebellum, 20*(2), 266–281.

Handforth, A., Tse, W., & Elble, R. J. (2020). A pilot double-blind randomized trial of perampanel for essential tremor. *Movement Disorders Clinical Practice*, 7(4), 399–404.

Hanson, T. L., Fuller, A. M., Lebedev, M. A., Turner, D. A., & Nicolelis, M. A. (2012). Subcortical neuronal ensembles: An analysis of motor task association, tremor, oscillations, and synchrony in human patients. *The Journal of Neuroscience, 32*(25), 8620–8632.

Hoebeek, F. E., Witter, L., Ruigrok, T. J., & De Zeeuw, C. I. (2010). Differential olivo-cerebellar cortical control of rebound activity in the cerebellar nuclei. *Proceedings of the National Academy of Sciences of the United States of America, 107*(18), 8410–8415.

Homberg, V., Hefter, H., Reiners, K., & Freund, H. J. (1987). Differential effects of changes in mechanical limb properties on physiological and pathological tremor. *Journal of Neurology, Neurosurgery, and Psychiatry, 50*(5), 568–579.

Horsley, V., & Schafer, E. A. (1886). Experiments on the character of the muscular contractions which are evoked by excitation of the various parts of the motor tract. *The Journal of Physiology*, 7(2), 96–110.

Hortnagl, H., Tasan, R. O., Wieselthaler, A., Kirchmair, E., Sieghart, W., & Sperk, G. (2013). Patterns of mRNA and protein expression for 12 GABAA receptor subunits in the mouse brain. *Neuroscience, 236*, 345–372.

Hurlock, E. C., McMahon, A., & Joho, R. H. (2008). Purkinje-cell-restricted restoration of Kv3.3 function restores complex spikes and rescues motor coordination in Kcnc3 mutants. *The Journal of Neuroscience, 28*(18), 4640–4648.

Jenkins, I. H., Bain, P. G., Colebatch, J. G., Thompson, P. D., Findley, L. J., Frackowiak, R. S., et al. (1993). A positron emission tomography study of essential tremor: Evidence for overactivity of cerebellar connections. *Annals of Neurology, 34*(1), 82–90.

Kato, K. (1990). Novel GABAA receptor alpha subunit is expressed only in cerebellar granule cells. *Journal of Molecular Biology, 214*(3), 619–624.

Kelly, R. M., & Strick, P. L. (2003). Cerebellar loops with motor cortex and prefrontal cortex of a nonhuman primate. *The Journal of Neuroscience, 23*(23), 8432–8444.

Kobayashi, K., Katayama, Y., Kasai, M., Oshima, H., Fukaya, C., & Yamamoto, T. (2003). Localization of thalamic cells with tremor-frequency activity in Parkinson's disease and essential tremor. *Acta Neurochirurgica. Supplement, 87*, 137–139.

Kosmowska, B., Ossowska, K., Glowacka, U., & Wardas, J. (2017). Tremorolytic effect of 5′-chloro-5′-deoxy-(+/−)-ENBA, a potent and selective adenosine A1 receptor agonist, evaluated in the harmaline-induced model in rats. *CNS Neuroscience & Therapeutics, 23*(5), 438–446.

Kosmowska, B., Ossowska, K., Konieczny, J., Lenda, T., Berghauzen-Maciejewska, K., & Wardas, J. (2020). Inhibition of excessive glutamatergic transmission in the ventral thalamic nuclei by a selective adenosine A1 receptor agonist, 5′-chloro-5′-deoxy-(+/−)-ENBA underlies its tremorolytic effect in the harmaline-induced model of essential tremor. *Neuroscience*, *429*, 106–118.

Kuo, S. H., Tang, G., Louis, E. D., Ma, K., Babji, R., Balatbat, M., et al. (2013). Lingo-1 expression is increased in essential tremor cerebellum and is present in the basket cell pinceau. *Acta Neuropathologica*, *125*(6), 879–889.

Lakie, M., Walsh, E. G., Arblaster, L. A., Villagra, F., & Roberts, R. C. (1994). Limb temperature and human tremors. *Journal of Neurology, Neurosurgery, and Psychiatry*, *57*(1), 35–42.

Lamarre, Y., & Mercier, L. A. (1971). Neurophysiological studies of harmaline-induced tremor in the cat. *Canadian Journal of Physiology and Pharmacology*, *49*(12), 1049–1058.

Lang, E. J. (2001). Organization of olivocerebellar activity in the absence of excitatory glutamatergic input. *The Journal of Neuroscience*, *21*(5), 1663–1675.

Lang, E. J. (2002). GABAergic and glutamatergic modulation of spontaneous and motor-cortex-evoked complex spike activity. *Journal of Neurophysiology*, *87*(4), 1993–2008.

Lang, E. J., & Blenkinsop, T. A. (2011). Control of cerebellar nuclear cells: A direct role for complex spikes? *Cerebellum*, *10*(4), 694–701.

Lang, E. J., Sugihara, I., & Llinas, R. (1996). GABAergic modulation of complex spike activity by the cerebellar nucleoolivary pathway in rat. *Journal of Neurophysiology*, *76*(1), 255–275.

Lang, E. J., Sugihara, I., & Llinas, R. (2006). Olivocerebellar modulation of motor cortex ability to generate vibrissal movements in rat. *The Journal of Physiology*, *571*(Pt 1), 101–120.

Lang, E. J., Sugihara, I., Welsh, J. P., & Llinas, R. (1999). Patterns of spontaneous purkinje cell complex spike activity in the awake rat. *The Journal of Neuroscience*, *19*(7), 2728–2739.

Laurie, D. J., Seeburg, P. H., & Wisden, W. (1992). The distribution of 13 GABAA receptor subunit mRNAs in the rat brain. II. Olfactory bulb and cerebellum. *The Journal of Neuroscience*, *12*(3), 1063–1076.

LeDoux, M. S., & Lorden, J. F. (2002). Abnormal spontaneous and harmaline-stimulated Purkinje cell activity in the awake genetically dystonic rat. *Experimental Brain Research*, *145*(4), 457–467.

Lee, J., & Chang, S. Y. (2019). Altered primary motor cortex neuronal activity in a rat model of harmaline-induced tremor during thalamic deep brain stimulation. *Frontiers in Cellular Neuroscience*, *13*, 448.

Lee, D., Gan, S. R., Faust, P. L., Louis, E. D., & Kuo, S. H. (2018). Climbing fiber-Purkinje cell synaptic pathology across essential tremor subtypes. *Parkinsonism & Related Disorders*, *51*, 24–29.

Leznik, E., & Llinas, R. (2005). Role of gap junctions in synchronized neuronal oscillations in the inferior olive. *Journal of Neurophysiology*, *94*(4), 2447–2456.

Leznik, E., Makarenko, V., & Llinas, R. (2002). Electrotonically mediated oscillatory patterns in neuronal ensembles: An in vitro voltage-dependent dye-imaging study in the inferior olive. *The Journal of Neuroscience*, *22*(7), 2804–2815.

Lin, C. Y., Louis, E. D., Faust, P. L., Koeppen, A. H., Vonsattel, J. P., & Kuo, S. H. (2014). Abnormal climbing fibre-Purkinje cell synaptic connections in the essential tremor cerebellum. *Brain*, *137*(Pt 12), 3149–3159.

Llinas, R., Baker, R., & Sotelo, C. (1974). Electrotonic coupling between neurons in cat inferior olive. *Journal of Neurophysiology*, *37*(3), 560–571.

Llinas, R., Bloedel, J. R., & Hillman, D. E. (1969). Functional characterization of neuronal circuitry of frog cerebellar cortex. *Journal of Neurophysiology*, *32*(6), 847–870.

Llinas, R., & Muhlethaler, M. (1988). Electrophysiology of guinea-pig cerebellar nuclear cells in the in vitro brain stem-cerebellar preparation. *The Journal of Physiology*, *404*, 241–258.
Llinas, R., & Volkind, R. A. (1973). The olivo-cerebellar system: Functional properties as revealed by harmaline-induced tremor. *Experimental Brain Research*, *18*(1), 69–87.
Llinas, R., & Yarom, Y. (1981a). Electrophysiology of mammalian inferior olivary neurones in vitro. Different types of voltage-dependent ionic conductances. *The Journal of Physiology*, *315*, 549–567.
Llinas, R., & Yarom, Y. (1981b). Properties and distribution of ionic conductances generating electroresponsiveness of mammalian inferior olivary neurones in vitro. *The Journal of Physiology*, *315*, 569–584.
Llinas, R., & Yarom, Y. (1986). Oscillatory properties of guinea-pig inferior olivary neurones and their pharmacological modulation: An in vitro study. *The Journal of Physiology*, *376*, 163–182.
Lolova, I., & Davidoff, M. (1990). Immuno- and histochemical data on changed GABA transmission in aged rat cerebellum. *Journal für Hirnforschung*, *31*(4), 423–428.
Long, M. A., Deans, M. R., Paul, D. L., & Connors, B. W. (2002). Rhythmicity without synchrony in the electrically uncoupled inferior olive. *The Journal of Neuroscience*, *22*(24), 10898–10905.
Lorden, J. F., Oltmans, G. A., McKeon, T. W., Lutes, J., & Beales, M. (1985). Decreased cerebellar 3′,5′-cyclic guanosine monophosphate levels and insensitivity to harmaline in the genetically dystonic rat (dt). *The Journal of Neuroscience*, *5*(10), 2618–2625.
Lorden, J. F., Stratton, S. E., Mays, L. E., & Oltmans, G. A. (1988). Purkinje cell activity in rats following chronic treatment with harmaline. *Neuroscience*, *27*(2), 465–472.
Louis, E. D., Babij, R., Lee, M., Cortes, E., & Vonsattel, J. P. (2013). Quantification of cerebellar hemispheric purkinje cell linear density: 32 ET cases versus 16 controls. *Movement Disorders*, *28*(13), 1854–1859.
Louis, E. D., Diaz, D. T., Kuo, S. H., Gan, S. R., Cortes, E. P., Vonsattel, J. P. G., et al. (2018). Inferior olivary nucleus degeneration does not lessen tremor in essential tremor. *Cerebellum Ataxias*, *5*, 1.
Louis, E. D., Hernandez, N., Dyke, J. P., Ma, R. E., & Dydak, U. (2018). In vivo dentate nucleus gamma-aminobutyric acid concentration in essential tremor vs. controls. *Cerebellum*, *17*(2), 165–172.
Louis, E. D., Lee, M., Cortes, E., Vonsattel, J. P., & Faust, P. L. (2014). Matching asymmetry of tremor with asymmetry of postmortem cerebellar hemispheric changes in essential tremor. *Cerebellum*, *13*(4), 462–470.
Louis, E. D., & Lenka, A. (2017). The olivary hypothesis of essential tremor: Time to lay this model to rest? *Tremor and Other Hyperkinetic Movements*, *7*, 473.
Marshall, S. P., & Lang, E. J. (2009). Local changes in the excitability of the cerebellar cortex produce spatially restricted changes in complex spike synchrony. *The Journal of Neuroscience*, *29*(45), 14352–14362.
Marshall, S. P., van der Giessen, R. S., de Zeeuw, C. I., & Lang, E. J. (2007). Altered olivocerebellar activity patterns in the connexin36 knockout mouse. *Cerebellum*, *6*(4), 287–299.
Martin, F. C., & Handforth, A. (2006). Carbenoxolone and mefloquine suppress tremor in the harmaline mouse model of essential tremor. *Movement Disorders*, *21*(10), 1641–1649.
Martin, F. C., Thu Le, A., & Handforth, A. (2005). Harmaline-induced tremor as a potential preclinical screening method for essential tremor medications. *Movement Disorders*, *20*(3), 298–305.
Masugi-Tokita, M., Tarusawa, E., Watanabe, M., Molnar, E., Fujimoto, K., & Shigemoto, R. (2007). Number and density of AMPA receptors in individual synapses in the rat cerebellum as revealed by SDS-digested freeze-fracture replica labeling. *The Journal of Neuroscience*, *27*(8), 2135–2144.

McMahon, A., Fowler, S. C., Perney, T. M., Akemann, W., Knopfel, T., & Joho, R. H. (2004). Allele-dependent changes of olivocerebellar circuit properties in the absence of the voltage-gated potassium channels Kv3.1 and Kv3.3. *The European Journal of Neuroscience*, *19*(12), 3317–3327.

Meera, P., Wallner, M., & Otis, T. S. (2011). Molecular basis for the high THIP/gaboxadol sensitivity of extrasynaptic GABA(A) receptors. *Journal of Neurophysiology*, *106*(4), 2057–2064.

Mehta, H., Saravanan, K. S., & Mohanakumar, K. P. (2003). Serotonin synthesis inhibition in olivo-cerebellar system attenuates harmaline-induced tremor in Swiss albino mice. *Behavioural Brain Research*, *145*(1–2), 31–36.

Mignani, S., Bohme, G. A., Birraux, G., Boireau, A., Jimonet, P., Damour, D., et al. (2002). 9-Carboxymethyl-5H,10H-imidazo[1,2-a]indeno[1,2-e]pyrazin-4-one-2-carbocylic acid (RPR117824): Selective anticonvulsive and neuroprotective AMPA antagonist. *Bioorganic & Medicinal Chemistry*, *10*(5), 1627–1637.

Milner, T. E., Cadoret, G., Lessard, L., & Smith, A. M. (1995). EMG analysis of harmaline-induced tremor in normal and three strains of mutant mice with Purkinje cell degeneration and the role of the inferior olive. *Journal of Neurophysiology*, *73*(6), 2568–2577.

Mishina, M., Uemura, T., Yasumura, M., & Yoshida, T. (2012). Molecular mechanism of parallel fiber-Purkinje cell synapse formation. *Frontiers in Neural Circuits*, *6*, 90.

Miwa, H., Nishi, K., Fuwa, T., & Mizuno, Y. (2000). Differential expression of c-fos following administration of two tremorgenic agents: Harmaline and oxotremorine. *Neuroreport*, *11*(11), 2385–2390.

Miyazaki, T., Yamasaki, M., Takeuchi, T., Sakimura, K., Mishina, M., & Watanabe, M. (2010). Ablation of glutamate receptor GluRdelta2 in adult Purkinje cells causes multiple innervation of climbing fibers by inducing aberrant invasion to parallel fiber innervation territory. *The Journal of Neuroscience*, *30*(45), 15196–15209.

Mukamel, E. A., Nimmerjahn, A., & Schnitzer, M. J. (2009). Automated analysis of cellular signals from large-scale calcium imaging data. *Neuron*, *63*(6), 747–760.

Nahab, F. B., Wittevrongel, L., Ippolito, D., Toro, C., Grimes, G. J., Starling, J., et al. (2011). An open-label, single-dose, crossover study of the pharmacokinetics and metabolism of two oral formulations of 1-octanol in patients with essential tremor. *Neurotherapeutics*, *8*(4), 753–762.

Najac, M., & Raman, I. M. (2015a). Integration of Purkinje cell inhibition by cerebellar nucleo-olivary neurons. *The Journal of Neuroscience*, *35*(2), 544–549.

Najac, M., & Raman, I. M. (2015b). Integration of Purkinje cell inhibition by cerebellar nucleo-olivary neurons. *Journal of Neuroscience*, *35*(2), 544–549.

Nandy, K. (1981). Morphological changes in the cerebellar cortex of aging Macaca nemestrina. *Neurobiology of Aging*, *2*(1), 61–64.

Nelson, B. J., & Mugnaini, E. (1988). The rat inferior olive as seen with immunostaining for glutamate decarboxylase. *Anatomy and Embryology*, *179*, 109–127.

Nelson, B. J., & Mugnaini, E. (1989). Origins of GABAergic inputs to the inferior olive. In P. Strata (Ed.), *Vol. 17. The olivocerebellar system in motor control*. Berlin: Springer-Verlag.

Ondo, W. G. (2020). Current and emerging treatments of essential tremor. *Neurologic Clinics*, *38*(2), 309–323.

Pahapill, P. A., Levy, R., Dostrovsky, J. O., Davis, K. D., Rezai, A. R., Tasker, R. R., et al. (1999). Tremor arrest with thalamic microinjections of muscimol in patients with essential tremor. *Annals of Neurology*, *46*(2), 249–252.

Palkovits, M., Mezey, E., Hamori, J., & Szentagothai, J. (1977). Quantitative histological analysis of the cerebellar nuclei in the cat. I. Numerical data on cells and on synapses. *Experimental Brain Research*, *28*, 189–209.

Pan, M. K., Li, Y. S., Wong, S. B., Ni, C. L., Wang, Y. M., Liu, W. C., et al. (2020). Cerebellar oscillations driven by synaptic pruning deficits of cerebellar climbing fibers contribute to tremor pathophysiology. *Science Translational Medicine*, *12*(526).

Papapetropoulos, S., Lee, M. S., Versavel, S., Newbold, E., Jinnah, H. A., Pahwa, R., et al. (2021). A phase 2 proof-of-concept, randomized, placebo-controlled trial of CX-8998 in essential tremor. *Movement Disorders*, *36*(8), 1944–1949.

Park, Y. G., Park, H. Y., Lee, C. J., Choi, S., Jo, S., Choi, H., et al. (2010). Ca(V)3.1 is a tremor rhythm pacemaker in the inferior olive. *Proceedings of the National Academy of Sciences of the United States of America*, *107*(23), 10731–10736.

Paterson, N. E., Malekiani, S. A., Foreman, M. M., Olivier, B., & Hanania, T. (2009). Pharmacological characterization of harmaline-induced tremor activity in mice. *European Journal of Pharmacology*, *616*(1–3), 73–80.

Pedrosa, D. J., Nelles, C., Brown, P., Volz, L. J., Pelzer, E. A., Tittgemeyer, M., et al. (2017). The differentiated networks related to essential tremor onset and its amplitude modulation after alcohol intake. *Experimental Neurology*, *297*, 50–61.

Persohn, E., Malherbe, P., & Richards, J. G. (1992). Comparative molecular neuroanatomy of cloned GABAA receptor subunits in the rat CNS. *The Journal of Comparative Neurology*, *326*(2), 193–216.

Pozos, R. S., & Iaizzo, P. A. (1992). Effects of topical anesthesia on essential tremor. *Electromyography and Clinical Neurophysiology*, *32*(7–8), 369–372.

Rappaport, M. S., Gentry, R. T., Schneider, D. R., & Dole, V. P. (1984). Ethanol effects on harmaline-induced tremor and increase of cerebellar cyclic GMP. *Life Sciences*, *34*(1), 49–56.

Roffler-Tarlov, S., Beart, P. M., O'Gorman, S., & Sidman, R. L. (1979). Neurochemical and morphological consequences of axon terminal degeneration in cerebellar deep nuclei of mice with inherited Purkinje cell degeneration. *Brain Research*, *168*(1), 75–95.

Roffler-Tarlov, S., & Turey, M. (1982). The content of amino acids in the developing cerebellar cortex and deep cerebellar nuclei of granule cell deficient mutant mice. *Brain Research*, *247*(1), 65–73.

Sasaki, K., Bower, J. M., & Llinas, R. (1989). Multiple Purkinje cell recording in rodent cerebellar cortex. *The European Journal of Neuroscience*, *1*(6), 572–586.

Schnitzler, A., Munks, C., Butz, M., Timmermann, L., & Gross, J. (2009). Synchronized brain network associated with essential tremor as revealed by magnetoencephalography. *Movement Disorders*, *24*(11), 1629–1635.

Schnitzler, A., Timmermann, L., & Gross, J. (2006). Physiological and pathological oscillatory networks in the human motor system. *Journal of Physiology, Paris*, *99*(1), 3–7.

Shaffer, C. L., Hurst, R. S., Scialis, R. J., Osgood, S. M., Bryce, D. K., Hoffmann, W. E., et al. (2013). Positive allosteric modulation of AMPA receptors from efficacy to toxicity: The interspecies exposure-response continuum of the novel potentiator PF-4778574. *The Journal of Pharmacology and Experimental Therapeutics*, *347*(1), 212–224.

Sharabi, S., Daniels, D., Last, D., Guez, D., Zivli, Z., Castel, D., et al. (2019). Non-thermal focused ultrasound induced reversible reduction of essential tremor in a rat model. *Brain Stimulation*, *12*(1), 1–8.

Simantov, R., Snyder, S. H., & Oster-Granite, M. L. (1976). Harmaline-induced tremor in the rat: Abolition by 3-acetylpyridine destruction of cerebellar climbing fibers. *Brain Research*, *114*(1), 144–151.

Sinclair, J. G., Lo, G. F., & Harris, D. P. (1982). Ethanol effects on the olivocerebellar system. *Canadian Journal of Physiology and Pharmacology*, *60*(5), 610–614.

Sinton, C. M., Krosser, B. I., Walton, K. D., & Llinas, R. R. (1989). The effectiveness of different isomers of octanol as blockers of harmaline-induced tremor. *Pflügers Archiv*, *414*(1), 31–36.

Sotelo, C., Llinas, R., & Baker, R. (1974). Structural study of inferior olivary nucleus of the cat: Morphological correlates of electrotonic coupling. *Journal of Neurophysiology*, *37*(3), 541–559.

Stratton, S. E., & Lorden, J. F. (1991). Effect of harmaline on cells of the inferior olive in the absence of tremor: Differential response of genetically dystonic and harmaline-tolerant rats. *Neuroscience*, *41*(2–3), 543–549.

Sturrock, R. R. (1989). Age related changes in Purkinje cell number in the cerebellar nodulus of the mouse. *Journal für Hirnforschung*, *30*(6), 757–760.

Sugihara, I. (2011). Compartmentalization of the deep cerebellar nuclei based on afferent projections and aldolase C expression. *Cerebellum*, *10*(3), 449–463.

Sugihara, I., Lang, E. J., & Llinas, R. (1993). Uniform olivocerebellar conduction time underlies Purkinje cell complex spike synchronicity in the rat cerebellum. *The Journal of Physiology*, *470*, 243–271.

Sugihara, I., Lang, E. J., & Llinas, R. (1995). Serotonin modulation of inferior olivary oscillations and synchronicity: A multiple-electrode study in the rat cerebellum. *The European Journal of Neuroscience*, 7(4), 521–534.

Sugihara, I., Marshall, S. P., & Lang, E. J. (2007). Relationship of complex spike synchrony bands and climbing fiber projection determined by reference to aldolase C compartments in crus IIa of the rat cerebellar cortex. *The Journal of Comparative Neurology*, *501*(1), 13–29.

Szapiro, G., & Barbour, B. (2007). Multiple climbing fibers signal to molecular layer interneurons exclusively via glutamate spillover. *Nature Neuroscience*, *10*(6), 735–742.

Tang, T., Blenkinsop, T. A., & Lang, E. J. (2019). Complex spike synchrony dependent modulation of rat deep cerebellar nuclear activity. *eLife*, *8*.

Tang, T., Suh, C. Y., Blenkinsop, T. A., & Lang, E. J. (2016). Synchrony is key: Complex spike inhibition of the deep cerebellar nuclei. *Cerebellum*, *15*(1), 10–13.

Tian, J. B., & Bishop, G. A. (2002). Stimulus-dependent activation of c-Fos in neurons and glia in the rat cerebellum. *Journal of Chemical Neuroanatomy*, *23*(3), 157–170.

Tomiyama, M., Palacios, J. M., Cortes, R., & Mengod, G. (1999). Flip and flop variants of AMPA receptor subunits in the human cerebellum: Implication for the selective vulnerability of Purkinje cells. *Synapse*, *31*(2), 163–167.

Turecek, J., Yuen, G. S., Han, V. Z., Zeng, X. H., Bayer, K. U., & Welsh, J. P. (2014). NMDA receptor activation strengthens weak electrical coupling in mammalian brain. *Neuron*, *81*(6), 1375–1388.

Uusisaari, M., Obata, K., & Knopfel, T. (2007). Morphological and electrophysiological properties of GABAergic and non-GABAergic cells in the deep cerebellar nuclei. *Journal of Neurophysiology*, *97*(1), 901–911.

Vallbo, A. B., & Wessberg, J. (1993). Organization of motor output in slow finger movements in man. *The Journal of Physiology*, *469*, 673–691.

Verhaart, W. J., & Voogd, J. (1962). Hypertrophy of the inferior olives in the cat. *Journal of Neuropathology and Experimental Neurology*, *21*, 92–104.

Vilis, T., & Hore, J. (1977). Effects of changes in mechanical state of limb on cerebellar intention tremor. *Journal of Neurophysiology*, *40*(5), 1214–1224.

Wallner, M., Hanchar, H. J., & Olsen, R. W. (2003). Ethanol enhances alpha 4 beta 3 delta and alpha 6 beta 3 delta gamma-aminobutyric acid type a receptors at low concentrations known to affect humans. *Proceedings of the National Academy of Sciences of the United States of America*, *100*(25), 15218–15223.

Watanabe, M. (2008). Molecular mechanisms governing competitive synaptic wiring in cerebellar Purkinje cells. *The Tohoku Journal of Experimental Medicine*, *214*(3), 175–190.

Welsh, J. P., Lang, E. J., Suglhara, I., & Llinas, R. (1995). Dynamic organization of motor control within the olivocerebellar system. *Nature*, *374*(6521), 453–457.

Wiklund, L., Sjolund, B., & Bjorklund, A. (1981). Morphological and functional studies on the serotoninergic innervation of the inferior olive. *Journal of Physiology, Paris*, 77(2–3), 183–186.
Wills, A. J., Jenkins, I. H., Thompson, P. D., Findley, L. J., & Brooks, D. J. (1994). Red nuclear and cerebellar but no olivary activation associated with essential tremor: A positron emission tomographic study. *Annals of Neurology*, *36*(4), 636–642.
Wisden, W., Laurie, D. J., Monyer, H., & Seeburg, P. H. (1992). The distribution of 13 GABAA receptor subunit mRNAs in the rat brain. I. Telencephalon, diencephalon, mesencephalon. *The Journal of Neuroscience*, *12*(3), 1040–1062.
Wu, Y. C., Louis, E. D., Gionco, J., Pan, M. K., Faust, P. L., & Kuo, S. H. (2021). Increased climbing fiber lateral crossings on Purkinje cell dendrites in the cerebellar hemisphere in essential tremor. *Movement Disorders*, *36*(6), 1440–1445.
Zagha, E., Lang, E. J., & Rudy, B. (2008). Kv3.3 channels at the Purkinje cell soma are necessary for generation of the classical complex spike waveform. *The Journal of Neuroscience*, *28*(6), 1291–1300.

CHAPTER SIX

Is inferior olive central to the pathophysiology of essential tremor? No

Mario Manto[a,b,*]

[a]Unité des Ataxies Cérébelleuses, Service de Neurologie, CHU-Charleroi, Belgium
[b]Service des Neurosciences, Université de Mons, Belgium
*Corresponding author: e-mail address: mario.manto@ulb.be

Contents

Abstract

Essential tremor (ET) represents one of the commonest movement disorder worldwide and is the most common tremor disorder. ET manifests with various combinations of motor and nonmotor symptoms. The clinical hallmark is a kinetic tremor of upper limbs. Historically, the pathogenesis of ET has been based on the hypothesis of an overactivity of the inferior olive (inferior olive hypothesis: IOH) where the inferior olive would act as the central pace-maker of ET, resulting in impaired electrophysiological discharges of the olivo-cerebellar tract. The absence of structural alterations in post-mortem studies of the inferior olive is a striking argument against the IOH. Furthermore, neuroimaging studies point towards the implication of the cerebello-thalamo-cerebral pathway rather than the IO, and the harmaline model which has been considered as an animal model of ET presents important weaknesses. By contrast, a series of experiments by Louis et al. have provided convincing evidence of impaired wiring of the Purkinje cell microcircuitry

International Review of Neurobiology, Volume 163
ISSN 0074-7742
https://doi.org/10.1016/bs.irn.2022.02.010

and progressive neurodegeneration of the cerebellar cortex. The Purkinje neuron appears as the primary culprit (Purkinjopathy). The cerebellar cortex hypothesis (CCH) has solid neuropathological signatures, unlike the purely physiological IOH. Rather than a dysregulatory electrophysiological disorder suggested by IOH, ET is a clinical-pathological entity similar to late onset neurodegenerative disorders such as Parkinson's disease or Alzheimer's disease. The CCH emphasizes the need to develop novel therapeutic strategies in order to maintain or promote the cerebellar reserve. The modern reconceptualization of ET in a genuine cerebellar disorder is cleaning the IOH to the light of histopathological studies. ET falls in the large basket of the neurodegenerative diseases and we have entered into a novel formulation of the disease pathogenesis with direct impacts on future therapies.

1. Epidemiology

Essential tremor (ET) represents one of the commonest movement disorder worldwide, with up to 7 million cases in USA (Louis, 2016). The prevalence is around 0.4% across all ages, with a marked increase with age since it reaches 4.6% in patients older than 65 years and more than 20% above 90 years according to studies published between 1998 and 2010 (Louis & Ferreira, 2010). In a recent meta-analysis, prevalence in patients older than 65 years is 5.79% and in descriptive analyses, median crude prevalence (age above 60 years) is 5.9% and mean is 8.0% (Louis & McCreary, 2021). The prevalence increases by 74% for every decade increase in age, without gender effect. There is a small peak from 20 to 30 years and a larger peak in the sixties (Brin & Koller, 1998), but an ascertainment bias cannot be excluded due to the referral to tertiary centers of young-onset cases (Louis, 2013a, 2013b). In terms of incidence, the adjusted incidence is 619 per 100.000 persons aged 65 or more (Benito-León, Bermejo-Pareja, & Louis, 2005). These studies have underlined that age is a risk factor, similarly to other neurodegenerative diseases. Nevertheless, ET onset may occur at any age. ET is a disabling disorder with up to 25% of patients being forced to retire prematurely because of the severity of shaking (Louis, 2013a, 2013b).

2. Clinical presentation

ET patients are diagnosed and followed not only by neurologists, but also by general practitioners, internists and geriatricians in many countries (Louis, 2013a, 2013b). ET is often considered as a disorder which is "easy to diagnose," but misdiagnosis is very common, especially because of the clinical heterogeneity (Amlang, Trujillo Diaz, & Louis, 2020; Louis, 2013a, 2013b). In particular, ET may be confounded with Parkinson's disease and dystonia (Amlang et al., 2020).

The traditional view of considering ET as a benign monosymptomatic tremorous condition has evolved with time. Indeed, ET is phenotypically a heterogeneous condition, manifesting with various combinations of a core of motor symptoms associated with nonmotor symptoms. Motor symptoms are dominated by tremor in the upper extremities. The central feature is kinetic upper limb tremor visible during tasks of daily life such as eating, drinking or writing. There is often a postural tremor elicited by the Barré manoeuver (holding the arms outstretched in front of the body). The kinetic tremor has an amplitude which is usually greater as compared to the postural tremor (Brennan, Jurewicz, Ford, Pullman, & Louis, 2002). The kinetic tremor score (scale from 0 to 3) is usually 1 point above the postural tremor score (Louis, 2013a, 2013b). The intention component is observed in 58% of patients (Deuschl, Wenzelburger, Löffler, Raethjen, & Stolze, 2000). The amplitude of tremor increases slowly with time, suggestive of a progressive brain disorder. Later age of onset tends to be associated with a more rapid progression. Tremor may also appear in other regions of the body especially head and neck with a prevalence of 15–55%. Head tremor often develops several years after the onset of upper limb tremor and is strongly associated with female gender (Hubble, Busenbark, Pahwa, Lyons, & Koller, 1997). Jaw tremor tends to be observed in patients with clinically severe ET. Leg tremor occurs in up to half of patients. Small doses of alcohol improve tremor in up to half of the patients. Dysmetric movements are common when the patients perform fast movements of the upper limbs, as observed also in many cerebellar disorders (Manto & Pandolfo, 2002). Some patients develop a rest tremor at the level of the arm (Cohen, Pullman, Jurewicz, Watner, & Louis, 2003; Louis, Hernandez, & Michalec, 2015). In these patients, the disease duration is longer and tremor is often more severe. Isolated rest tremor may develop in absence of bradykinesia or rigidity, and there is no evidence of Lewy body pathology in the mesencephalon in these patients (Louis, 2013a, 2013b). Nevertheless, ET patients can subsequently develop a Parkinson's disease (Yahr, Orosz, & Purohit, 2003). The risk would be multiplied by a factor of four to fivefold (Benito-Leon, Louis, & Bermejo-Pareja, 2008).

Non motor symptoms include mild cognitive changes including a dysexecutive syndrome, anxiety, changes in personality, social phobia, depressive symptoms (Zesiewicz, Chari, Jahan, Miller, & Sullivan, 2010;

Schneier, Barnes, Albert, & Louis, 2001). The dynamic course of the cognitive changes is still unclear (Louis, Joyce, & Cosentino, 2019). Sensory abnormalities include problems with hearing and olfaction. Non motor symptoms might be the primary manifestation of the disease. These cognitive/neuropsychological deficits observed in ET patients share similar features to those reported in the Schmahmann's syndrome, including impaired executive functions, altered visuospatial cognition and abnormal control of emotion-affect, but we lack a direct evaluation of the Schmahmann's syndrome in ET (Argyropoulos, van Dun, Adamaszek, et al., 2020). The motor dysmetria/dysmetria of thought applies also to ET.

3. Genetics

Twin studies have shown elevated concordance amongst monozygotic twins (Lorenz et al., 2004). Assessment of ET families suggests that the disorder is familial in 40–50% of cases, often inherited in a manner indicating an autosomal dominant transmission with incomplete penetrance (Coppede, 2013). First-degree relatives of ET patients have a fivefold increased risk to develop the disease (Louis, et al., 2001). ET with childhood onset is familial in most cases (Louis & Dogu, 2007).

The causative gene has not been discovered, although linkage studies have shown several foci (in particular ETM1 on 3q13, ETM2 on 2p24.1 and ETM3 on 6p23) in Iceland and North America. LINGO1 (LINGO1 is part of the complex Nogo-66 receptor which is expressed mainly in neurons and is involved in the inhibition of axonal regeneration) might be a susceptibility gene (Their, Lorenz, Nothnagel, et al., 2010). According to a recent review, LINGO1 rs9652490 and STK32B rs10937625 appear to influence ET susceptibility (Siokas et al., 2020). However, results are conflicting and there is a lack of replication for many candidate genes.

4. Diagnosis

Diagnosis of ET relies on medical history and neurological examination, with laboratory testing being used to exclude ET-mimics including hyperthyroidism and Wilson's disease. Numerous medications may enhance a physiological tremor (especially bronchodilators, prednisone, theophylline, lithium salts, valproid acid). Caffeine/tobacco also trigger tremor. Several sets of criteria have been proposed and revised over the recent years (Bhatia et al., 2018; Deuschl, Bain, et al., 1998; Elble, 2000).

5. Pathogenesis

5.1 The inferior olive hypothesis

Historically, the pathogenesis of ET has been based on the hypothesis of an overactivity of the inferior olive (inferior olive hypothesis: IOH) where the inferior olive would act as the central pace-maker (central oscillator) of ET, resulting in impaired electrophysiological discharges of the olivo-cerebellar tract. Physiological studies starting in the 1970s have put forward the importance of the olivary-cerebellar network, arguing that the electrophysiological derangement within the inferior olive is the prime mover of ET. The absence of neuropathological impairment in the inferior olive has been viewed as an argument for ET as a disorder at the electrophysiological level (Rajput, Rozdilsky, Ang, & Rajput, 1991).

5.1.1 The inferior olive as the pace-maker of ET

Each inferior olivary neuron is coupled with a variable number of neighboring neurons (Hoge et al., 2011). Neurons of the inferior olive show a oscillatory behavior (Llinás & Yarom, 1981). Subthreshold 3–6 Hz sinusoidal oscillations of the resting membrane potential are observed (Llinás & Yarom, 1986). In the hyperpolarized state, these sinusoidal oscillations increase the propensity to generate action potentials in a rhythmic manner (Shaikh, Optican, & Zee, 2013). The gap junctions between adjacent neurons of the inferior olive facilitate the synchronization of patches in the inferior olive. Excitatory inputs to the inferior olive arise from the somatosensory system, the vestibular system, spinal cord and midbrain nuclei (Luo & Sugihara, 2016). These inputs reset the oscillatory rhythm to evoke firings, whose frequency is around 1 Hz (Eccles, Llinás, & Sasaki, 1966). It is presumed that enhanced synchronization will lead to oscillations within the cerebellar circuit via the climbing fibers. Climbing fibers establish a powerful one-to-one synaptic connection with target Purkinje cells, following a topographic arrangement with a triangular loop of neuronal connection inferior olive-cerebellar cortex-cerebellar nuclei, also called the cerebellar module (Apps et al., 2018; Luo & Sugihara, 2016). The matching olivocorticonuclear connectivity has to be considered together with the chemical diversity of Purkinje cells belonging to stripes, the diversity of information processing between cerebellar modules and the modular plasticity (Apps et al., 2018). Stripes correspond to chemically identified banding patterns of cerebellar cortex (usually identifying zebrin II/aldolase c bands) (Apps et al., 2018). There is generally a good correspondence between the chemical signature of Purkinje cells and their anatomical connections, but the biochemical complexity is very high. It is likely that modules are involved in diverse motor and nonmotor functions. Still, the inferior olivary neurons are not the sole neurons with pacemaker properties in the CNS. For instance, thalamus, Purkinje neurons, cerebellar nuclei, locus ceruleus and pyramidal neurons of the cerebral cortex also show striking pacemaking features (de Oliveira, Howlett, Gravina, et al., 2010; Steriade & Llinás, 1988; Raman, Gustafson, & Padgett, 2000; Llinás & Sugimori, 1980; Silva, Amitai, & Connors, 1991: Louis & Lenka, 2017).

5.1.2 The harmaline model

The group of beta-carboline alkaloids (BCAs) includes harmaline, harmine and harmane. It has been proposed that the chronic environmental exposure to BCAs via food intake contributes to the pathogenesis of ET (Laviţă, Aro, Kiss, Manto, & Duez, 2016). BCAs might be generated through Maillard-type reaction. BCAs are members of a large group of heterocyclic amines, products of cooking meat.

Harmaline-induced tremor is commonly used to produce a generalized acute tremor in animals (Miwa, 2013). Administration of harmaline by the intraperitoneal route causes an action tremor in rodents in the minutes following the administration. Harmaline increases the membrane excitability to trigger 3–6 Hz spike trains of action potentials (Llinás & Yarom, 1986). It has been postulated that the harmaline-induced tremor is a reliable model of ET. The tremor frequencies range from 8–10 Hz in monkeys, 10–12 Hz in rats and 11–14 Hz in mice (Milner, Cadoret, Lessard, & Smith, 1995). Tremor is synchronous between agonist and antagonist muscles. The local application of harmaline in the inferior olive triggers a rhythmic activity in loco (De Montigny & Lamarre, 1975). Harmaline provokes sustained rhythmic activity in the inferior olive of decerebrate cats. Harmaline modulates the rhythmic coupling of neurons in the inferior olive. The rhythmic discharges are transmitted to Purkinje neurons via the crossed olivo-cerebellar pathway. The harmaline model has been closely linked with the inferior olivary discharges for historical reasons. In the work of Lamarre and Mercier, harmaline triggered rhythmic discharges in the inferior olive, the Purkinje neurons, cerebellar nuclei and reticular formation (Lamarre & Mercier, 1971). It was presumed that tremor was relayed to muscles by the brainstem nuclei and their spinal projections including the reticulospinal tracts. The studies of Llinas' group pointed towards the inferior olive, since rhythmic activity persisted despite decerebration and cerebellectomy (Llinás & Volkind, 1973).

In mutant mice with Purkinje cell degeneration, harmaline does not trigger tremor, suggesting that the Purkinje neurons are a key-element to generate oscillatory activity in muscles (Milner et al., 1995). Milner et al. investigated wild-type mice (+/+), lurcher (lc/+: apoptotic death of cerebellar Purkinje cells during post-natal development), nervous (nr/nr), and Purkinje cell degeneration (pcd/pcd) mutants before and after intraperitoneal injections of harmaline. Spectral analysis was used to quantify the amplitude and frequency of tremor found in the EMGs following harmaline administration. Normal mice responded to harmaline with strong, continuous 11- to 14-Hz tremor. Mutants from the pcd/pcd strain showed continuous tremor, but of lower amplitude and frequency. By contrast, nr/nr mutants exhibited intermittent paroxysmal tremor lasting for only a few seconds, and lc/+ mutants showed no evidence of tremor. Knock-down of the $Ca_v3.1$ gene in the inferior olive suppresses the harmaline-induced tremor, emphasizing the importance of T-type calcium channels in tremorgenesis (Park et al., 2010). Harmaline could potentiate Ca(V)3.1 channels by changing voltage-dependent responsiveness in the hyperpolarizing direction. Another argument which has been used in favor of the model is that the harmaline-induced tremor is reduced by ethanol, as observed in ET (Rappaport, Gentry, Schneider, & Dole, 1984). Neurophysiological and neurochemical evidence indicates that both ethanol and benzodiazepines enhance the inhibitory neurotransmission mediated by GABA.

There are several criticisms against the harmaline model as a robust animal model of ET. First, ET is not an acute generalized tremor as observed in animals receiving an acute dose of harmaline, but is a chronic tremor disorder. Second, the repeated administration of harmaline results in a long-lasting tolerance in rodents (Lutes, Lorden, Beales, & Oltmans, 1988). It is likely that the site of tolerance to the tremorogenic effects of harmaline lies within the olivocerebellar system. If ET is the result of a regular exposure to BCAs in the lifetime, this is a major weakness of the model, which is still used to develop anti-tremor drugs in human. Third, harmaline is aversive in animals (Miwa, 2013). The avoid movement is unlike what is observed in ET. Fourth, harmaline induces a patchy loss of Purkinje neurons (band-like necrosis) which is likely mediated by the inferior olive by a mechanism of excitotoxicity (excessive exposure to glutamate) since chemical ablation of the olivocerebellar fibers by 3-acetylpyridine blocks the harmaline-induced Purkinje cell degeneration. Fifth, there are species-specific differences in response to harmaline. Sixth, some drugs improve the harmaline-induced tremor whereas they clearly worsen ET, for instance valproic acid (Handforth, 2012).

In summary, the harmaline-induced tremor is a weak model for ET. Overall, it does not bring more relevant information on the pathogenesis of ET than the MPTP model for Parkinson's disease. The physiological findings in animals, especially rodents, have been extrapolated to human without evidence that they represent neurophysiological signatures in ET.

5.1.3 Post-mortem studies

The absence of structural alterations in post-mortem studies of the inferior olive is a striking argument against the IOH. A detailed postmortem study of the inferior olivary nucleus of 14 essential tremor cases versus 15 age-matched controls has failed to show any difference in terms of microscopic neuronal and glial changes in the inferior olivary nucleus. Olivary linear neuronal density is unaffected in ET (Louis, Babij, Cortés, Vonsattel, & Faust, 2013).

The observation of Ekouzi et al. provides another argument against the IOH (Elkouzi, Kattah, & Elble, 2015). The authors have reported a 77-year-old patient with long-standing ET (more than 30 years) unaffected by bilateral olivary hypertrophy with palatal tremor. The patient developed loss of balance and falls. MRI revealed a bilateral olivary hypertrophy and a 2 Hz palatal tremor was observed. The patient presented a syndrome of progressive ataxia with palatal tremor (PAPT) at least 30 years after the onset of ET. His ET remained unchanged despite 3 years of progressive ataxia, bilateral olivary hypertrophy, and persistent palatal tremor. The authors concluded that the inferior olive is not the source of ET. In other words, the olivary degeneration did not influence the course of ET.

Louis et al. have reported a 94-years-old patient with longstanding severe familial ET (Louis et al., 2018; Louis et al., 2018). The tremor had started at the age of 10 years and was clinically progressive until death. Post-mortem examination showed olivary degeneration. The neuronal linear density of the inferior olive demonstrated a marked neuronal loss. Cerebellar degenerative changes were identified. There was moderate to severe atrophy in the cerebellar vermis bilaterally and cerebellar cortical folial atrophy bilaterally in the neocerebellum. Dentate nuclei were unremarkable. A marked Purkinje cell loss was found in association with severe Bergmann gliosis and a patchy loss of granule cells (see also next section). Torpedoes were increased and a marked hypertrophy of basket cell axonal processes was found. The nucleo-olivary tract were relatively well preserved.

If the inferior olive was the culprit in the 2 cases reported above (the first with pseudo-hypertrophic olivary degeneration on neuroimaging and the second with olivary degeneration demonstrated pathologically), one would have expected a clear disappearance of ET. By contrast, tremor continued to worsen during the last decade of life. ET evolved independently of the inferior olive in these cases. This suggests that another pathway than the olivo-cerebellar pathway should be considered.

5.1.4 Neuroimaging studies

Neuroimaging studies point towards the implication of the cerebello-thalamo-cerebral pathway in the pathogenesis of ET rather than the IO. The abnormal activation is observed in the red nucleus, cerebellum and thalamus (Wills, Jenkins, Thompson, Findley, & Brooks, 1995). At rest, PET studies show a cerebellar blood flow which is significantly increased bilaterally in ET, but no increased olivary activity is found. ET during arm extension is associated with further abnormal increases in bilateral cerebellar and abnormal red nuclear activation. Again, no olivary overactivity is observed. Voluntary wrist oscillation in control subjects induce only an ipsilateral cerebellar activation.

fMRI studies show abnormal activities of the cerebellar outflow tract (Lenka et al., 2017). Resting state fMRI of 30 patients with ET and 30 age- and gender-matched healthy controls have been compared by Lenka et al. The functional connectivity of the two groups was compared using whole-brain seed-to-voxel-based analysis. ET patients showed decreased connectivity of several cortical regions especially of the primary motor cortex and the primary somatosensory cortex with several right cerebellar lobules. The thalamus on both hemispheres had increased connectivity with several posterior cerebellar lobules and vermis. Connectivity of several right cerebellar seeds with the cortical and thalamic seeds showed significant correlation with an overall score of Fahn-Tolosa-Marin tremor rating scale (FTM-TRS) as well as the subscores for head tremor and limb tremor. The results of the connectivity analysis point towards alterations in the cerebello-thalamo-cortical network in patients with ET. These alterations are correlated with the FTM scores and the subscores for limb tremor and head tremor in patients, supporting the hypothesis of a cerebellar pathology in ET. MRS studies also indicate that the defects occur in the cerebellar cortex (Louis et al., 2002). Indeed, cerebellar cortical NAA/tCR ratio is reduced in ET patients, with an inverse association between cerebellar cortical NAA/tCR and dominant arm tremor severity. In other words, these results point towards a cerebellar cortical damage.

There are controversies regarding whether PET studies in patients with ET show a hypermetabolism or not in the IO, as discussed above (Wills, Jenkins, Thompson, Findley, & Brooks, 1994). ET has been associated with bilateral cerebellar activation including the cerebellar vermis (Boecker et al., 1996). Ethanol ingestion decreases cerebellar blood flow and this is associated with suppression of tremor in the patients. Alcohol-associated increases of regional cerebral blood flow have been observed in the inferior olivary nuclei. It has been suggested that alcohol-induced suppression of ET is mediated via a reduction of cerebellar synaptic overactivity resulting in increased afferent input to the inferior olivary nuclei. However, the blood flow in the inferior olive is not different in ET and in control subjects if one considers the baseline (before ethanol intake).

Overall, there is so far no clear evidence of a key-role of the inferior olive in the pathogenesis of ET. The problems related to the olivary model have been discussed by Louis and Lenka (2017). This does not exclude at all the strong evidence that the inferior olive show oscillatory activities, but oscillatory activities in the inferior olive does not imply rhythmic contractions in muscles in ET.

5.2 The cerebellar cortex hypothesis

5.2.1 Clinical evidence that ET is a cerebellar disorder

Clinically, ET patients show a spectrum of clinical deficits, which are very suggestive of a genuine cerebellar disorder. Intention tremor may involve the upper limbs as well as the head, and patients show impaired gait/balance and oculomotor deficits (Helmchen et al., 2003; Parisi, Héroux, Culham, & Norman, 2006). ET patients exhibit a deficit of pursuit initiation with a correlation with the intensity of tremor. The oculomotor deficits indicate an impairment of the caudal vermis in ET. Both motor and nonmotor symptoms are typical of cerebellar disorders (Manto & Pandolfo, 2002).

5.2.2 Neuropathological findings

A series of experiments by Louis et al. have provided convincing evidence of impaired wiring of the Purkinje cell microcircuitry and progressive neurodegeneration of the cerebellar cortex. The cerebellar cortex hypothesis (CCH) has solid neuropathological signatures, unlike the IOH:

- increase in "torpedoes" (swellings of the Purkinje cell axons in their proximal portions) by six- to sevenfold (Louis, Yi, Erickson-Davis, Vonsattel, & Faust, 2009). A single Purkinje neuron may contain several torpedoes. These most likely sign an damaged Purkinje neuron

(Mizushima, 1976). Torpedoes contain a massive accumulation of disorganized neurofilaments which impair both anterograde and retrograde axonal transport and lead finally to neuronal death by a phenomenon of strangulation (Louis et al., 2009). Torpedoes may also be observed in other diseases of the cerebellum.

- increased number of thickened axons, arciform axons and recurrent collaterals (Babij et al., 2013). Axonal branching is exaggerated. Axons show multiple bifurcations and terminal sprouting is hypertrophied. It should be kept in mind that the Purkinje neurons are well known to reorganize themselves facing an injury, especially at the axonal level probably to get access to trophic factors or a more favorable micro-environment (Dusart & Sotelo, 1994). Cell bodies are particularly prone to maintain themselves whereas the axonal compartment will show changes. For instance, in the model of axotomy neither the Purkinje cells which have been axotomized close to their cell bodies nor those which have been axotomized in the white matter degenerate (Dusart & Sotelo, 1994). The major changes occur in the axonal compartment: axonal alterations, namely the presence of torpedoes and hypertrophy of the recurrent collateral system, are already observed 24 h after the lesion, although they later differ in time course and spatial distribution.
- the dendritic tree of the Purkinje neuron is characterized by increased dendritic swellings, abnormal pruning and loss of spines (Yu et al., 2012; Louis et al., 2014). Pruning is considered as a stage preceding neuronal death.
- decrease in the cerebellar reserve of Purkinje neurons. Interestingly, the brains with a high number of torpedoes show a greater loss in the reserve, in agreement with the hypothesis that the torpedoes are associated with a mechanism of neuronal death. The fact that neuronal loss of Purkinje neurons is mild in some cases does not exclude at all the concept that ET is a disorder of the Purkinje neuron because the neuronal death might occur at a late stage. This would be the final stage of longstanding process and therefore might not be grasped on all the pathological samples. Furthermore, one has to consider the attempts of the neuron to survive and reorganize its close environment. Post-mortem pictures are an assembly of a complex mixture of degenerative events and attempts of the neuron to regenerate, reorganize (collaterals, sprouting) and survive.
- heterotopic Purkinje neurons outside the Purkinje cell layer. Purkinje cells are found in the granular layer or the molecular layer. In some cases,

the dendritic arbor is directed towards the granular layer, with the axon in direction of the molecular layer (Louis, Diaz, et al., 2018; Louis, Kuo, et al., 2018).

- abnormal connectivity between the climbing fibers and Purkinje neurons (Lin et al., 2014). There is a decrease in the climbing fiber-Purkinje cell synaptic density. Climbing fibers tend to extend to the outer portion of the molecular layer. In principle, climbing fibers form synapses mainly on the thick proximal Purkinje cell dendrites in the inner portion of the molecular layer, whereas parallel fibers form synapses on the thin, Purkinje cell spiny branchlets distally. The increased distribution of climbing fiber-Purkinje cell synapses on the thin Purkinje cell branchlets is inversely correlated with clinical tremor severity. Recent studies show that climbing fibers cross a greater number of Purkinje neurons somas/dendrites as compared to control cases, raising the possibility that there is abnormal wiring onto the Purkinje cells. Interestingly, the increase in crossings is also positively correlated with tremor severity (Wu et al., 2021).
- hypertrophic changes in basket cell axonal processes ("hairy baskets") surrounding the Purkinje cell body, pointing towards a reorganization of the local network around the Purkinje neurons (Erickson-Davis et al., 2010).

In addition, Lewy bodies have been found in clear excess in the locus ceruleus on alpha-synuclein-stained sections (Louis, 2010). The locus ceruleus is a major source of norepinephrine in the CNS and its neurons target directly the Purkinje neurons with a neuromodulation of the inhibitory output and an effect upon cytoskeletal proteins (Hoffer, Siggins, Oliver, & Bloom, 1973, Rogers, Zornetzer, & Bloom, 1981; Mavridis, Degryse, Lategan, Marien, & Colpaert, 1991). The presence of brainstem Lewy bodies is not documented in all pathological series.

Immunohistochemical studies have demonstrated a decrease in GABA-R in dentate nuclei and a reduction in the EAAT2 protein levels in the cerebellar cortex (Lee et al., 2014; Paris-Robidas, Brochu, Sintes, et al., 2012). A post-mortem decrease in GABA(A) (35% reduction) and GABA(B) (22–31% reduction) receptors in the dentate nucleus of ET patients has been reported. Concentrations of GABA(B) receptors are inversely correlated with the duration of symptoms, suggesting that the loss of GABA(B) receptors follows the progression of the disease. It has been suggested that the decrease in GABA receptors in the dentate nucleus causes a disinhibition of cerebellar pacemaker output activity, propagating along the cerebello-thalamo-cortical pathways. EAAT2 is expressed in glial cells and clears glutamate from the synaptic cleft. EAAT2 is expressed in astrocytic processes surrounding directly the Purkinje cell axon initial segment. Therefore, reduction of EAAT2 might lead to excitotoxicity at the level of the cerebellar cortex where glutamate is the neurotransmitter of climbing fibers terminals and parallel fibers crossing the dendritic trees of the Purkinje neurons (Lee et al., 2014; Ozawa, 2007).

The abnormal histopathological findings of the cerebellar cortex contrast with the sparing of the inferior olive, the red nucleus, the thalamus and sensorimotor cortex. This is a major pathological argument for a primary disorder of the cerebellar cortex.

5.2.3 *Structural neuroimaging*

Gray matter loss has been documented in ET by VBM studies, especially in the vermis (Cerasa, Messina, Nicoletti, et al., 2009; Quattrone, Cerasa, Messina, et al., 2008). The atrophy has been confirmed using an automated volumetric method (FreeSurfer) to quantify subcortical atrophy. Atrophy of the vermis is correlated with cerebellar clinical signs, suggesting a link between the loss of neurons of the cerebellar cortex and the clinical syndrome (Shin, Lee, Lee, et al., 2016).

5.2.4 *Function of the cerebellar circuitry: the impact of the Purkinje neuron dysfunction*

Dysfunction of Purkinje neurons has major impacts on the function of cerebellar circuitry. Purkinje neurons are the sole output of the cerebellar cortex. They target cerebellar nuclei and vestibular nuclei. The inhibitory control exerted by the Purkinje neurons tunes the nuclear discharges of the cerebellum circuitry by a mechanism of inhibition/disinhibition (Gruol et al., 2016). The Purkinje cell layer is central is the remote control of thalamic nuclei and cerebral cortex. Purkinje neuron dysfunction impacts on the prediction performed by the whole cerebellar circuitry (see below).

6. Re-considering our physiological conceptualization of cerebellar tremor

One of the main features of the cerebellum is its capacity for compensation and restoration after pathological injury (Mitoma, Kakei, Yamaguchi, & Manto, 2021). This is attributed to two unique morphological and physiological features of the cerebellum: (1) mossy fibers that convey peripheral and central information run mediolaterally over a wide area of the cerebellum, innervating multiple microzones; (2) the circuitry shows a high degree of synaptic plasticity including long-term depression (LTD). Cerebellum appears now as a modular and potentially reconfiguring brain structure. The multiple pathological changes observed in the cerebellar cortex of ET might reflect the attempts of the circuitry to compensate the initial Purkinje neuron insult.

Lesions in the Guillain-Mollaret triangle (dentato-rubro-olivary tract) cause various forms of tremors or tremor-like movements (Kakei, Manto, Tanaka, & Mitoma, 2021). The inferior olive receives afferents via the central tegmental tract from both the magnocellular portion (linked with the precentral motor cortex) and the parvicellular portion (linked with the supplementary motor cortex, the post-central gyrus and BA5) of the red nucleus (Schmahmann, 2016). The inferior olive is also the target of zona incerta fibers. Cerebellar nuclei have two main outputs: (1) the primary excitatory path to the thalamus, the red nucleus (RN), and other brain stem nuclei, and (2) the secondary inhibitory projection to the inferior olive (nucleo-olivary tract), which contributes to the dentato-olivo-cerebellar loop. The first path is involved in a long loop. The second path is involved in a short loop. Impairment of the long loop for instance as a result of a Purkinjopathy would impair the cerebellar forward model and affect its accuracy of prediction and compensation of the feedback delay, resulting in irregular instability of voluntary motor control and shaking. This type of tremor called *forward model-related tremor*, such as kinetic tremor, is associated with dysmetria. Evidence supporting the cerebellar forward model hypothesis has been provided (Bastian, 2006). A series of studies has confirmed the impaired predictive control in movements of patients with degenerative cerebellar ataxia (Kakei et al., 2019). Such studies should be performed in ET. The second type of tremor (*inferior olive-oscillation-related tremor*) is correlated with synchronized oscillation of inferior olive neurons due, at least in animal models, to reduced degrees of freedom in inferior olive activities. Although this type of regular tremor does not necessarily accompany ataxia, the aberrant activities of the inferior olive may induce secondary a maladaptation of cerebellar forward models through aberrant patterns of long-term depression (LTD) and/or long-term potentiation (LTP) of the cerebellar circuitry. In other words, the Purkinjopathy of ET would cause tremor due to errors in predictions carried out by the cerebellar circuitry with a participation of the inferior olive for learning purposes. The inferior olive would be involved in palatal tremor.

The massive reciprocal interconnections of the cerebellum with the cerebral cortex are intricated with the numerous afferent/efferent pathways to the brainstem and spinal cord. The numerous loops involving the cerebral circuitry involve also the recently reported connections with the basal ganglia (Bostan & Strick, 2018). The basal ganglia and the cerebellum are distinct subcortical systems influencing many of the same cortical areas but their projections to thalamic nuclei is different. The subthalamic nucleus is the source of a dense disynaptic projection to the cerebellar cortex and the dentate nucleus in the cerebellum is the source of a dense disynaptic projection to the striatum. A great challenge is to define how the unique learning mechanisms at each network node interact to improve performance (Schmahmann, 2016). One can speculate that the inferior olive is a master-piece for learning mechanisms but the evidence that it is the master-piece in the pathogenesis of ET is missing.

7. Conclusion

The CCH is taking the lead or even cancelling progressively the IOH. The CCH is based upon neuropathological signatures which cannot be ignored anymore. ET is no longer viewed as a monosymptomatic illness but can be considered as a progressive heterogeneous neurodegenerative disorder with a clinical-neuropathological profile, by analogy with other neurodegenerative disorders such as Parkinson's disease or Alzheimer's disease. It can be stated in 2022 that ET is the commonest cerebellar degenerative disorder, characterized by a constellation of neuropathological changes at the level of the Purkinje neuron and its adjacent neurons. The Purkinje neuron is at the center of the pathogenesis, hence the terminology of Purkinjopathy (Grimaldi & Manto, 2013). Tremor is likely the consequence of the abnormal activity of a complex network originating in the cerebellar cortex and passing through thalamic nuclei and sensorimotor cortex (cerebello-thalamo-cortical pathway). The CCH opens the door for a novel set of therapies aiming at a neuroprotective approach of the cerebellar reserve or its potentiation (Mitoma et al., 2020). Efforts should be made to better understand how the Purkinjopathy impacts on the feedforward and feedback limbs of the cerebrocerebellar system.

References

Amlang, C. J., Trujillo Diaz, D., & Louis, E. D. (2020). Essential tremor as a "waste basket" diagnosis: Diagnosing essential tremor remains a challenge. *Frontiers in Neurology*, *25*(11), 172.

Apps, R., Hawkes, R., Aoki, S., Bengtsson, F., Brown, A. M., Chen, G., et al. (2018). Cerebellar modules and their role as operational cerebellar processing units: A consensus paper. *Cerebellum*, *17*(5), 654–682.

Argyropoulos, G., van Dun, K., Adamaszek, M., et al. (2020). The cerebellar cognitive affective/schmahmann syndrome: A task force paper. *Cerebellum*, *19*(1), 102–125. https://doi.org/10.1007/s12311-019-01068-8.

Babij, R., Lee, Cortés, E., JPG, V., Faust, P. L., & Louis, E. D. (2013). Purkinje cell axonal anatomy: Quantifying morphometric changes in essential tremor versus control brains. *Brain*, *136*(Pt 10), 3051–3061. https://doi.org/10.1093/brain/awt238. Epub 2013 Sep 12.

Bastian, A. J. (2006). Learning to predict the future: The cerebellum adapts feedforward movement control. *Current Opinion in Neurobiology*, *16*, 645–649.

Benito-León, J., Bermejo-Pareja, F., & Louis, E. D. (2005). Neurological Disorders in Central Spain (NEDICES) Study Group. Incidence of essential tremor in three elderly populations of central Spain. *Neurology*, *64*(10), 1721–1725.

Benito-Leon, J., Louis, E. D., & Bermejo-Pareja, F. (2008). Risk of incident Parkinson's disease and Parkinsonism in essential tremor: A population-based study. *Neurology*, *70*(Suppl 1), A191.

Bhatia, K. P., Bain, P., Bajaj, N., Elble, R. J., Hallett, M., Louis, E. D., et al. (2018). Consensus Statement on the classification of tremors. from the task force on tremor of the International Parkinson and Movement Disorder Society. Tremor Task Force of the International Parkinson and Movement Disorder Society. *Movement Disorders*, *33*(1), 75–87.

Boecker, H., Wills, A. J., Ceballos-Baumann, A., Samuel, M., Thompson, P. D., Findley, L. J., et al. (1996). The effect of ethanol on alcohol-responsive essential tremor: A positron emission tomography study. *Annals of Neurology*, *39*(5), 650–658. https://doi.org/10.1002/ana.410390515.

Bostan, A. C., & Strick, P. L. (2018). The basal ganglia and the cerebellum: Nodes in an integrated network. *Nature Reviews. Neuroscience*, *19*(6), 338–350. https://doi.org/10.1038/s41583-018-0002-7.

Brennan, K. C., Jurewicz, E. C., Ford, B., Pullman, S. L., & Louis, E. D. (2002). Is essential tremor predominantly a kinetic or a postural tremor? A clinical and electrophysiological study. *Movement Disorders*, *17*(2), 313–316. https://doi.org/10.1002/mds.10003.

Brin, M. F., & Koller, W. (1998). Epidemiology and genetics of essential tremor. *Movement Disorders*, *13*(Suppl 3), 55–63.

Cerasa, A., Messina, D., Nicoletti, G., et al. (2009). Cerebellar atrophy in essential tremor using an automated segmentation method. *AJNR. American Journal of Neuroradiology*, *30*(6), 1240–1243. https://doi.org/10.3174/ajnr.A1544. Epub 2009 Apr 2.

Cohen, O., Pullman, S., Jurewicz, E., Watner, D., & Louis, E. D. (2003). Rest tremor in patients with essential tremor: Prevalence, clinical correlates, and electrophysiologic characteristics. *Archives of Neurology*, *60*(3), 405–410. https://doi.org/10.1001/archneur.60.3.405.

Coppede, F. (2013). Advances in the genetics of human tremor. In G. Grimaldi, & M. Manto (Eds.), *Mechanisms and Emerging Therapies in Tremor Disorders* (pp. 53–78). New York: Springer.

De Montigny, C., & Lamarre, Y. (1975). Effects produced by local applications of harmaline in the inferior olive. *Canadian Journal of Physiology and Pharmacology*, *53*(5), 845–849. https://doi.org/10.1139/y75-116.

de Oliveira, R. B., Howlett, M. C. H., Gravina, F. S., et al. (2010). Pacemaker currents in mouse locus coeruleus neurons. *Neuroscience*, *170*(1), 166–177. https://doi.org/10.1016/j.neuroscience.2010.06.028. Epub 2010 Jul 8.

Deuschl, G., Bain, P., et al. (1998). Consensus statement of the Movement Disorder Society on Tremor. Ad Hoc Scientific Committee. *Movement Disorders*, *13*(Suppl 3), 2–23.

Deuschl, G., Wenzelburger, R., Löffler, K., Raethjen, J., & Stolze, H. (2000). Essential tremor and cerebellar dysfunction clinical and kinematic analysis of intention tremor. *Brain, 123*(Pt 8), 1568–1580.

Dusart, I., & Sotelo, C. (1994). Lack of Purkinje cell loss in adult rat cerebellum following protracted axotomy: Degenerative changes and regenerative attempts of the severed axons. *The Journal of Comparative Neurology, 347*(2), 211–232. https://doi.org/10.1002/cne.903470206.

Eccles, J. C., Llinás, R., & Sasaki, K. (1966). The excitatory synaptic action of climbing fibres on the Purkinje cells of the cerebellum. *The Journal of Physiology, 182*(2), 268–296.

Elble, R. J. (2000). Diagnostic criteria for essential tremor and differential diagnosis. *Neurology, 54*(11 Suppl 4), S2–S6.

Elkouzi, A., Kattah, J. C., & Elble, R. J. (2015). Hypertrophic Olivary Degeneration Does Not Reduce Essential Tremor. *Movement Disorders Clinical Practice, 3*(2), 209–211. https://doi.org/10.1002/mdc3.12275. eCollection Mar-Apr 2016.

Erickson-Davis, C. R., Faust, P. L., Vonsattel, J. P., Gupta, S., Honig, L. S., & Louis, E. D. (2010). "Hairy baskets" associated with degenerative Purkinje cell changes in essential tremor. *Journal of Neuropathology and Experimental Neurology, 69*(3), 262–271.

Grimaldi, G., & Manto, M. (2013). Is essential tremor a Purkinjopathy? The role of the cerebellar cortex in its pathogenesis. *Movement Disorders, 28*, 1759–1761.

Gruol, D. L., Koibuchi, N., Manto, M., Molinari, M., Schmahmann, J. D., & Shen, Y. (2016). *Essentials of Cerebellum and Cerebellar Disorders*. Switzerland: Springer.

Handforth, A. (2012). Harmaline tremor: Underlying mechanisms in a potential animal model of essential tremor. *Tremor and Other Hyperkinetic Movements, 2*. https://doi.org/10.7916/D8TD9W2P. 02-92-769-1. Epub 2012 Sep 12.

Helmchen, C., Hagenow, A., Miesner, J., Sprenger, A., Rambold, H., Wenzelburger, R., et al. (2003). Eye movement abnormalities in essential tremor may indicate cerebellar dysfunction. *Brain, 126*(Pt 6), 1319–1332.

Hoffer, B. J., Siggins, G. R., Oliver, A. P., & Bloom, F. E. (1973). Activation of the pathway from locus coeruleus to rat cerebellar Purkinje neurons: Pharmacological evidence of noradrenergic central inhibition. *The Journal of Pharmacology and Experimental Therapeutics, 184*(3), 553–569.

Hoge, G. H., Davidson, K., Yasumura, T., Castillo, P. E., Rash, J. E., & Pereda, A. E. (2011). The extent and strength of electrical coupling between inferior olivary neurons is heterogeneous. *Journal of Neurophysiology, 105*(3), 1089–1101. https://doi.org/10.1152/jn.00789.2010. Epub 2010 Dec 22.

Hubble, J. P., Busenbark, K. L., Pahwa, R., Lyons, K., & Koller, W. C. (1997). Clinical expression of essential tremor: Effects of gender and age. *Movement Disorders, 12*(6), 969–972.

Kakei, S., Lee, J., Mitoma, H., Tanaka, H., Manto, M., & Hampe, C. S. (2019). Contribution of the cerebellum to predictive motor control and its evaluation in ataxic patients. *Frontiers in Human Neuroscience, 13*, 216.

Kakei, S., Manto, M., Tanaka, H., & Mitoma, H. (2021). Pathophysiology of cerebellar tremor: The forward model-related tremor and the inferior olive oscillation-related tremor. *Frontiers in Neurology, 28*(12). https://doi.org/10.3389/fneur.2021.694653, 694653. eCollection 2021.

Lamarre, Y., & Mercier, L. A. (1971). Neurophysiological studies of harmaline-induced tremor in the cat. *Canadian Journal of Physiology and Pharmacology, 49*(12), 1049–1058. https://doi.org/10.1139/y71-149.

Laviţă, S. I., Aro, R., Kiss, B., Manto, M., & Duez, P. (2016). The Role of beta-Carboline Alkaloids in the Pathogenesis of Essential Tremor. *Cerebellum, 15*(3), 276–284.

Lee, M., Cheng, M. M., Lin, C. Y., Louis, E. D., Faust, P. L., & Kuo, S. H. (2014). Decreased EAAT2 protein expression in the essential tremor cerebellar cortex. *Acta Neuropathologica Communications*, *13*(2), 157. https://doi.org/10.1186/s40478-014-0157-z.

Lenka, A., Bhalsing, K. S., Panda, R., Jhunjhunwala, K., Naduthota, R. M., Saini, J., et al. (2017). Role of altered cerebello-thalamo-cortical network in the neurobiology of essential tremor. *Neuroradiology*, *59*(2), 157–168.

Lin, C. Y., Louis, E. D., Faust, P. L., Koeppen, A. H., Vonsattel, J. P. G., & Kuo, S. H. (2014). Abnormal climbing fibre-Purkinje cell synaptic connections in the essential tremor cerebellum. *Brain*, *137*(Pt 12), 3149–3159. https://doi.org/10.1093/brain/awu281. Epub 2014 Oct 1.

Llinás, R., & Sugimori, M. (1980). Electrophysiological properties of in vitro Purkinje cell somata in mammalian cerebellar slices. *The Journal of Physiology*, *305*, 171–195. https://doi.org/10.1113/jphysiol.1980.sp013357.

Llinás, R., & Volkind, R. A. (1973). The olivo-cerebellar system: Functional properties as revealed by harmaline-induced tremor. *Experimental Brain Research*, *18*(1), 69–87. https://doi.org/10.1007/BF00236557.

Llinás, R., & Yarom, Y. (1981). Electrophysiology of mammalian inferior olivary neurones in vitro. Different types of voltage-dependent ionic conductances. *The Journal of Physiology*, *315*, 549–567.

Llinás, R., & Yarom, Y. (1986). Oscillatory properties of guinea-pig inferior olivary neurones and their pharmacological modulation: An in vitro study. *The Journal of Physiology*, *376*, 163–182. https://doi.org/10.1113/jphysiol.1986.sp016147.

Lorenz, D., Frederiksen, H., Moises, H., Kopper, F., Deuschl, G., & Christensen, K. (2004). High concordance for essential tremor in monozygotic twins of old age. *Neurology*, *62*(2), 208–211.

Louis, E. D. (2010). Essential tremor: Evolving clinicopathological concepts in an era of intensive post-mortem enquiry. *Lancet Neurology*, *9*(6), 613–622. https://doi.org/10.1016/S1474-4422(10)70090-9. Epub 2010 May 5.

Louis, E. D. (2013a). Essential tremor and other forms of kinetic tremor. In G. Grimaldi, & M. Manto (Eds.), *Mechanisms and Emerging Therapies in Tremor Disorders* (pp. 167–201). New York: Springer.

Louis, E. D. (2013b). The primary type of tremor in essential tremor is kinetic rather than postural: Cross-sectional observation of tremor phenomenology in 369 cases. *European Journal of Neurology*, *20*(4), 725–727.

Louis, E. D. (2016). Essential Tremor: A common disorder of purkinje neurons? *The Neuroscientist*, *22*, 108–118.

Louis, E. D., Babij, R., Cortés, E., Vonsattel, J. P. G., & Faust, P. L. (2013). The inferior olivary nucleus: A postmortem study of essential tremor cases versus controls. *Movement Disorders*, *28*(6), 779–786. https://doi.org/10.1002/mds.25400. Epub 2013 Mar 11.

Louis, E. D., Diaz, D. T., Kuo, S. H., Gan, S. R., Cortes, E. P., Vonsattel, J. P. G., et al. (2018). Inferior Olivary nucleus degeneration does not lessen tremor in essential tremor. *Cerebellum Ataxias*, *15*(5), 1.

Louis, E. D., & Dogu, O. (2007). Does age of onset in essential tremor have a bimodal distribution? Data from a tertiary referral setting and a population-based study. *Neuroepidemiology*, *29*(3–4), 208–212. https://doi.org/10.1159/000111584. Epub 2007 Nov 27.

Louis, E. D., & Ferreira, J. J. (2010). How common is the most common adult movement disorder? Update on the worldwide prevalence of essential tremor. *Movement Disorders*, *25*(5), 534–541.

Louis, E. D., Ford, B., Frucht, S., Barnes, L. F., X-Tang, M., & Ottman, R. (2001). Risk of tremor and impairment from tremor in relatives of patients with essential tremor: A community-based family study. *Annals of Neurology*, *49*(6), 761–769.

Louis, E. D., Hernandez, N., & Michalec, M. (2015). Prevalence and correlates of rest tremor in essential tremor: Cross-sectional survey of 831 patients across four distinct cohorts. *European Journal of Neurology*, *22*(6), 927–932.

Louis, E. D., Joyce, J. L., & Cosentino, S. (2019). Mind the gaps: What we don't know about cognitive impairment in essential tremor. *Parkinsonism & Related Disorders*, *63*, 10–19.

Louis, E. D., Kuo, S. H., Tate, W. J., Kelly, G. C., Gutierrez, J., Cortes, E. P., et al. (2018). Heterotopic Purkinje cells: A comparative postmortem study of essential tremor and spinocerebellar ataxias 1, 2, 3, and 6. *Cerebellum*, *17*(2), 104–110.

Louis, E. D., Lee, M., Babij, R., Ma, K., Cortés, E., Vonsattel, J. P., & Faust, P. L. (2014). Reduced Purkinje cell dendritic arborization and loss of dendritic spines in essential tremor. *Brain, Dec, 137*(Pt 12), 3142–3148.

Louis, E. D., & Lenka, A. (2017). The olivary hypothesis of Essential Tremor: Time to lay this model to rest? *Tremor and Other Hyperkinetic Movements*, *13*(7), 473. https://doi.org/10.7916/D8FF40RX. eCollection 2017.

Louis, E. D., & McCreary, M. (2021). How common is essential tremor? Update on the worldwide prevalence of essential tremor. *Tremor and Other Hyperkinetic Movements*, *9*(11), 28.

Louis, E. D., Shungu, D. C., Chan, S., Mao, X., Jurewicz, E. C., & Watner, D. (2002). Metabolic abnormality in the cerebellum in patients with essential tremor: A proton magnetic resonance spectroscopic imaging study. *Neuroscience Letters*, *333*(1), 17–20. https://doi.org/10.1016/s0304-3940(02)00966-7.

Louis, E. D., Yi, H., Erickson-Davis, C., Vonsattel, J. P. G., & Faust, P. L. (2009). Structural study of Purkinje cell axonal torpedoes in essential tremor. *Neuroscience Letters*, *450*(3), 287–291. https://doi.org/10.1016/j.neulet.2008.11.043. Epub 2008 Nov 24.

Luo, Y., & Sugihara, I. (2016). The olivocerebellar tract. In D. L. Gruol, N. Koibuchi, M. Manto, M. Molinari, J. D. Schmahmann, & Y. Shen (Eds.), *Essentials of Cerebellum and Cerebellar Disorders* (pp. 55–61). Springer: Switzerland.

Lutes, J., Lorden, J. F., Beales, M., & Oltmans, G. A. (1988). Tolerance to the tremorogenic effects of harmaline: Evidence for altered olivo-cerebellar function. *Neuropharmacology*, *27*(8), 849–855. https://doi.org/10.1016/0028-3908(88)90102-5.

Manto, M., & Pandolfo, M. (2002). *The Cerebellum and Its Disorders*. Cambridge, UK: Cambridge University Press.

Mavridis, M., Degryse, A. D., Lategan, A. J., Marien, M. R., & Colpaert, F. C. (1991). Effects of locus coeruleus lesions on parkinsonian signs, striatal dopamine and substantia nigra cell loss after 1-methyl-4-phenyl-1,2,3,6-tetrahydropyridine in monkeys: A possible role for the locus coeruleus in the progression of Parkinson's disease. *Neuroscience*, *41*(2-3), 507–523. https://doi.org/10.1016/0306-4522(91)90345-o.

Milner, T. E., Cadoret, G., Lessard, L., & Smith, A. M. (1995). EMG analysis of harmaline-induced tremor in normal and three strains of mutant mice with Purkinje cell degeneration and the role of the inferior olive. *Journal of Neurophysiology*, *73*(6), 2568–2577. https://doi.org/10.1152/jn.1995.73.6.2568.

Mitoma, H., Kakei, S., Yamaguchi, K., & Manto, M. (2021). Physiology of Cerebellar Reserve: Redundancy and Plasticity of a Modular Machine. *International Journal of Molecular Sciences*, *22*(9), 4777. https://doi.org/10.3390/ijms22094777.

Mitoma, H., Buffo, A., Gelfo, F., Guell, X., Fucà, E., Kakei, S., et al., (2020). Consensus paper. Cerebellar reserve: From cerebellar physiology to cerebellar disorders. *Cerebellum*, *19*(1), 131–153.

Miwa, H. (2013). Rodent models of tremor. In G. Grimaldi, & M. Manto (Eds.), *Mechanisms and Emerging Therapies in Tremor Disorders* (pp. 37–51). New York: Springer.

Mizushima, S. (1976). An ultrastructural observation of torpedoes in the human degenerative cerebellum. *Journal of Clinical Electron Microscopy*, *9*, 672–673.

Ozawa, S. (2007). Role of glutamate transporters in excitatory synapses in cerebellar Purkinje cells. *Brain and Nerve, 59*(7), 669–676.
Parisi, S. L., Héroux, M. E., Culham, E. G., & Norman, K. E. (2006). Functional mobility and postural control in essential tremor. *Archives of Physical Medicine and Rehabilitation, 87*(10), 1357–1364.
Paris-Robidas, S., Brochu, E., Sintes, M., et al. (2012). Defective dentate nucleus GABA receptors in essential tremor. *Brain, 135*(Pt 1), 105–116. https://doi.org/10.1093/brain/awr301. Epub 2011 Nov 26.
Park, Y. G., Park, H. Y., Lee, C. J., Choi, S., Jo, S., Choi, H., et al. (2010). Ca(V)3.1 is a tremor rhythm pacemaker in the inferior olive. *Proceedings of the National Academy of Sciences of the United States of America, 107*(23), 10731–10736.
Quattrone, A., Cerasa, A., Messina, D., et al. (2008). Essential head tremor is associated with cerebellar vermis atrophy: A volumetric and voxel-based morphometry MR imaging study. *AJNR. American Journal of Neuroradiology, 29*(9), 1692–1697. https://doi.org/10.3174/ajnr.A1190. Epub 2008 Jul 24.
Rajput, A. H., Rozdilsky, B., Ang, L., & Rajput, A. (1991). Clinicopathologic observations in essential tremor: Report of six cases. *Neurology, 41*(9), 1422–1424. https://doi.org/10.1212/wnl.41.9.1422.
Raman, I. M., Gustafson, A. E., & Padgett, D. (2000). Ionic currents and spontaneous firing in neurons isolated from the cerebellar nuclei. *The Journal of Neuroscience, 20*(24), 9004–9016. https://doi.org/10.1523/jneurosci.20-24-09004.2000.
Rappaport, M. S., Gentry, R. T., Schneider, D. R., & Dole, V. P. (1984). Ethanol effects on harmaline-induced tremor and increase of cerebellar cyclic GMP. *Life Sciences, 34*(1), 49–56. https://doi.org/10.1016/0024-3205(84)90329-1.
Rogers, J., Zornetzer, S. F., & Bloom, F. E. (1981). Senescent pathology of cerebellum: Purkinje neurons and their parallel fiber afferents. *Neurobiology of Aging, 2*(1), 15–25. https://doi.org/10.1016/0197-4580(81)90054-3.
Schmahmann, J. D. (2016). The cerebrocerebellar system. In D. L. Gruol, N. Koibuchi, M. Manto, M. Molinari, J. D. Schmahmann, & Y. Shen (Eds.), *Essentials of cerebellum and cerebellar disorders* (pp. 101–115). Switzerland: Springer.
Schmahmann, J. D. (2016). The cerebellar cognitive affective syndrome and the neuropsychiatry of the cerebellum. In D. L. Gruol, N. Koibuchi, M. Manto, M. Molinari, J. D. Schmahmann, & Y. Shen (Eds.), *Essentials of cerebellum and cerebellar disorders* (pp. 499–511). Switerland: Springer.
Schneier, F. R., Barnes, L. F., Albert, S. M., & Louis, E. D. (2001). Characteristics of social phobia among persons with essential tremor. *The Journal of Clinical Psychiatry, 62*(5), 367–372. https://doi.org/10.4088/jcp.v62n0511.
Shaikh, A. G., Optican, L. M., & Zee, D. S. (2013). Membrane mechanisms of tremor. In G. Grimaldi, & M. Manto (Eds.), *Mechanisms and Emerging Therapies in Tremor Disorders* (pp. 11–35). New York: Springer.
Shin, H., Lee, D. K., Lee, J. M., et al. (2016). Atrophy of the cerebellar vermis in essential tremor: Segmental volumetric MRI analysis. *Cerebellum, 15*(2), 174–181. https://doi.org/10.1007/s12311-015-0682-8.
Silva, L. R., Amitai, Y., & Connors, B. W. (1991). Intrinsic oscillations of neocortex generated by layer 5 pyramidal neurons. *Science, 251*(4992), 432–435. https://doi.org/10.1126/science.1824881.
Siokas, V., Aloizou, A. M., Tsouris, Z., Liampas, I., Aslanidou, P., Dastamani, M., et al. (2020). Genetic risk factors for essential tremor: A review. *Tremor and Other Hyperkinetic Movements, 11*(10), 4.
Steriade, M., & Llinás, R. (1988). The functional states of the thalamus and the associated neuronal interplay. *Physiological Reviews, 68*(3), 649–742. https://doi.org/10.1152/physrev.1988.68.3.649.

Their, S., Lorenz, D., Nothnagel, M., et al. (2010). LINGO1 polymorphisms are associated with essential tremor in Europeans. *Movement Disorders*, *25*(6), 717–723. https://doi.org/10.1002/mds.22887.

Wills, A. J., Jenkins, I. H., Thompson, P. D., Findley, L. J., & Brooks, D. J. (1994). Red nuclear and cerebellar but no olivary activation associated with essential tremor: A positron emission tomographic study. *Annals of Neurology*, *36*(4), 636–642. https://doi.org/10.1002/ana.410360413.

Wills, A. J., Jenkins, I. H., Thompson, P. D., Findley, L. J., & Brooks, D. J. (1995). A positron emission tomography study of cerebral activation associated with essential and writing tremor. *Archives of Neurology*, *52*(3), 299–305. https://doi.org/10.1001/archneur.1995.00540270095025.

Wu, Y. C., Louis, E. D., Gionco, J., Pan, M. K., Faust, P. L., & Kuo, S. H. (2021). Increased Climbing Fiber Lateral Crossings on Purkinje Cell Dendrites in the Cerebellar Hemisphere in Essential Tremor. *Movement Disorders*, *36*(6), 1440–1445.

Yahr, M. D., Orosz, D., & Purohit, D. P. (2003). Co-occurrence of essential tremor and Parkinson's disease: Clinical study of a large kindred with autopsy findings. *Parkinsonism & Related Disorders*, *9*(4), 225–231.

Yu, M., Ma, K., Faust, P. L., Honig, L. S., Cortés, E., Vonsattel, J. P. G., et al. (2012). Increased number of Purkinje cell dendritic swellings in essential tremor. *European Journal of Neurology*, *19*(4), 625–630. https://doi.org/10.1111/j.1468-1331.2011.03598.x. Epub 2011 Dec 5.

Zesiewicz, T. A., Chari, A., Jahan, I., Miller, A. M., & Sullivan, K. L. (2010). Overview of essential tremor. *Neuropsychiatric Disease and Treatment*, 7(6), 401–408. https://doi.org/10.2147/ndt.s4795.

SECTION IV

Essential tremor as a harbinger: What is the cognitive fallout of having essential tremor?

Elan D. Louis[a] and Sheng-Han Kuo[b,c]

[a]Department of Neurology, University of Texas Southwestern Medical Center, Dallas, TX, United States
[b]Department of Neurology, College of Physicians and Surgeons, Columbia University, New York, NY, United States
[c]Initiative for Columbia Ataxia and Tremor, Columbia University, New York, NY, United States

As part of the current series, which highlights a select group of controversies in the essential tremor (ET) field, we invited authors to submit papers with opposing viewpoints, one using published data to put forth the argument that ET is a risk factor for dementia and the other, to examine these data to put forth the argument that ET is not a risk factor for dementia (Cosentino & Shih, 2022; Rajput, 2022). We are grateful to both sets of authors for reviewing and synthesizing the relevant data and constructing their arguments. Here we provide additional commentary.

We should take note that both sets of authors readily acknowledge that numerous studies have documented poorer cognitive functioning among individuals with ET compared to age matched controls (Cosentino & Shih, 2022; Rajput, 2022). Already by 2013, a systematic review of studies of cognition in ET examined 12 studies spanning the period from 1999 to 2011 (Janicki, Cosentino, & Louis, 2013) and numerous additional studies have been published since that review.(Louis, Joyce, & Cosentino, 2019) Beyond the authors (Cosentino & Shih, 2022; Rajput, 2022), there is a general consensus in the field that there is a cognitive side to ET. One caveat noted in both pieces, however, is that the suggestion that there is an increased risk for cognitive impairment in ET is not synonymous with suggesting that ET is universally characterized by cognitive impairment. In other words, the situation in ET is somewhat parallel to that in Parkinson's disease (PD); *presence in some* should not be equated with *presence in all.*

The question that is the focal point of these two articles, however, is whether ET is associated with increased risk of dementia. That is, whether the cognitive side of ET includes the development in some patients of more marked changes in cognition above and beyond what one would expect in similarly aged controls.

The authors of these articles cite evidence from several population-based studies in which the odds and risk of dementia in ET cases have been reported (Cosentino & Shih, 2022; Rajput, 2022). In a population-based study in central Spain, the Neurological Disorders in Central Spain (NEDICES) Study, in an adjusted model, ET cases with tremor onset after age 65% were 70% more likely to be demented than were controls (odds ratio [OR] = 1.70, 95% confidence interval [CI] = 1.04 to 2.76, $P = 0.03$) (Benito-Leon, Louis, & Bermejo-Pareja, 2006). In a population-based study of community-dwelling elders in northern Manhattan, the adjusted odds of dementia among ET cases age 65 and older was elevated compared to controls: OR = 1.84, 95% CI = 1.13–2.98, $P = 0.01$) (Thawani, Schupf, & Louis, 2009). In addition to ET being associated with and increased odds of *prevalent* dementia, the authors of the two papers in our series (Cosentino & Shih, 2022; Rajput, 2022) note that ET seems to be associated with increased risk of *incident* dementia in both the Central Spain and northern Manhattan cohorts. In NEDICES, during follow-up, 7.8% of ET cases developed dementia versus 3.9% of non-ET controls, yielding an adjusted relative risk (RR) of 1.66 (95% CI = 0.99–2.80, $P = 0.054$) and in an adjusted model, ET cases with tremor onset after age 65 years were twice as likely to develop incident dementia than were controls (RR = 1.98, 95% CI = 1.14–3.45, $P = 0.01$) (Bermejo-Pareja, Louis, & Benito-Leon, 2007). Similarly, in the northern Manhattan cohort of community-dwelling elders (Thawani et al., 2009), 18.3% of ET cases vs. 8.7% of controls developed incident dementia, with an adjusted hazards ratio (HR) of 1.64, 0.99–2.72, 95% CI, $P = 0.055$). The authors of that report were not able to stratify ET cases based on age of onset.

Notably, however, a third study consisting of community-dwelling residents in Sun City, Arizona (the Arizona Study of Aging and Neurodegenerative Disorders, AZSAND) did not detect an increased incidence of dementia in ET (Shill, Hentz, Jacobson, et al., 2014). During follow up, 6% of ET cases versus 8% of non-ET controls developed dementia, resulting in a HR = 0.79 (95% CI = 0.33–1.85). However, the HR for the association with dementia for ET cases with tremor onset at or above age 65 versus onset before age 65, was 2.1, and the authors did not perform the central analysis, which would have been a comparison between older onset ET cases and controls. This being said, their data suggest that individuals with elderly onset might be at higher risk for dementia. One other concern that has been raised with this study is that the prevalence of ET was considerably higher than expected in prior population-based estimates of ET and this raises the question of whether a significant number of the cases labeled as ET represented non-ET cases with enhanced physiologic tremor, thereby biasing their estimates of the HR towards the null (Louis, Radler, Huey, & Cosentino, 2021).

Hence, while the data are somewhat mixed, and in some cases the statistical significance is marginally close to the conventional cutoff of $P < 0.05$, there is a signal across all studies to date that older onset ET is associated with an increased risk for dementia. That is, that ET is a harbinger for or risk factor for dementia, in the same way as PD is, although the HRs for ET are not as large as they are for PD.

Although not central to either piece, both authors put forth some initial thoughts as to why ET is associated with cognitive dysfunction (Cosentino & Shih, 2022; Rajput, 2022). Both sets of authors agreed that our understanding of the cognitive impairment in ET is still in its early stage and that numerous mechanistic questions remaining to be answered (Cosentino & Shih, 2022; Rajput, 2022). Both sets of authors raise the possibility that cerebellar dysfunction or cerebellar pathway dysfunction may give rise not only to motor impairments in ET, but to cognitive impairments as well. Yet, as argued by one of the groups, it is challenging to attribute the majority of mild cognitive impairment cases and dementia in ET to cerebellar dysfunction alone, and this invokes the likelihood that more marked levels of cognitive impairment in ET reflect the presence of pathology, likely degenerative, outside of the cerebellum. That is, some of the cognitive impairment in ET arises secondary to co-morbid degenerative pathology, and that based on the current state of knowledge, it is likely that people with ET are at increased risk for cognitive impairment and dementia based on a heterogeneous, and potentially additive or synergistic brain changes.

Regardless of the data present in scientific studies, there remains some resistance to the notion that ET could be a risk factor for dementia. We should recall a similar situation with PD. Classical neurological descriptions of the disorder did not recognize dementia as part of the disease (Lieberman, Dziatolowski, Kupersmith, et al., 1979), and dementia was infrequently discussed before the advent of levodopa therapy (Lang & Blair, 1984). Mental deterioration was noted but thought to occur infrequently and its association was viewed initially as an aging phenomenon rather than a disease-linked one (Lieberman et al., 1979). Systematic clinical studies of PD cases and controls in the early to mid-1970s began to call greater attention to the presence of this non-motor feature of PD and bring this into focus both for clinicians and researchers (Lieberman et al., 1979). Prior to that point, it had not been part of the clinical dialogue. Similarly, in ET, the prevailing perception and teaching among clinicians has been that ET is a monosymptomatic motor-only disorder. Clinically apparent cognitive impairment has not been widely taught or recognized as a feature of ET. The emergence of literature, noted above, is producing scientific data that directly challenges this notion. Moreover, there are several reasons inherent to the nature of the cognitive deficits in ET that may make detection and recognition by patients and clinicians difficult. First, the degree of cognitive impairment is less than that seen in PD and there may be a lack of patient awareness, concern or self-report. Second, impairment in certain cognitive domains outside of memory may be less noticeable to patients and their families. Finally, the presence of potentially confounding factors such as age, mood and medication may be perceived to account for cognitive deficits.

In summary, debate leads to further enquiry and additional studies. These are necessary in order to delineate the nuances of this disease. We are witnessing an expansion of the clinical spectrum of ET, and part of that process is bringing to light data that suggest that ET is not only a cognitive-motor disorder, but one that is a potential harbinger of or risk factor for dementia.

References

Benito-Leon, J., Louis, E. D., & Bermejo-Pareja, F. (2006). Elderly-onset essential tremor is associated with dementia. *Neurology, 66*, 1500–1505.

Bermejo-Pareja, F., Louis, E. D., & Benito-Leon, J. (2007). Neurological disorders in Central Spain study G. Risk of incident dementia in essential tremor: A population-based study. *Movement Disorders, 22*, 1573–1580.

Cosentino, S., & Shih, L. C. (2022). Does essential tremor increase risk of cognitive impairment and dementia? Yes. *International Review of Neurobiology, 163*, 195–231.

Janicki, S. C., Cosentino, S., & Louis, E. D. (2013). The cognitive side of essential tremor: What are the therapeutic implications? *Therapeutic Advances in Neurological Disorders, 6*, 353–368.

Lang, A. E., & Blair, R. D. (1984). Parkinson's disease in 1984: An update. *Canadian Medical Association Journal, 131*, 1031–1037.

Lieberman, A., Dziatolowski, M., Kupersmith, M., et al. (1979). Dementia in Parkinson disease. *Annals of Neurology, 6*, 355–359.

Louis, E. D., Joyce, J. L., & Cosentino, S. (2019). Mind the gaps: What we don't know about cognitive impairment in essential tremor. *Parkinsonism & Related Disorders.*

Louis, E. D., Radler, K. H., Huey, E. D., & Cosentino, S. (2021). Progression to dementia in patients with essential tremor. *Parkinsonism & Related Disorders, 83*, 123–124.

Rajput, A. (2022). ET and DEMENTIA. *International Review of Neurobiology, 163*, 233–253.

Shill, H. A., Hentz, J. G., Jacobson, S. A., et al. (2014). Essential tremor in the elderly and risk for dementia. *Journal of Neurodegenerative Disorders, 2014*, 328765.

Thawani, S. P., Schupf, N., & Louis, E. D. (2009). Essential tremor is associated with dementia: Prospective population-based study in New York. *Neurology, 73*, 621–625.

CHAPTER SEVEN

Does essential tremor increase risk of cognitive impairment and dementia? Yes

Stephanie Cosentino[a,b,*] and Ludy C. Shih[c,d,*]

[a]Cognitive Neuroscience Division of the Taub Institute for Research on Alzheimer's Disease and the Aging Brain, Columbia University Irving Medical Center, New York, United States
[b]Department of Neurology, Columbia University Irving Medical Center, New York, United States
[c]Department of Neurology, Parkinson's Disease and Movement Disorders Center, Boston Medical Center, Boston, Massachusetts, United States
[d]Department of Neurology, Boston University School of Medicine, Boston, Massachusetts, United States
*Corresponding authors: e-mail address: sc2460@cumc.columbia.edu; lshih@bu.edu

Contents

Abstract

Essential Tremor (ET), by definition, is a disorder of movement. Yet over the years, epidemiologic, clinical, pathologic, and neuroimaging studies have converged to reveal a cognitive side of ET. The cognitive symptoms in ET are heterogeneous and are likely to reflect heterogeneous underlying mechanisms. In this chapter, we review and synthesize a diverse set of studies from both population-based settings to cohorts with more detailed investigations into cognition to consider the various mechanisms by which cognitive symptoms may emerge in a subset of individuals with ET. As part of our analysis, we consider questions surrounding ET diagnosis and the possibility of comorbid disease as potential factors that, upon closer examination, appear to strengthen the argument in favor of ET as a risk factor for dementia. Importantly, we also consider the clinical relevance of cognitive impairment in ET. While ET is not universally characterized by significant cognitive deficits, the data from epidemiological, cognitive, neuroimaging, and postmortem neuropathologic studies converge to reveal an increased

International Review of Neurobiology, Volume 163
ISSN 0074-7742
https://doi.org/10.1016/bs.irn.2022.02.011

risk for cognitive impairment and dementia among individuals with ET. We conclude by offering directions for future research, and a neurocognitive framework with which to consider existing findings and to use in the design of novel studies dedicated to clarifying the basis, nature, and course of cognitive impairments in ET.

1. Introduction

Essential tremor (ET) predominantly causes hand, head and voice tremor. The overall prevalence of ET is estimated to be approximately 6% over the age of 60 (Louis & McCreary, 2021), or 2.2% (7 million) of the entire population in the US (Louis & Ottman, 2014). This makes ET one of the most common neurological disorders, approximately 7 times more prevalent than Parkinson's disease (PD) (Marras, Beck, Bower, et al., 2018). In its mildest stages, ET can be mistaken for enhanced physiologic tremor, which is tremor that may be temporarily caused by medications, hunger, caffeine, stress, or other medical conditions (Benito-Leon & Louis, 2006). However, continuous presence of tremor and worsening over time are key features of ET as a distinct and progressive disorder (Putzke, Whaley, Baba, Wszolek, & Uitti, 2006). Individuals with ET develop both physical and emotional difficulty with activities of daily living (Bain, Findley, Thompson, et al., 1994), and tremor can become disabling in nearly 2/3 of patients (Louis et al., 2001), with up to 15–25% of individuals taking early retirement (Bain et al., 1994). Moreover, a growing literature links ET to increased prevalence of cognitive deficits including mild cognitive impairment (MCI) (Benito-León, Louis, Mitchell, & Bermejo-Pareja, 2011), as well as increased incidence and prevalence of Alzheimer's dementia (AD) (Bermejo-Pareja, Louis, & Benito-León, 2007; Shill, Hentz, Jacobson, et al., 2014; Thawani, Schupf, & Louis, 2009). Whether and how these various cognitive impairments in ET manifest has important potential implications for physician counseling regarding dementia risk modification, prognosis, and how choices of medical and surgical therapy may impact future risk.

Despite the boundaries between movement and cognitive disorders that define clinical neurologic subspecialties, there are a multitude of mechanisms by which cognition and motor function are intimately intertwined. In fact, in the context of age-related neurodegenerative disease, the selective deterioration of cognitive versus motor abilities is the exception rather than the rule (Montero-Odasso, Pieruccini-Faria, Bartha, et al., 2017). The mechanisms by which cognitive and motor impairment jointly arise in age-related degenerative disease are myriad. From a broad perspective, both cognitive and motor symptoms may arise from a singular disease process (e.g., Lewy Body disease (LBD)). Alternatively, co-morbid disease processes may be present because one disease may be in the causal pathway of another (i.e., cerebrovascular disease and AD) (Lee, Zimmerman, Narkhede, et al., 2018), two diseases may share genetic pathways (e.g., Amyotrophic Lateral Sclerosis and behavioral variant frontotemporal dementia (FTD)) (DeJesus-Hernandez, Mackenzie, Boeve, et al., 2011), accumulation of abnormal proteins in one disease may initiate a cascade of processes that increase risk for developing a separate disease via promoting formation of pathologic inclusions (Giasson, Forman, Higuchi, et al., 2003), or because of a general proclivity to protein misfolding and aggregation (Gidalevitz, Ben-Zvi, Ho, Brignull, & Morimoto, 2006). Correspondingly, there are thus two overarching ways by which ET may confer risk for cognitive impairment. The first is that ET "itself" causes such impairment. The second is that ET is associated, or co-morbid, with a cognitively impairing disease such as AD or LBD, for example. In this article, we highlight a combination of epidemiologic, clinical, pathologic, and neuroimaging studies that converge to reveal the cognitive side of ET, and we consider the various mechanisms by which cognitive symptoms may emerge in ET.

Prior to reviewing the evidence for cognitive impairment in ET, there is an important point to establish. Suggesting that there is an increased risk for cognitive impairment in ET is not synonymous with suggesting that ET is universally characterized by cognitive impairment in all patients. As opposed to a condition defined by cognitive decline, ET is by definition, at its core, a movement disorder. An analogous situation applies to PD, in which cognitive impairment is not universally present, but the increased risk for cognitive impairment is well-accepted. Indeed, in all studies that will be reviewed below, there is clear heterogeneity in the cognitive functioning of patients and participants with ET, with many individuals with ET demonstrating cognitive performance within the normal range. Moreover, within the group of individuals diagnosed as cognitively normal, are individuals who perform consistently high on neuropsychological testing. In a cross-sectional study examining profiles of performance among cognitively normal individuals diagnosed with ET, cluster analysis identified three primary cognitive profiles including a group who performed in the high average range across all measured domains except for language which was average (Cersonsky et al., 2020), perhaps more a reflection of psychometric limitations of the language measures than a relative weakness in this domain. As such, regardless of the evidence to be reviewed for increased risk of cognitive impairment in ET, we do not suggest that all individuals diagnosed with ET have, or are destined to have, cognitive impairment. Rather, toward the end of the chapter, we argue for the importance of establishing the features of ET that accompany, and even foreshadow, cognitive change to ensure that the individuals most at risk for such changes are informed, monitored, guided, and treated optimally.

As patients with ET are not regularly assessed for cognitive changes and may not be aware of changes that occur (Azar, Bertrand, Louis, et al., 2017), it is important that clinicians appreciate the potential for such changes in their patients. There have been few clinical observations describing the systematic assessment of cognitive changes in ET. Existing studies largely have focused upon individuals with ET undergoing stereotactic brain surgery for treatment of their tremor as reviewed below. There have been no practice guidelines to suggest that clinicians should conduct cognitive screening in ET. In a 2017 study examining self-awareness for cognitive functioning, cognitively normal individuals with ET provided accurate self-ratings regarding performance in language and executive functioning, with mild over-estimation of memory performance as assessed by neuropsychological testing. In contrast, individuals with ET who were diagnosed with clinically significant cognitive impairment (i.e., MCI or dementia) overestimated their performance in all measured domains. Such metacognitive impairment has been observed in a range of neurologic conditions including AD, FTD, stroke, and traumatic brain injury, and has been linked to reduced decision-making capacity (Cosentino, Metcalfe, Cary, De Leon, & Karlawish, 2011; Kelleher, Tolea, & Galvin, 2016; Shaked, Sunderaraman, Piscitello, et al., 2019; Starkstein, 2014). Impaired self-awareness may reflect compromise to brain networks and cognitive processes important for accurate self-evaluation and which may deteriorate in the context of neuroanatomic changes (Bertrand, Azar, Rizvi, et al., 2018; Cosentino, Brickman, Griffith, et al., 2015; Miller et al., 2001; Shany-Ur et al., 2014; Sollberger, Rosen, Shany-Ur, et al., 2014).

2. Cognitive impairment in ET

The evidence for cognitive impairment in ET derives from both clinic-based and population-based samples, both relevant for different reasons. Clinic-based samples reflect a population that clinicians will most likely encounter in clinical care and also provide a context in which the nature of cognitive deficits and their potential mechanisms can be explored more precisely. However, patients who present to the clinic and enroll in studies of cognition may be more likely to have cognitive difficulties about which they are concerned. Conversely, population-based studies presumably avoid the risk of such bias, can be used to characterize the features of ET more broadly, and enable quantification of risk for adverse outcomes in the ET population at-large, features which are important for long-term trends and research funding priority-setting. Population- and community-based epidemiologic studies also allow the opportunity to test the reproducibility and generalizability of observations made in smaller clinic-based cohorts.

However, large cohort studies are often unable to carefully map the various brain-behavior associations that characterize ET, an exercise which requires comprehensive neuropsychological testing, neuroimaging, and ultimately post-mortem neuropathological examination. Indeed, the focus of large cohort studies is often to establish a statistical association between a particular variable or condition and dementia, rather than to characterize the specific clinical features of the dementia. Given the number of variables that are often measured as part of such studies, cognitive protocols are brief and broad enough to capture all cognitive domains rather than delving into patterns within specific domains (i.e., memory, executive function, language, processing speed, spatial abilities).

In this chapter, we attempt to synthesize various streams of evidence across studies to determine the extent to which ET is characterized not only by motor features, but cognitive impairments as well. We include a discussion of the evidence for a spectrum of cognitive impairments, including: (1) cognitive performance below that of a non-ET control group, or below that of clinically defined normative scores; (2) Mild Cognitive Impairment (MCI) defined as the presence of clinically significant cognitive impairment on neuropsychological testing in conjunction with subjective concerns of cognitive impairment either by the person with ET or a knowledgeable informant; and (3) diagnosis of dementia defined as clinically significant cognitive impairment in at least two domains, and which interferes with daily functioning. In cases of dementia diagnoses, impairments in everyday life are required to result from cognitive difficulty rather than motor dysfunction.

It is important to emphasize that the staging of cognitive impairments (i.e., MCI, dementia) is independent from the underlying pathogenesis, or mechanism of impairments (e.g., cerebellar dysfunction, AD, PD or some combination). It is frequently the case that studies which examine the staging of cognitive impairment (i.e., epidemiological investigations) are not fully equipped to examine its pathogenesis (i.e., postmortem neuropathological studies), and vice versa. In certain cases, epidemiologic studies can inform the pathogenesis through the assignment of dementia subtypes (e.g., AD, DLB) and MCI categories (e.g., amnestic versus non-amnestic, and single versus multi-domain). Below we highlight a series of clinic-based studies which speak to different elements of cognitive impairment in ET, and then turn to data from three large epidemiologic studies. In the subsequent sections, we then address: (1) the potential effect that ET misdiagnosis may have on rates of cognitive impairment, (2) possible mechanisms of cognitive impairment in ET; and (3) clinical relevance and future directions.

3. Clinic based studies

A number of clinic-based studies have documented poorer cognitive functioning among individuals with ET as compared to age matched controls (Duane & Vermilion, 2002; Frisina, Tse, Hälbig, & Libow, 2009; Gasparini, Bonifati, Fabrizio, et al., 2001; Higginson Christopher et al., 2008; Lacritz, Dewey, Giller, & Cullum, 2002; Lombardi, Woolston, Roberts, & Gross, 2001; Passamonti, Novellino, Cerasa, et al., 2011; Şahin et al., 2006; Schwartz, Badarny, Gofman, & Hocherman, 1999; Tröster, Woods, Fields, et al., 2002). The results of many early studies, reviewed by Janicki, Cosentino, and Louis (2013)) consistently converge to reflect elements of executive dysfunction on tests of working memory, verbal fluency, and abstract reasoning, for example. Also documented in the 2013 review were studies demonstrating deficits in aspects of language, visuospatial abilities, and episodic memory among individuals with ET compared to controls. Indeed, in this systematic examination of 12 studies spanning 1999 through 2011, impairment was present in each study in at least one measured domain among individuals diagnosed with ET as compared to controls.

Since 2011, additional studies have reported on either objective or subjective cognitive symptoms among patients with ET. As a whole, review of existing clinic-based studies did not document any studies revealing comparable performance between an ET group and a non-ET group. In a recent study by Peng and colleagues conducted in 199 patients with ET and 132 healthy controls, the ET group reported more memory difficulty than controls (Peng et al., 2020). A brief report by Pedrosa and colleagues (2016) described specific deficits in the processing of time intervals among 24 ET patients as compared to 24 controls (Pedrosa, Nelles, Maier, et al., 2016). This deficit was accompanied by deficits in immediate word list recall and verbal fluency. Shalash and colleagues (2019) described significantly impaired global cognition in 30 participants with ET compared to 30 controls (Shalash, Mohamed, Mansour, et al., 2019). Pauletti and colleagues (2015) reported deficits in attentional abilities including alerting and executive efficiency among 21 non-demented participants with ET versus controls (Pauletti, Mannarelli, De Lucia, et al., 2015). Lee and colleagues (2015) examined 60 ET participants in comparison to 22 normal controls, documenting significant cognitive dysfunction (Lee, Kim, Lee, Kwon, & Koh, 2015). Sengul and colleagues (2015) also reported lower global cognitive scores among 45 young individuals with ET as compared to 35 controls (Sengul, Sengul, Yucekaya, et al., 2015). Finally, Mameli and colleagues (2013) conducted an investigation focused on the Guilty Knowledge Task in which individuals are asked to produce lies, an ability previously linked to executive functioning and shown to be impaired in patients with PD (Mameli, Tomasini, Scelzo, et al., 2013). In this task, participants are asked to select 5 pictures from a set of 10 and then directed to answer truthfully or untruthfully to the question, "Do you have this picture?" Interestingly, while no differences were found on other cognitive measures in this study, participants with ET had significantly longer reaction times than both the PD group and the NC group for all answer types. Moreover, when asked to produce a deceitful response, participants with ET and PD had lower lie accuracy scores than NCs.

Other studies have focused on diagnoses of MCI in ET (Collins et al., 2017; Park, Oh, Lee, et al., 2015; Sinoff & Badarny, 2014). An initial study by Park and colleagues described primarily non-amnestic cases of MCI in ET (Park et al., 2015). However, a later study by Collins and colleagues reported a majority of amnestic as opposed to non-amnestic MCI cases, a somewhat surprising finding given the traditional conceptualization of cognitive deficits in ET as primarily dysexecutive (Collins et al., 2017). Furthermore, participants did not meet criteria for amnestic MCI based only on the free-recall components of the memory test (i.e., the components which may tap into an executively mediated memory impairment). Rather, participants also had difficulty on the storage component of the memory testing, measured by testing participants' ability to recognize words (i.e., discriminate between recently learned and novel information). The presence of memory deficits in ET, and particularly memory deficits that implicate hippocampal compromise, raises interesting questions about the mechanisms underlying cognitive impairment in ET. Indeed, using the newly developed Cerebellar Cognitive and Affective Syndrome (CCAS)

Scale, (Hoche, Guell, Vangel, Sherman, & Schmahmann, 2018) has found that patients with cerebellar lesions have difficulty retrieving but not recognizing newly learned words. We will return to discussing mechanisms of cognitive impairment (Hoche et al., 2018) after reviewing epidemiologic evidence for increased rates of prevalent and incident dementia in ET, and also considering the role that ET misdiagnosis may play in determining the cognitive features of ET.

4. Population based studies

Forty-two population-based studies have estimated the prevalence of ET, but only three of these cohorts have included cognitive outcomes. Case ascertainment of ET in large population-based studies is challenging, but there have been significant developments in methodological approaches to address this. The methodological differences across the different population-based studies have been extensively reviewed (Louis & Ottman, 2014). Among the higher quality studies, several important methods were identified (Louis & Ferreira, 2010): (1) explicit definition of the case definition of ET; (2) age and gender stratification of prevalence estimates; and (3) either a direct neurological examination by a movement disorders specialist, or a screening questionnaire or drawing followed by neurologic examination, with sensitivity statistics allowing an estimate of the overall true prevalence. The three population-based cohorts which examine cognitive outcomes among individuals with ET include the Neurological Disorders in Central Spain (NEDICES) cohort, the Washington Heights and Inwood Columbia Aging Project (WHICAP) cohort, and the Arizona Study of Aging and Neurodegenerative Disorders (AZSAND) cohort. Notably, the first two cohorts included only individuals older than age 65.

First, investigations stemming from these population-based cohorts suggest poorer cognitive performance among ET participants on formal testing. In the Central Spain cohort, 232 ET cases demonstrated lower scores on the Mini-Mental Status Examination (MMSE), and tests of executive function, with more errors on the Trail Making Test compared to 696 age-matched controls, adjusting for age, gender, education level, presence of medications that affect cognition, and depression (Benito-Leon, Louis, & Bermejo-Pareja, 2006a). Notably, neither tremor duration nor tremor severity was associated with neuropsychological test performance. In a comparison study within the same cohort of 2212 healthy controls, 46 individuals with PD, and 180 individuals with ET, PD cases performed least well on cognitive tests of global cognition, memory and verbal fluency, followed by ET cases, who performed better than PD cases, but also performed worse than healthy controls (Sánchez-Ferro, Benito-León, Louis, et al., 2017). Notably, PD and ET differed on verbal fluency, with the latter not significantly impaired in ET relative to healthy controls. Finally, in a subgroup analysis of ET participants grouped by age of onset (< or >65 years old), there were no differences in cognitive performance.

Furthermore, population-based cohorts suggest a higher risk of dementia in ET compared to similarly aged individuals without ET. Odds and risk of dementia have been characterized in several studies. In a study characterizing prevalence of dementia in the Central Spain cohort, 11.4% of the 273 ET cases met criteria for dementia based on the Diagnostic and Statistical Manual of Mental Disorders (4th edition [DSM-IV]) criteria, which included a Clinical Dementia Rating (CDR) of 1 or more (indicating cognitively based functional impairment in everyday life), and the National Institute of Neurological and Communication Disorders Disease and Stroke/ Alzheimer's Disease and Related Disorders Association (NINCDS-ADRDA) criteria (Benito-Leon, Louis, & Bermejo-Pareja, 2006b). By comparison, only 6.0% of 3382 non-ET controls had dementia, producing an OR of 1.38–1.35, adjusting for age, history of stroke, educational level, ethanol consumption, smoking and depressive symptoms.

Age of onset of ET was associated with prevalent dementia in this study. There was an even stronger risk of dementia, when specifically looking at the subgroup of older-onset ET cases (age of onset >65 years of age), OR 1.70 (1.04–2.76 95% CI), compared to a smaller risk seen in younger-onset ET cases (age of onset <65 years of age, OR 0.38 (0.09–1.73 95% CI) compared to controls. This study finding suggests that older-onset ET might harbor pathologic mechanisms distinct from younger-onset ET. With regard to dementia subtype, AD predominated in both ET and non-ET controls (74.9% of dementia cases overall) with vascular dementia being the second most common subtype (14.9%). Of the 31 ET cases with dementia, 84% were of older-onset ET while 16% consisted of younger-onset ET cases, and the older-onset ET cases were highly associated with AD (OR 2.28, 95% CI 1.47–3.52). The dementia classification from the younger-onset ET cases with dementia was not mentioned. Presumably, the number of younger-onset ET cases was too small to draw any conclusions about differences in dementia phenotypes between older-onset and younger-onset ET cases. Therefore, the question remains unresolved as to whether potentially different dementia phenotypes emerge depending upon age of onset.

In the WHICAP cohort of community-dwelling elders age ≥65 in Manhattan, New York, an even higher prevalence of dementia was observed compared to the central Spain study. Fully a *quarter* of the 124 ET cases had prevalent dementia compared to 9.2% of 2161 controls (OR 1.84, 95% CI 1.13–2.98) adjusting for age, education and ethnicity (Thawani et al., 2009). As in the central Spain study, dementia diagnoses were based on clinical evaluation using DSM criteria as well as neuropsychological testing. The participants were of similar age (mean = 77) in the Manhattan study as the Spanish study, but information about age of onset was not collected. In all other respects, the two studies arrived at similar results.

In addition to ET being associated with prevalent dementia, ET is also associated with increased risk of incident dementia in both the Central Spain and WHICAP cohorts, despite relatively short follow up. In Central Spain, after 3.2 years, 7.8% of 206 ET cases developed dementia versus 3.9% of non-ET controls, yielding an adjusted relative risk of 1.66 (0.99–2.80 95%CI) (Bermejo-Pareja et al., 2007). Here, age of onset greater than 65 conferred increased dementia risk as compared to individuals with age of onset less than 65. Additional factors associated with increased risk of incident dementia included family history of ET. In the Manhattan cohort, mean follow up was 3.8 years, with incident dementia rates of 18.3% in ET cases versus 8.7% in non-ET controls, yielding an adjusted HR of 1.64, 0.99–2.72 95% CI. Both studies therefore showed very similar and consistently elevated risks of developing incident dementia over a similar time period. It remains to be seen whether over longer periods of follow-up the risks are even more substantial compared to a non-ET control sample.

Consistent with the idea that ET is associated with progression to dementia, a faster rate of cognitive decline has also been observed in cognitively normal ET cases, excluding individuals who went onto develop dementia during the follow up period. Specifically, compared to non-ET controls ($n = 2184$), ET cases ($n = 135$) declined more quickly on a measure of global cognition over a mean follow up of 3.4 years as measured by a 37-item Mini-Mental Status Examination. Though the difference in change was small (mean change 0.70 ± 3.2 points in cases vs. 0.11 ± 3.8 points in controls ($P = 0.03$) (Louis, Benito-León, Vega-Quiroga, & Bermejo-Pareja, 2010a), it was seven times faster than that of normal controls, a statistically significant difference. Further information over longer periods of time is necessary to confirm this trajectory, and detailed neuropsychological testing collected longitudinally will enable a clearer understanding of the specific aspects of cognition that decline.

Notably, however, a third population-based study consisting of community-dwelling residents in Sun City, Arizona (AZSAND) did not find an increased incidence of dementia in ET. After excluding cases of prevalent dementia, the authors identified 83 subjects with ET and 424 non-ET controls (Shill et al., 2014). After mean follow up of 5.4 years, 6% of ET cases versus 8% of non-ET controls developed dementia, according to DSM-IV criteria applied during consensus conference in conjunction with available data from neuropsychological testing. However, the authors did report that individuals with elderly onset and shorter duration tremor had double the risk for dementia as their younger onset and longer tremor duration counterparts.

The broad discrepancy between findings in the AZSAND study and those of the other cohorts has been a topic of recent discussion (Louis, Radler, Huey, & Cosentino, 2021; Shill, Mehta, & Adler, 2021), and may reflect important methodologic differences across cohorts particularly as relate to the diagnosis of ET. Prevalence of ET in the AZSAND study (16.4%) is significantly higher than prior population-based estimates of age-based ET prevalence. This increased rate raises the question of whether some of the cases labeled as ET represented non-ET cases with enhanced physiologic tremor (Louis, Radler, et al., 2021; Shill et al., 2021). Among the AZSAND cohort, ET was diagnosed based on: (1) previous clinical diagnosis of ET and an exam consistent with that diagnosis; (2) isolated voice or head tremor without dystonia; (3) presence of postural or kinetic hand tremor score of $>=2$ on the Fahn Tolosa Marin scale (moderate amplitude, may be intermittent, no measurement anchor specified) without secondary cause; and (4) postural or terminal tremor of the hands <2 for more than 3 years without secondary cause. In contrast, diagnoses of ET in the WHICAP study, for example, required moderate or greater amplitude (>1–2 cm) kinetic tremor during three or more tests of hand tremor in the absence of PD, dystonia, or other causes. These tests consisted of pouring, cup drinking, spoon feeding, finger-to-nose movements and spiral drawing. The amplitude requirement as well as tremor during manual tasks likely eliminates many cases of enhanced physiologic tremor which often produces only a postural tremor, while minimally affecting manual tasks. Therefore, differences in the classification of ET likely have direct relevance for the extent to which ET diagnosis is associated with an increased risk for dementia.

5. Question of misdiagnosis

If ET cohorts are enriched with cases of physiologic or dystonic tremor, comparable rates of dementia incidence across ET and non-ET cases may be seen, reflecting an admixed ET population. By the same token, it may be reasonable to consider the theoretical argument that increased rates of cognitive impairment in ET could indicate that in some subset of cases with significant cognitive impairment, individuals had been misdiagnosed with ET when in fact the emergence of tremor is a feature of a different degenerative disease. Critically, however, prospective longitudinal studies of incident dementia among individuals with longstanding ET provide an additional level of evidence that what was labeled "ET" was not merely a non-specific action tremor in an individual with dementia. Moreover, in population-based studies, the prevalence of ET dwarfs that of PD and even further dwarfs that of multiple system atrophy (MSA) (Bower, Maraganore, McDonnell, & Rocca, 1997; Schrag, Ben-Shlomo, & Quinn, 1999) and fragile X-associated tremor/ataxia syndrome (Hall and O'keefe, 2012). In some population-based studies, the prevalence of ET is 20 × greater than that of PD. Thus, while some "ET" cases could be "PD", the proportion is likely to be exceedingly low in such population-based settings and such theoretic diagnostic misclassification is not likely to account for the observed association between ET and incident dementia. However, this does not fully preclude the possibility that tremor emerges as an early feature of a different disease, particularly in late onset ET cases. Below we discuss this consideration at a theoretical level while highlighting the features of the clinical presentation and course that increase confidence in ET diagnoses.

Although ET has a prototypical clinical presentation, misdiagnosis of ET is common. The diagnostic misclassification of chief concern for the inflation of cognitive impairment in individuals with ET would likely be PD. The potential for misdiagnosis between these conditions arises due to the overlap in clinical phenomenology. Indeed, it is generally assumed that some rest tremor is an acceptable finding in ET cases with moderate to severe postural and kinetic tremor. Conversely, the appearance of kinetic and postural tremor in PD has been well documented (Josephs, Matsumoto, & Ahlskog, 2006; Kraus, Lemke, & Reichmann, 2006). Though often readily distinguished based on marked asymmetry or even unilateral presence of upper limb tremor, which is characteristic of PD, there have been instances in which bilateral postural kinetic tremor is then followed later on by resting tremor in the upper limbs among individuals who are ultimately diagnosed with PD. In such cases, the eventual development of bradykinesia is inevitable even in tremor-predominant PD cases, which are initially characterized by milder symptoms of bradykinesia and less severe nigral neuronal loss (Selikhova, Kempster, Revesz, Holton, & Lees, 2013). Thus, ET cases with long disease duration (for example, over 10 years) (Tarakad & Jankovic, 2018), and without development of unequivocal bradykinesia, can be confidently confirmed as ET diagnoses even if overflow resting tremor happens to be present.

The implication of these findings for the question of cognitive impairment in ET is that PD is unlikely to be the underlying etiology if: 1) there is no cross-sectional correlation between rest tremor and cognition; and 2) individuals with longstanding ET develop incident cognitive impairment. Recent work from the The Clinical-Pathological Study of Cognitive Impairment in Essential Tremor (COGNET) study examined correlations between motor features in ET including severity of the postural tremor, gait, presence of head or voice tremor, and presence of rest tremor. Correlations were conducted among cognitively normal adults, as well as a more heterogenous group including those with MCI. In neither analysis was rest tremor associated with cognitive scores (Peterson, Chapman, Radler, et al., 2021).

In addition to PD, atypical parkinsonism syndromes such as multiple system atrophy (MSA) and progressive supranuclear palsy (PSP) might initially be mistaken for ET if postural and/or action hand tremor is an early prominent feature. Tremor in MSA can appear as a jerky, irregular tremor-like movement which has been suggested to represent polyminimyoclonus, but can appear similar to a bilateral postural hand tremor if the irregularity of tremor is not observed (Gilman, Wenning, Low, et al., 2008; Köllensperger, Geser, Seppi, et al., 2008; Miki, Foti, Asi, et al., 2019). However, this misclassification does not occur in cases where tremor is carefully assessed among MSA, PSP and LBD patients (Miki et al., 2019). A jerky postural tremor has also been purported in PSP, and as in PD or MSA, progressive development of other striking neurologic signs (e.g., oculomotor gaze palsy, postural instability and frequent falls, autonomic dysfunction, respiratory stridor) clearly distinguish the course of PSP from ET. As such, while other forms of parkinsonism could conceivably be mistaken for ET, it would likely be only for a short time, given the typically more rapidly progressive nature of atypical parkinsonism. Other diagnostic entities that could be mistaken as ET include dystonia, orthostatic tremor or certain ataxic syndromes, namely Fragile X tremor-ataxia syndrome, or even some spinocerebellar ataxias, as in spinocerebellar ataxia 12. However, at some point such conditions would be suspected when ataxic signs accrue beyond the acceptable threshold for ET.

There have been two studies of note that have retrospectively reviewed diagnoses of ET. Schrag and colleagues (2000) re-evaluated 50 charts randomly selected from 200 patients previously coded as having ET between 1994 and 1998 (Schrag, Münchau, Bhatia, Quinn, & Marsden, 2000). The authors applied diagnostic criteria, outlined by Bain and colleagues, which had been developed to define classical hereditary ET based on descriptions of ET among families in which at least three generations were similarly affected. Using these criteria for "classical ET" as bilateral postural hand tremor, with or without kinetic tremor, and age of onset before 65 with the upper limbs as the site of tremor onset, 50% of individuals diagnosed with ET were re-classified as "atypical ET." ET was re-classified as "atypical" based on: isolated tremor onset in a body part other than the arms, presence of other movement abnormalities such as dystonia and myoclonus, or clear presence of resting tremor in a fully relaxed state. In this study, younger age of onset was significantly associated with being diagnosed with classical ET.

Additionally, Jain et al. (2006) later reviewed 71 cases of patients presenting to the Neurologic Institute of New York who received diagnoses of ET over a 5-year period from 2000 to 2005 (Jain, Lo, & Louis, 2006). The group re-evaluated the presence of ET according to the Movement Disorders Society consensus statement requiring the presence of bilateral, predominantly symmetric postural or kinetic tremor in the upper extremities with possible head tremor in the absence of abnormal posturing. Of the 71 cases, 37% were re-diagnosed, with PD being the most common re-diagnosis. *Re*-diagnosed cases tended to have shorter tremor duration and, in contrast to the results of Schrag and colleagues, were marginally *younger* than those with confirmed diagnoses of ET, suggesting that an ET diagnosis was more often correct when made in older patients with longer duration tremor.

The significance of "atypical ET" and similarly, ET with subtle neurologic signs of dystonia, and whether they represent a different diagnostic entity is still debated. In recent years, the clinicopathologic significance of additional neurologic signs seen among individuals classified as having ET has come under greater scrutiny and discussion. The extent of the findings has led to proposals to distinguish ET from so-called "ET plus" as a diagnostic classification for both research and clinical purposes. However, we are just at the beginning stages of defining how various neurologic features may arise at different points throughout the disease course, and there is controversy surrounding use of "ET plus" as a diagnostic label (Louis, Bares, Benito-Leon, et al., 2020). Careful longitudinal studies of the evolution of signs may help resolve ET case ascertainment issues related to defining what ET and "ET plus" represent. In particular, whether "ET plus" is a disease subtype per se, or instead a more advanced disease stage than "ET" is yet to be determined. There is evidence that features of "ET plus" are both age- and stage-dependent (Iglesias-Hernandez et al., 2021; Louis, Huey, & Cosentino, 2021). Thus, while it is possible that cognitive impairment may be more common among individuals who meet criteria for "ET plus" (Huang, Yang, Zhao, et al., 2020); the implications of this finding for the mechanism underlying cognitive impairment are uncertain.

Finally, given the variability in ET diagnostic criteria that characterize existing studies, and which may continue to characterize future studies, it is important not only to examine cognition in ET versus non-ET groups, but to examine cognitive function in direct relation to specific motor features of ET (see Fig. 1). In this manner, more precise associations can be

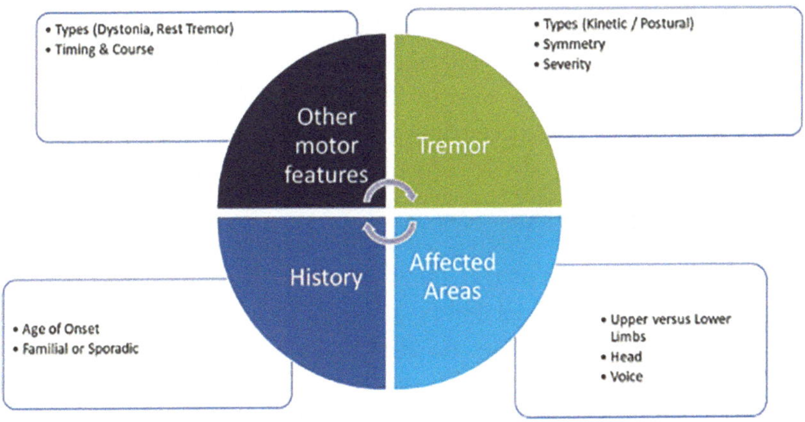

Fig. 1 Features of ET to be examined in relation to neuroanatomy and cognition. Features listed are those which may vary among existing cohorts of individuals diagnosed with ET, but which may be considered by movement disorders specialists to argue against a diagnosis of ET (e.g., asymmetric tremor, cranial tremor in the absence of upper limb tremor). Such features are included here to draw attention to dimensions of the ET diagnosis which may have varied across studies to date and may have led to variable results regarding the cognitive profile(s) of ET. Understanding how variability in these features relates to cognitive performance and neuroanatomic structure and function will likely provide greater clarity on the nature and basis of cognitive impairment in ET.

established and potential discrepancies in findings across studies may be reduced. Moreover, attempts to harmonize motor assessments and establish clear adjudication procedures across studies may allow for better comparison and resolution of issues surrounding ET diagnosis in the future.

6. Mechanisms of cognitive impairment in ET

Against this backdrop, our understanding of the cognitive impairment in ET is still nascent with many mechanistic questions remaining to be answered. First, we consider the possibility that there is a singular pathologic disease process which is ET. This scenario, of course, depends on what ET actually is (Louis, 2014), and various forms of evidence suggest that ET would be better described as the "ETs" given the clear heterogeneity in both clinical, genetic, and pathologic symptoms. A challenge of establishing the cognitive features of ET is that the neuropathologic landscape of ET is only beginning to be mapped. Though not universal, there is general agreement, based on converging evidence from neuroimaging, clinical, and postmortem studies that ET pathobiology resides primarily in the cerebellum (Filip, Lungu, Manto, & Bareš, 2016; Hanajima et al., 2016; Shill, Adler, & Beach, 2012; Shill, Adler, Sabbagh, et al., 2008). This pathobiology appears to center around Purkinje cells, specialized neurons found in the cerebellar cortex receiving massive inputs via their dendritic branches (Paul & Limaiem, 2021). Cerebellar neuropathology from post-mortem ET brains has been reviewed in detail previously (Louis & Faust, 2020) and includes reduced Purkinje cell density (Axelrad, Louis, Honig, et al., 2008; Choe et al., 2016; Louis, Faust, Vonsattel, et al., 2007), abnormal protein filled accumulations including neurofilament-filled torpedoes (Louis et al., 2007), dendritic changes (Louis, Lee, Babij, et al., 2014; Yu, Ma, Faust, et al., 2012), axonal injury and remodeling (Babij et al., 2013; Louis, Kerridge, Chatterjee, et al., 2019) and changes to neighboring cells (i.e., Basket cells).

7. Mechanisms: Cerebellar dysfunction

Cerebellar dysfunction may give rise not only to motor impairments in ET, but to cognitive impairments as well. With case studies, functional imaging, and tract tracing in non-human primates, decades of research have established what has been referred to as the "cerebellar sensorimotor-cognitive" dichotomy at both the behavioral and anatomic level.

Whereas the anterior lobe of the cerebellum receives spinal afferents through the spinocerebellar tracts and reciprocal connections with the motor cortices, the posterior cerebellar lobe shares reciprocal connections with association areas of the cerebral cortex including the prefrontal, posterior parietal, and superior temporal cortex as well as the cingulate gyrus and parahippocampal area (see Schmahmann, 2019, for a review). Task-based functional MRI studies reveal that cognitive and motor paradigms activate topographically distinct regions, and that regions of the posterior cerebellar lobe are involved in a wide variety of cognitive tasks spanning language, spatial, and executive functions (Stoodley & Schmahmann, 2009; Stoodley, Valera, & Schmahmann, 2010, 2012). Consistent with the crossed nature of cerebro-cerebellar projections, language tasks activate right sided cerebellar areas whereas spatial tasks map onto left-sided regions.

While multiple models have been put forth to describe the specific manner in which the cerebellum subserves cognition, a broad theme is that the cerebellum performs similar functions in the cognitive domain as it does in the motor domain, translating conscious, planned and controlled processes into those that are automatic and routinized (Koziol, Budding, Andreasen, et al., 2014). In this formulation, the cerebral cortex draws on the cerebellum to carry out skillfully executed movements as well as cognitive operations. In support of this theory, functional neuroimaging work has suggested that the rate at which new rules can be learned and automatized is associated with the rate of excitability changes which signal cerebellar plasticity (Balsters & Ramnani, 2011). The conceptualization of the cerebellum as critical for the optimization of performance across both cognitive and motor domains, as well as affective and autonomic functioning, is the foundation of the Universal Cerebellar Transform (UCT) model proposed by Schmahmann (2000) in which the cerebellum is postulated to automatically integrate internal representations with external stimuli and self-generated responses (Schmahmann, 2000). Correspondingly, damage to the posterior lobe of the cerebellum results not in movement abnormalities, but in "dysmetria of thought" (Schmahmann, 1998), or impairment in the coordination of cognitive processes. Analogous to impaired control of the distance, speed, and motions necessary to execute an action, dysmetria of thought disrupts an individual's ability to efficiently apply previously automatized rules, and interferes with higher-order problem solving, multi-tasking, and organization of behavior, whether verbal or nonverbal. Schmahmann (1998) has described a cerebellar cognitive affective syndrome (CCAS) which is characterized by deficits in executive function, linguistic processing, spatial cognition and affect regulation (Schmahmann & Sherman, 1998). The effects of posterior cerebellar lobe damage on cognition can thus be far-reaching, presumably encompassing a wide range of cognitive functions, particularly those considered to be executive functions and which rely heavily on pre-frontal regions.

Cerebellar volume loss and morphological changes have been observed with cognitive aging in the general population (Bernard and Seidler, 2014) and are highly associated with that of prefrontal cortical regions (Schmahmann, Guell, Stoodley, & Halko, 2019) . In particular, the volume of the anterior lobes of the cerebellum decline with age, while posterior cerebellum correlates with prefrontal cognitive function (Bernard & Seidler, 2013). The extent to which cerebellar gray matter is reduced in ET is not yet established. Of the studies that examined cerebellar volume, 14 studies reported significant reductions in cerebellar gray matter, while another 14 studies did not find changes. However, there were several limitations among these studies. All studies included clinic-based samples and were based upon cross-sectional analyses, ranging from 20 subjects to 110 subjects (largest sample 60 ET and 50 healthy controls). These findings do, however, raise the possibility that cerebellar degeneration is consistently present to the degree that gross changes can be observed radiologically and might therefore contribute to cognitive impairment in ET.

It is therefore conceivable that cerebellar dysfunction alone could produce both relative cognitive deficits as well as clinically significant deficits sufficient for a diagnosis of MCI, or perhaps even dementia. However, overlaid on the known cerebellar features of ET, are a range of imaging and pathologic studies documenting more widespread cortical and subcortical changes among individuals with ET in regions including the corpus callosum, thalamus, and hippocampus. Moreover, even the staunchest proponents of the cognitive cerebellum cast doubt on its potential role in memory deficits. While the cognitive profile of ET shares features of the CCAS, primarily executive dysfunction, there are notable exceptions which will be discussed below. It is therefore challenging to attribute the majority of MCI cases in ET to cerebellar dysfunction alone, invoking the likelihood that cognitive impairment in ET reflects the presence of pathology outside of the cerebellum. Another formulation regarding the cognitive impairments in ET is that this disease encompasses a distributed set of extra-cerebellar regions, and that depending on which regions are affected in a given individual, a certain type and degree of cognitive impairment will emerge.

8. Mechanisms: Extra-cerebellar dysfunction

If we expand our schema of ET to include a range of extra-cerebellar dysfunctions, there is greater room for considering the basis by which MCI and dementia may evolve with greater frequency in such patients. Structural MRI brain scans have implicated both cortical and subcortical cerebral and cerebellar regions affected in ET, typically using voxel-based morphometry (VBM) or diffusion tensor imaging (DTI), however, the findings have not been consistent across studies. These have been extensively reviewed (Holtbernd & Shah, 2021) and summarized in a recent meta-analysis (Luo, Pan, Xu, & Chen, 2019). The key findings are discussed here across several regions of interest with respect to cognitive function in ET.

Subcortical structural changes have been characterized in ET in relation to cognitive performance. Pelzer et al. reported on structural MRI correlates of verbal fluency, a language-based measure which relies heavily on executive functioning, in 19 individuals with ET versus 23 age- and gender-matched controls. Both verbal fluency and word list recall were reduced among ET participants compared to controls. Using VBM and DTI, the investigators documented a positive correlation between fractional anisotropy of the corpus callosum and verbal fluency. Interestingly, no cerebellar white or gray matter regions were found to correlate with verbal fluency. Notably, corpus callosal integrity is also implicated in other conditions in which verbal fluency is reduced, including in AD (Rodríguez-Aranda, Waterloo, Johnsen, et al., 2016; Sánchez, Duarte-Abritta, Abulafia, et al., 2020), semantic dementia (Agosta, Henry, Migliaccio, et al., 2010), multiple sclerosis (Manca, Stabile, Bevilacqua, et al., 2019) and cerebral small vessel disease (Jokinen, Ryberg, Kalska, et al., 2007). Therefore, the findings do not implicate the corpus callosum as an ET-specific imaging marker of cognitive impairment, but rather potentially as a mechanistic pathway of cognitive impairment shared in common with other diseases.

In another study by Prasad et al. (2019) examining subcortical volumes in 40 subjects with ET, based upon 3T MRI and VBM, reduced volumes in bilateral thalami, hippocampi and ventral diencephalon were observed in comparison to 40 healthy controls (Prasad, Shah, Bhalsing, Ingalhalikar, et al., 2019). Interestingly, increases in regional volumes were also noted, including caudate, pallidum, amygdala, putamen and nucleus accumbens, but there were no associations between brain structure and cognition.

However, cognition was measured using the MMSE, likely limiting the degree to which variability in cognition could be adequately detected.

Cortical structures have been examined in detail in a few studies with some correlations between regional gray matter volume and cognition. In another study using 3 T MRI, Bhalsing and colleagues (2014) demonstrated volumetric losses in medial frontal gyrus, post-central gyrus, anterior cingulate and insula in 17 cognitively impaired ET patients compared with 8 cognitively intact individuals with ET (Bhalsing, Upadhyay, Kumar, et al., 2014). Neuropsychological test performance correlated with gray matter volume (GMV) in numerous regions. These included visuospatial scores correlating with GMV in left post-central gyrus, middle temporal gyrus and occipital lobe, and working memory with superior frontal gyrus, insula and posterior lobe of the cerebellum. Immediate visual memory was correlated with medial frontal, inferior occipital and superior parietal gyri GMV, while delayed visual memory correlated with medial frontal, superior frontal as well as inferior occipital gyrus GMV. Immediate verbal memory correlated with insula and lentiform nucleus GMV, and delayed verbal memory correlated with lentiform nucleus, middle frontal gyrus, and superior parietal lobe. In this study, there were no significant associations between cognitive performance and regional GMV in the group of ET cases without cognitive impairment nor in non-ET controls, suggesting the GMV correlations seen were specific to ET cognitive impairment. However, this study employed simple correlation analyses to test associations between cognitive domains and gray matter regions. Neuroimaging-cognitive performance analyses may benefit from multivariate covariance-based approaches, given between-subject variability and collinearity of regional volume measures and cognitive measures (Habeck & Stern, 2010).

In addition to a number of neuroimaging studies in ET suggestive of extra-cerebellar morphological changes, postmortem investigations have begun to provide insight into the nature of extra-cerebellar pathologic changes that occur in ET. A 2014 study systematically assessing hippocampus, limbic regions, brainstem, cerebellum, mesencephalon and cerebral cortices using a semi-quantitative rating system documented higher Braak neocortical neurofibrillary tangle (NFT) stage in 40 ET brains as compared to 32 controls in the context of comparable neuritic plaque scores (Pan et al., 2014). More recently, a quantitative and more precise examination of NFTs was undertaken among 26 ET cases versus 73 control cases, all of whom had detailed cognitive assessments and clinical diagnoses available. ET and non-ET controls were carefully matched on a 1:3 basis, within each cognitive diagnostic category (normal, MCI, dementia), according to age, gender, ethnicity, education, duration between cognitive testing and death, and performance on six cognitive tests. Overall, ET and control brains were characterized by similar levels of beta amyloid and Lewy bodies, as well as medial temporal NFT-positive neurons. However, among the cognitively normal and MCI groups, there was a higher number of neocortical NFTs in the ET cases than among control brains. NFT counts were correlated with cognition in both ET and non-ET cases. The higher level of neocortical tau among ET cases suggests that ET is accompanied by an increased predisposition to accumulate tau outside the medial temporal lobe, and therefore distinct from the pattern of tau pathology observed nearly ubiquitously in aging (Primary Age Related Tauopathy; PART) (Crary, Trojanowski, Schneider, et al., 2014).

Moreover, emerging work suggests that the expression profile of tau protein isoforms diverges from that observed in three other tauopathies, PSP, AD, and Pick's disease (Kim, Farrell, Cosentino, et al., 2021). An exploratory 2021 study assessed immunohistochemical AT8-positive phosphorylated-tau (p-tau) burden in the brains of 13 patients with ET, 6 of whom were clinically demented in life, 4 of whom carried diagnoses of MCI, and three of whom were cognitively normal at the time of death. Age of death was 90.2 years old on average, ranging from 83 to 99. ET cases, including those with dementia, had variable, and significantly lower p-tau than the AD and Pick's cases. Most striking, however, were differences in tau isoforms derived from alternative pre-mRNA splicing of exons 2, 3, and 10, reflecting divergent underlying pathology. Specifically, while levels of 3R isoforms were comparable in ET, Pick's, and PSP, and levels of 4R were comparable across ET and all other tauopathies, ET brains demonstrated differences in N-terminal exons of tau. Specifically, levels of 0N tau were markedly lower in the ET group than in AD and PSP. The authors concluded that ET patients might exhibit an expression signature of tau isoforms that is unique from other degenerative diseases.

9. Mechanisms: Co-morbid degenerative disease

A second conceptual framework for explaining the presence of cognitive impairments in ET (which is not mutually exclusive from the first), is that cognitive impairment in ET arises secondary to co-morbid degenerative pathology. The possibility that multiple diseases produce a combination of motor and cognitive symptoms in ET is in line with idea that the development of one disorder is a marker of propensity to neurologic illness in general, and that individuals with a degenerative disease often have multiple pathologies on autopsy. In other words, patients with ET may simply be at greater risk for developing another neurodegenerative disease by virtue of having ET (Braak, Rüb, Schultz, & Del Tredici, 2006). Several genetic and postmortem studies have pointed to tauopathic features in ET, raising the possibility that ET is comorbid with a tauopathy such as PSP, AD, or FTD (Clark, Liu, Parmalee, Hernandez, & Louis, 2014; Crary et al., 2014; García-Martín, Martínez, Alonso-Navarro, et al., 2012; Louis et al., 2014; Pan et al., 2014; Vilariño-Güell, Soto-Ortolaza, Rajput, et al., 2011). Perhaps most attention has been given to the possibility that AD is a comorbid condition, given findings of epidemiologic studies.

As hippocampal volume is a biomarker of AD risk and pathology, based on the framework for neurodegeneration for AD (Jack, Bennett, Blennow, et al., 2016, 2018; Jack, Knopman, Weigand, et al., 2012), at least two studies have specifically examined the integrity of the hippocampus in ET. Building on work documenting more amnestic than non-amnestic cases of MCI among a cohort of ET cases, Novellino and colleagues (2020) examined the basis of memory storage deficits in ET (Novellino, Vasta, Saccà, et al., 2020). The authors examined 60 individuals with ET in comparison to 50 non-ET controls across a number of different neuropsychological assessments, with particular interest in understanding a potential role for hippocampal alterations in cognitive changes. The ET group performed within the normal range on memory and other tests; however, group level differences in neuropsychological test performance were found in composite measures of frontal function using the Frontal Assessment Battery, working memory (Digit Span backward), and short-term verbal and episodic memory using the Rey Auditory Verbal Learning Test (RAVLT)–Immediate and Delayed Recall, and the Free and Cued Selective Reminding Test (FCSRT)–Immediate Free Recall. Moreover, while total hippocampal volume on brain MRI was not found to be different, using DTI, the authors reported several alterations, including increased mean diffusivity (MD) values in bilateral hippocampi among ET patients as compared to controls, with hippocampal MD correlated with memory scores in ET. Specifically, greater mean diffusivity, reflective of reduced white matter integrity, was correlated with worse recall performance on RAVLT and FCSRT. In summary, these findings suggest that subtle microstructural hippocampal deficits are related to memory function in ET.

Prasad and colleagues also analyzed hippocampal volume in ET across 12 hippocampal subfields in 40 ET cases and 40 non-ET controls. Generally speaking, the subfields most affected in ET versus non-ET controls were parasubiculum, the granular cell layer of the dentate gyrus, the hippocampal amygdala transition area, and hippocampal fissure and tail regions. The ET group was further subdivided into ET with and without cognitive impairment (25 and 15 cases, respectively) based upon having three or more neuropsychological test scores 1.5 SD below the normative means. Between these two groups, there was larger fimbria volume in the ET without cognitive impairment group, compared to ET with cognitive impairment and also relative to non-ET healthy controls. Among individuals with ET, there were positive correlations between fimbria volume and verbal memory, and negative correlations between hippocampal tail volume and MMSE, Wisconsin Card Sort and Wechsler Memory Scale spatial span test performance (Prasad, Shah, Bhalsing, et al., 2019). These findings, however, were not compared to individuals in a non-ET group, so it remains unknown whether these MR biomarker correlates of cognitive function are specific to ET-related cognitive impairment.

The involvement of the hippocampus in ET is not equivalent to comorbid AD. Indeed, Prasad, Shah, Bhalsing, Ingalhalikar, et al. (2019) speculated that among the ET sample, tremor may perturb feedback regarding self-motion, leading to adaptive re-organization and increased demands on relevant structures including the retrosplenial cortex which provides input to the parasubiculum of the hippocampus. Moreover, there are numerous age-related conditions which include hippocampal compromise (Huang, Chen, Hu, & Guo, 2020; Pletnikova, Sloane, Renton, et al., 2014; Woodworth et al., 2021). However, when considered in light of the increased AD diagnoses in the Central Spain and Manhattan studies, there is a compelling argument to be made for co-morbid AD as a basis for increased cognitive impairment in ET. In fact, though amyloid levels are not always found to be higher in ET brains, Beliveau and colleagues (2015) quantified soluble and insoluble amyloid beta in cerebellar and parietal cortices of 9 ET brains, 16 control brains, and 10 PD brains and found that levels of insoluble amyloid beta were "many fold higher" in the cerebellum and parietal cortex of ET brains as compared to PD and control brains (Béliveau, Tremblay, Aubry-Lafontaine, et al., 2015). In contrast, these groups had similar levels of phosphorylated tau. A drawback of this study was that although it was reported that none of the brains met neuropathological criteria for AD, the cognitive status of participants during life was not known. Ideally, postmortem examinations will compare brains matched for cognitive status to ensure that, for example, the control group does not include only cognitively healthy elders whereas the ET group includes individuals with MCI and dementia (or vice versa).

10. Summary and conclusions

In total, the weight of the evidence examined here in both population-based settings and dedicated cohort studies, suggest impaired cognitive performance and longitudinal cognitive changes over time. In large, well-controlled epidemiologic studies, ET has been clearly demonstrated to be associated with increased incidence and prevalence of dementia. We also reviewed several neuroimaging studies which demonstrate associations between cognitive changes and structural neuroimaging changes in both hippocampal regions as well as other cerebral cortical and subcortical regions, while data on cerebellar structural changes in relation to cognitive changes are still lacking. The emerging neuropathologic data on ET brains also point toward considerable tau-related pathology in ET brains, with potentially different isoforms of pathologic tau in comparison to other tauopathies such as AD, FTD and PSP, the significance of which has still yet to be understood. The range of evidence examined suggests possible heterogeneity in the pathogenesis of cognitive deficits and dementia in ET. Contributions from cerebellar pathology and functional derangement as well evidence supporting a strong basis for ET as a possible trigger for accrual of additional neuropathologic changes in non-cerebellar regions sufficient to cause the increased risk of dementia seen at a population level.

Based on the current state of knowledge, it is likely the case that individuals with ET are at increased risk for cognitive impairment and dementia based on heterogeneous, and potentially additive or synergistic brain changes. This perspective on cognitive impairment in ET is in line with the idea that ET may be a family of diseases, defined at their core by kinetic tremor of the arms, and perhaps more accurately termed *the essential tremors*. Viewed through such a lens, we would expect significant heterogeneity in the motor as well as cognitive features of ET, and we would anticipate that certain presentations of ET might have little to no effect on cognition, whereas other presentations of ET may evolve to include a dementia syndrome. Indeed, while ET is not universally characterized by significant cognitive deficits; the bulk of data from epidemiological, cognitive, and postmortem studies converge to reveal an increased risk for cognitive impairment and dementia among individuals with ET.

11. Clinical relevance and future directions

Prevalence and severity of ET increase with aging (Louis & Ferreira, 2010; Louis, Gerbin, & Galecki, 2013), as does the prevalence of AD and other neurodegenerative diseases which may be comorbid with ET. Understanding cognitive impairment in ET will thus have greater and greater public health impact as the aging population increases. Additionally, recognizing these issues has important implications for quality of life among individuals living with ET who may experience impaired instrumental activities of daily living as a function of tremor and of cognitive impairment. Notably, in comparison to non-ET controls (Louis, Benito-León, Vega-Quiroga, & Bermejo-Pareja, 2010b), older adults with ET reported greater difficulty in six cognitive-only daily functional activities (e.g., shopping or going outside) as well as five daily functional activities that combine cognitive and motor (having steady hands). As might be expected, worse global cognitive score, as measured by MMSE, was correlated with worse functional impairment in individuals with ET. It is thus clear that cognitive impairment in ET adds to the burden of disease, and future studies are needed to flesh out the everyday relevance of specific neuropsychological test performance deficits observed in ET. Moving forward, it will be particularly useful to determine the extent to which routine assessment with the CCAS scale is feasible, clinically informative, and how it compares to screening instruments such as the MMSE and Montreal Cognitive Assessment.

Another critical issue is that although the evidence supports cognitive deficits in ET, a prevailing perception and teaching among clinicians has been that ET is a monosymptomatic motor-only disorder. The key features and concepts taught consist of an isolated tremor syndrome of bilateral upper limb action tremor, typically consisting of a postural and action tremor of the hands, head and voice with additional other motor features allowable (Bhatia, Bain, Bajaj, et al., 2018). Clinically apparent cognitive impairment has not been widely taught or recognized as a common feature associated with ET. There may be several reasons for this. First, most neurologists caring for ET patients frequently see PD patients, for whom the risk for cognitive impairment including MCI and dementia is demonstrably higher. The longer time course of care under the neurologist for a PD patient may reveal more readily apparent patterns of progression in cognitive impairment during face-to-face encounters over years and decades long follow-up. Cognitive impairment may also not be widely recognized as a feature of ET because of challenges detecting and recognizing deficits by patients, families, and clinicians: (1) lack of patient awareness, concern or self-report; (2) impairment in certain cognitive domains outside of memory may be less noticeable by the patient or families; and (3) the presence of potentially confounding factors such as age, mood and presence of medication that may impair cognition may be perceived to account for cognitive deficits. These issues are relevant for clinicians caring for patients with ET who need to discern whether their patients need additional evaluation, care or supervision.

With regard to future directions for research, studies are needed to identify risk factors for developing cognitive impairment in ET as well as predictors of specific cognitive trajectories (i.e., static, mildly progressive, rapidly progressive). It is clear that the cognitive profiles of ET are heterogenous. A natural question is whether or not the motor features of ET correlate with the cognitive features. To date, examination of cognitive variability among individuals with ET has generally suggested that tremor characteristics (i.e., total tremor score, duration and disability; presence of voice or head tremor) are not associated with cognitive scores (Cersonsky et al., 2020). In contrast, confidence in balance has been reported to be lower among those individuals with relatively weak memory scores compared to those whose cognition was consistently high. Moreover, emerging work is linking cognitive function to tandem gait (Peterson et al., 2021). Examination of motor-cognitive correlations across different cognitive groups (i.e., normal cognition, MCI, dementia) will help to further clarify the circumstances under which cognitive abilities may covary with different motor features (Fig. 1).

Non-motor correlates of cognitive impairment include later age of onset and possibly family history (Bermejo-Pareja et al., 2007). It is noteworthy perhaps that, at odds with this are studies of individuals seeking neurosurgical treatment of severe ET, who are much more likely to have earlier age of onset and longer tremor duration (Louis et al., 2013) but worse cognitive function. In one series of 201 individuals seeking evaluation for deep brain stimulation (DBS) for medication-refractory ET, 27% of patients were identified with MCI based upon a formal neuropsychological battery carried out prior to deep brain stimulation (DBS) surgery (Ratajska, Lopez, Kenney, et al., 2021). In another clinical cohort of 101 consecutively seen individuals with medication-refractory ET referred for thalamotomy or thalamic DBS, 34–60% of individuals with ET were found to have cognitive deficits, scoring greater than one standard deviation below the normative mean on measures of attention, verbal fluency, word list recall, semantic encoding and facial matching. There is little information on the long-term cognitive outcomes of individuals with ET who undergo thalamotomy or thalamic DBS, but short-term cognitive outcomes have been thought to be generally unchanged, with the exception of verbal fluency declines (Tröster, Fields, Pahwa, et al., 1999). Additional studies examining the association between long-term cognitive outcomes following stereotactic brain surgery for ET and presurgical cognitive status and imaging characteristics may elucidate the relative contributions of disease state to post-surgical cognitive changes.

In addition to work clarifying the correlates of cognitive impairment in ET, work is needed to identify early predictors of cognitive decline. Establishing the predictors of different cognitive outcomes will be important from a clinical prognosis and counseling standpoint, as well as from a treatment standpoint. A recent review addressing gaps in knowledge regarding cognitive change in ET details a range of demographic, genetic, psychiatric, physical, sensory, and other variables which may be relevant for predicting cognitive trajectories in ET based on their predictive utility in the general population of older adults as well as among individuals with PD (Louis, Joyce, & Cosentino, 2019). The COGNET study was designed to provide the first prospective, longitudinal characterization of cognitive function in ET and to examine both the neuropathological correlates as well as predictors of cognitive impairment.[132] Ongoing work is examining the extent to which mood, lifestyle, and motor features have implications for future cognitive course. In addition to establishing the anatomic and pathologic basis of cognitive heterogeneity in ET, prospective, longitudinal studies will provide much needed information regarding heterogeneity in the course of cognitive change over time.

With regard to cognitive course in ET, further study is particularly needed to better understand the temporal dynamics of MCI. Notably, in contrast to higher rates of both prevalent and incident dementia in ET, MCI prevalence was not higher among ET than non-ET cases in either the Central Spain or Manhattan cohorts. Considering age of onset in the Central Spain cohort did show that while rates of MCI were similar to controls overall (OR 1.32, 95% CI 0.93–1.89), MCI rates were higher in older-onset ET compared to controls (OR 1.57, 95% CI 1.03–2.37) in adjusted models (age, gender, educational level, smoking, ethanol consumption, diabetes, hypertension, heart disease, depressive symptoms, and use of antidepressants and medications that could potentially affect cognition).

The Manhattan cohort also did not show increased incident MCI, either with or without memory impairment. Furthermore, age of onset information was not collected, therefore, a subgroup analysis similar to the Central Spain cohort could not be done. One issue, particularly in the Manhattan cohort, was the low number of ET cases, such that the number of incident MCI cases numbered in the single digits. Another consideration is that it is not clear whether individuals with ET may possibly spend less time in a pre-dementia state of milder impairment but progress more quickly to a dementia than their non-ET counterparts. Annual conversion rate from MCI to dementia among individuals with ET has been reported to be approximately 12.5%,[48,132] higher than the 2.6–6.3% conversion rate described in historical control cohorts. This finding may in part explain the weak association between ET and MCI observed in the Central Spain and Manhattan cohorts. Questions surrounding rates of MCI in ET remain to be further clarified through other longitudinal studies of cognition in both ET and non-ET individuals. However, given the known prevalence of ET, this would require either a large population-based sample followed longitudinally or a dedicated clinic-based cohort to efficiently recruit enough individuals at risk for MCI or dementia.

In closing, a clearer conceptualization of the cognitive side of ET will require investigation of specific cognitive deficits, diagnoses, and trajectories alongside mapping of cerebellar and extra-cerebellar neuroanatomic changes measured via structural and functional imaging, in vivo biomarkers of neurodegenerative pathologies disease including amyloid, tau, and dopamine transporter (DaT) scans, and postmortem neuropathologic examination. As outlined in Table 1, characterization of cognitive function and neuroanatomic findings along at least three dimensions will advance the field's understanding of when, how and why cognitive changes occur in ET. In the first respect, it is important to consider the nature of cognitive impairment in a given individual, characterizing the level of severity, domains involved and temporal course. We then need to ask: How do these features map onto different neuroanatomic signatures? It will be necessary to examine the specific proteinopathies in the brains of individuals with ET, achieved through molecular imaging or neuropathologic examination to complete a second piece of the puzzle. Finally, it is critical to understand if the neuroanatomic changes are restricted to the cerebellum, or if they extend to either cortical or subcortical regions of the brain. Moreover, if the pathologic changes occur in cerebellar tissue only, it is critical to understand whether such changes are located in the posterior versus anterior lobe.

Table 1 Dimensions and mechanisms of cognitive impairment in ET.

	I. Nature of cognitive impairment			
II. Type of neuropathology	**Clinical status (NC, MCI, dementia)**	**Course (static, progressive)**	**Domains**	**Not available**
A. ET-specific				
1. Tau*				
2. Other yet unknown				
B. Extra-cerebellar				
1. Amyloid				
2. Synuclein				
3. Tau*				
4. Other				
C. Not available				
III. Location of neuropathology				
A. Cerebellar				
1. Anterior				
2. Posterior				
B. Extra-cerebellar				
1. Cortical				
2. Subcortical				
C. Not available				

Note: This table outlines three dimensions along which existing studies can be coded, and future studies can be designed, to organize and advance understanding of the mechanisms underlying cognitive impairment in ET. For display purposes, Location and Type of Neuropathology are presented. A given study could in theory fill a single cell addressing all three dimensions of the framework such that, for example, an investigation may determine that among individuals with Mild Cognitive Impairment (Dimension I), cognitive performance is linked to tauopathic changes (Dimension II) in cortical regions (Dimension III). *Tau is listed under ET-specific pathology as well as ET Non-Specific given its various subspecies which may be found uniquely in ET, or also in AD, PSP, FTD.

Within this schema, it would also be important to determine the extent to which cognitive deficits are attributable to focal cerebellar changes, or to the distal effects of such changes on areas such as the prefrontal cortex, for example. That is, are cognitive changes in ET part and parcel of ET, or evidence of a comorbid disease? Taken together, information acquired along each of these dimensions will answer fundamental questions about the cognitive side of ET.

References

Agosta, F., Henry, R. G., Migliaccio, R., et al. (2010). Language networks in semantic dementia. *Brain, 133*(Pt. 1), 286–299. https://doi.org/10.1093/brain/awp233.

Axelrad, J. E., Louis, E. D., Honig, L. S., et al. (2008). Reduced Purkinje cell number in essential tremor. *Archives of Neurology, 65*(1), 101–107. https://doi.org/10.1001/archneurol.2007.8.

Azar, M., Bertrand, E., Louis, E. D., et al. (2017). Awareness of cognitive impairment in individuals with essential tremor. *Journal of the Neurological Sciences, 377*, 155–160. https://doi.org/10.1016/j.jns.2017.04.009.

Babij, R., Lee, M., Cortés, E., Vonsattel, J. P. G., Faust, P. L., & Louis, E. D. (2013). Purkinje cell axonal anatomy: Quantifying morphometric changes in essential tremor versus control brains. *Brain, 136*(10), 3051–3061. https://doi.org/10.1093/brain/awt238.

Bain, P. G., Findley, L. J., Thompson, P. D., et al. (1994). A study of hereditary essential tremor. *Brain, 117*(4), 805–824. https://doi.org/10.1093/brain/117.4.805.

Balsters, J. H., & Ramnani, N. (2011). Cerebellar plasticity and the automation of first-order rules. *The Journal of Neuroscience, 31*(6), 2305–2312. https://doi.org/10.1523/JNEUROSCI.4358-10.2011.

Béliveau, E., Tremblay, C., Aubry-Lafontaine, É., et al. (2015). Accumulation of amyloid-β in the cerebellar cortex of essential tremor patients. *Neurobiology of Disease, 82*, 397–408. https://doi.org/10.1016/j.nbd.2015.07.016.

Benito-Leon, J., & Louis, E. D. (2006). Essential tremor: Emerging views of a common disorder. *Nature Clinical Practice. Neurology, 2*(12), 666–679. https://doi.org/10.1038/ncpneuro0347.

Benito-Leon, J., Louis, E. D., & Bermejo-Pareja, F. (2006a). On behalf of the Neurological Disorders in Central Spain (NEDICES) Study Group*. Population-based case-control study of cognitive function in essential tremor. *Neurology, 66*(1), 69–74. https://doi.org/10.1212/01.wnl.0000192393.05850.ec.

Benito-Leon, J., Louis, E. D., & Bermejo-Pareja, F. (2006b). On behalf of the Neurological Disorders in Central Spain (NEDICES) Study Group. Elderly-onset essential tremor is associated with dementia. *Neurology, 66*(10), 1500–1505. https://doi.org/10.1212/01.wnl.0000216134.88617.de.

Benito-León, J., Louis, E. D., Mitchell, A. J., & Bermejo-Pareja, F. (2011). Elderly-onset essential tremor and mild cognitive impairment: A population-based study (NEDICES). *Journal of Alzheimer's Disease, 23*(4), 727–735. https://doi.org/10.3233/JAD-2011-101572.

Bermejo-Pareja, F., Louis, E. D., & Benito-León, J. (2007). Risk of incident dementia in essential tremor: A population-based study. *Movement Disorders, 22*(11), 1573–1580. https://doi.org/10.1002/mds.21553.

Bernard, J. A., & Seidler, R. D. (2013). Relationships between regional cerebellar volume and sensorimotor and cognitive function in young and older adults. *Cerebellum, 12*(5), 721–737. https://doi.org/10.1007/s12311-013-0481-z.

Bernard, J. A., & Seidler, R. D. (2014). Moving forward: Age effects on the cerebellum underlie cognitive and motor declines. *Neuroscience and Biobehavioral Reviews, 42*, 193–207. https://doi.org/10.1016/j.neubiorev.2014.02.011.

Bertrand, E., Azar, M., Rizvi, B., et al. (2018). Cortical thickness and metacognition in cognitively diverse older adults. *Neuropsychology, 32*(6), 700–710. https://doi.org/10.1037/neu0000458.

Bhalsing, K. S., Upadhyay, N., Kumar, K. J., et al. (2014). Association between cortical volume loss and cognitive impairments in essential tremor. *European Journal of Neurology*, *21*(6), 874–883. https://doi.org/10.1111/ene.12399.

Bhatia, K. P., Bain, P., Bajaj, N., et al. (2018). Consensus statement on the classification of tremors. From the task force on tremor of the International Parkinson and Movement Disorder Society: IPMDS Task Force on Tremor Consensus Statement. *Movement Disorders*, *33*(1), 75–87. https://doi.org/10.1002/mds.27121.

Bower, J. H., Maraganore, D. M., McDonnell, S. K., & Rocca, W. A. (1997). Incidence of progressive supranuclear palsy and multiple system atrophy in Olmsted County, Minnesota, 1976 to 1990. *Neurology*, *49*(5), 1284–1288. https://doi.org/10.1212/wnl.49.5.1284.

Braak, H., Rüb, U., Schultz, C., & Del Tredici, K. (2006). Vulnerability of cortical neurons to Alzheimer's and Parkinson's diseases. *Journal of Alzheimer's Disease*, *9*(3 Suppl), 35–44. https://doi.org/10.3233/jad-2006-9s305.

Cersonsky, T. E. K., Kellner, S., Chapman, S., Huey, E. D., Louis, E. D., & Cosentino, S. (2020). Profiles of normal cognition in essential tremor. *Journal of the International Neuropsychological Society*, *26*(2), 197–209. https://doi.org/10.1017/S1355617719001140.

Choe, M., Cortés, E., Vonsattel, J. P. G., Kuo, S. H., Faust, P. L., & Louis, E. D. (2016). Purkinje cell loss in essential tremor: Random sampling quantification and nearest neighbor analysis. *Movement Disorders*, *31*(3), 393–401. https://doi.org/10.1002/mds.26490.

Clark, L. N., Liu, X., Parmalee, N. L., Hernandez, N., & Louis, E. D. (2014). The microtubule associated protein tau H1 haplotype and risk of essential tremor. *European Journal of Neurology*, *21*(7), 1044–1048. https://doi.org/10.1111/ene.12335.

Collins, K., Rohl, B., Morgan, S., Huey, E. D., Louis, E. D., & Cosentino, S. (2017). Mild cognitive impairment subtypes in a cohort of elderly essential tremor cases. *Journal of the International Neuropsychological Society*, *23*(5), 390–399. https://doi.org/10.1017/S1355617717000170.

Cosentino, S., Brickman, A. M., Griffith, E., et al. (2015). The right insula contributes to memory awareness in cognitively diverse older adults. *Neuropsychologia*, *75*, 163–169. https://doi.org/10.1016/j.neuropsychologia.2015.05.032.

Cosentino, S., Metcalfe, J., Cary, M. S., De Leon, J., & Karlawish, J. (2011). Memory awareness influences everyday decision making capacity about medication management in Alzheimer's disease. *International Journal of Alzheimer's Disease*, *2011*, 483897. https://doi.org/10.4061/2011/483897.

Crary, J. F., Trojanowski, J. Q., Schneider, J. A., et al. (2014). Primary age-related tauopathy (PART): A common pathology associated with human aging. *Acta Neuropathologica*, *128*(6), 755–766. https://doi.org/10.1007/s00401-014-1349-0.

DeJesus-Hernandez, M., Mackenzie, I. R., Boeve, B. F., et al. (2011). Expanded GGGGCC hexanucleotide repeat in non-coding region of C9ORF72 causes chromosome 9p-linked frontotemporal dementia and amyotrophic lateral sclerosis. *Neuron*, *72*(2), 245–256. https://doi.org/10.1016/j.neuron.2011.09.011.

Duane, D. D., & Vermilion, K. J. (2002). Cognitive deficits in patients with essential tremor. *Neurology*, *58*(11), 1706–1707. https://doi.org/10.1212/WNL.58.11.1706.

Filip, P., Lungu, O. V., Manto, M. U., & Bareš, M. (2016). Linking essential tremor to the cerebellum: Physiological evidence. *Cerebellum*, *15*(6), 774–780. https://doi.org/10.1007/s12311-015-0740-2.

Frisina, P. G., Tse, W., Hälbig, T. D., & Libow, L. S. (2009). The pattern of cognitive-functional decline in elderly essential tremor patients: An exploratory-comparative study with Parkinson's and Alzheimer's disease patients. *Journal of the American Medical Directors Association*, *10*(4), 238–242. https://doi.org/10.1016/j.jamda.2008.10.013.

García-Martín, E., Martínez, C., Alonso-Navarro, H., et al. (2012). H1-MAPT and the risk for familial essential tremor. *PLoS One*, 7(7), e41581. https://doi.org/10.1371/journal.pone.0041581.

Gasparini, M., Bonifati, V., Fabrizio, E., et al. (2001). Frontal lobe dysfunction in essential tremor. *Journal of Neurology*, *248*(5), 399–402. https://doi.org/10.1007/s004150170181.

Giasson, B. I., Forman, M. S., Higuchi, M., et al. (2003). Initiation and synergistic fibrillization of tau and alpha-synuclein. *Science*, *300*(5619), 636–640. https://doi.org/10.1126/science.1082324.

Gidalevitz, T., Ben-Zvi, A., Ho, K. H., Brignull, H. R., & Morimoto, R. I. (2006). Progressive disruption of cellular protein folding in models of polyglutamine diseases. *Science*, *311*(5766), 1471–1474. https://doi.org/10.1126/science.1124514.

Gilman, S., Wenning, G. K., Low, P. A., et al. (2008). Second consensus statement on the diagnosis of multiple system atrophy. *Neurology*, *71*(9), 670–676. https://doi.org/10.1212/01.wnl.0000324625.00404.15.

Habeck, C., & Stern, Y. (2010). Alzheimer's disease neuroimaging initiative. Multivariate data analysis for neuroimaging data: Overview and application to Alzheimer's disease. *Cell Biochemistry and Biophysics*, *58*(2), 53–67. https://doi.org/10.1007/s12013-010-9093-0.

Hall, D. A., & O'keefe, J. A. (2012). Fragile X-associated tremor ataxia syndrome: The expanding clinical picture, pathophysiology, epidemiology, and update on treatment. *Tremor and Other Hyperkinetic Movements (New York, NY)*, *2*. tre-02-56-352-1.

Hanajima, R., Tsutsumi, R., Shirota, Y., Shimizu, T., Tanaka, N., & Ugawa, Y. (2016). Cerebellar dysfunction in essential tremor. *Movement Disorders*, *31*(8), 1230–1234. https://doi.org/10.1002/mds.26629.

Higginson Christopher, I., Wheelock Vicki, L., Levine, D., King David, S., Pappas Conrad, T. E., & Sigvardt, K. A. (2008). Cognitive deficits in essential tremor consistent with frontosubcortical dysfunction. *Journal of Clinical and Experimental Neuropsychology*, *30*(7), 760–765. https://doi.org/10.1080/13803390701754738.

Hoche, F., Guell, X., Vangel, M. G., Sherman, J. C., & Schmahmann, J. D. (2018). The cerebellar cognitive affective/Schmahmann syndrome scale. *Brain*, *141*(1), 248–270. https://doi.org/10.1093/brain/awx317.

Holtbernd, F., & Shah, N. J. (2021). Imaging the pathophysiology of essential tremor—A systematic review. *Frontiers in Neurology*, *12*, 680254. https://doi.org/10.3389/fneur.2021.680254.

Huang, L., Chen, K., Hu, X., & Guo, Q. (2020). Differential atrophy in the hippocampal subfield volumes in four types of mild dementia. *Frontiers in Neuroscience*, *14*, 699. https://doi.org/10.3389/fnins.2020.00699.

Huang, H., Yang, X., Zhao, Q., et al. (2020). Clinical characteristics of patients with essential tremor or essential tremor plus. *Acta Neurologica Scandinavica*, *141*(4), 335–341. https://doi.org/10.1111/ane.13209.

Iglesias-Hernandez, D., Delgado, N., McGurn, M., Huey, E. D., Cosentino, S., & Louis, E. D. (2021). "ET plus": Instability of the diagnosis during prospective longitudinal follow-up of essential tremor cases. *Frontiers in Neurology*, *12*, 782694. https://doi.org/10.3389/fneur.2021.782694.

Jack, C. R., Bennett, D. A., Blennow, K., et al. (2016). A/T/N: An unbiased descriptive classification scheme for Alzheimer disease biomarkers. *Neurology*, *87*(5), 539–547. https://doi.org/10.1212/WNL.0000000000002923.

Jack, C. R., Bennett, D. A., Blennow, K., et al. (2018). NIA-AA research framework: Toward a biological definition of Alzheimer's disease. *Alzheimer's & Dementia*, *14*(4), 535–562. https://doi.org/10.1016/j.jalz.2018.02.018.

Jack, C. R., Knopman, D. S., Weigand, S. D., et al. (2012). An operational approach to National Institute on Aging-Alzheimer's Association criteria for preclinical Alzheimer disease. *Annals of Neurology*, *71*(6), 765–775. https://doi.org/10.1002/ana.22628.

Jain, S., Lo, S. E., & Louis, E. D. (2006). Common misdiagnosis of a common neurological disorder: How are we misdiagnosing essential tremor? *Archives of Neurology*, *63*(8), 1100. https://doi.org/10.1001/archneur.63.8.1100.

Janicki, S. C., Cosentino, S., & Louis, E. D. (2013). The cognitive side of essential tremor: What are the therapeutic implications? *Therapeutic Advances in Neurological Disorders*, *6*(6), 353–368. https://doi.org/10.1177/1756285613489591.

Jokinen, H., Ryberg, C., Kalska, H., et al. (2007). Corpus callosum atrophy is associated with mental slowing and executive deficits in subjects with age-related white matter hyperintensities: The LADIS study. *Journal of Neurology, Neurosurgery, and Psychiatry*, *78*(5), 491–496. https://doi.org/10.1136/jnnp.2006.096792.

Josephs, K. A., Matsumoto, J. Y., & Ahlskog, J. E. (2006). Benign tremulous parkinsonism. *Archives of Neurology*, *63*(3), 354–357. https://doi.org/10.1001/archneur.63.3.354.

Kelleher, M., Tolea, M. I., & Galvin, J. E. (2016). Anosognosia increases caregiver burden in mild cognitive impairment. *International Journal of Geriatric Psychiatry*, *31*(7), 799–808. https://doi.org/10.1002/gps.4394.

Kim, S. H., Farrell, K., Cosentino, S., et al. (2021). Tau isoform profile in essential tremor diverges from other tauopathies. *Journal of Neuropathology and Experimental Neurology*, *80*(9), 835–843. https://doi.org/10.1093/jnen/nlab073.

Köllensperger, M., Geser, F., Seppi, K., et al. (2008). Red flags for multiple system atrophy. *Movement Disorders*, *23*(8), 1093–1099. https://doi.org/10.1002/mds.21992.

Koziol, L. F., Budding, D., Andreasen, N., et al. (2014). Consensus paper: The cerebellum's role in movement and cognition. *Cerebellum*, *13*(1), 151–177. https://doi.org/10.1007/s12311-013-0511-x.

Kraus, P. H., Lemke, M. R., & Reichmann, H. (2006). Kinetic tremor in Parkinson's disease—An underrated symptom. *Journal of Neural Transmission (Vienna)*, *113*(7), 845–853. https://doi.org/10.1007/s00702-005-0354-9.

Lacritz, L. H., Dewey, R., Giller, C., & Cullum, C. M. (2002). Cognitive functioning in individuals with "benign" essential tremor. *Journal of the International Neuropsychological Society*, *8*(1), 125–129. https://doi.org/10.1017/S1355617702001121.

Lee, S. M., Kim, M., Lee, H. M., Kwon, K. Y., & Koh, S. B. (2015). Nonmotor symptoms in essential tremor: Comparison with Parkinson's disease and normal control. *Journal of the Neurological Sciences*, *349*(1–2), 168–173. https://doi.org/10.1016/j.jns.2015.01.012.

Lee, S., Zimmerman, M. E., Narkhede, A., et al. (2018). White matter hyperintensities and the mediating role of cerebral amyloid angiopathy in dominantly-inherited Alzheimer's disease. *PLoS One*, *13*(5), e0195838. https://doi.org/10.1371/journal.pone.0195838.

Lombardi, W. J., Woolston, D. J., Roberts, J. W., & Gross, R. E. (2001). Cognitive deficits in patients with essential tremor. *Neurology*, *57*(5), 785–790. https://doi.org/10.1212/wnl.57.5.785.

Louis, E. D. (2014). "Essential tremor" or "the essential tremors": Is this one disease or a family of diseases? *Neuroepidemiology*, *42*(2), 81–89. https://doi.org/10.1159/000356351.

Louis, E. D., Bares, M., Benito-Leon, J., et al. (2020). Essential tremor-plus: A controversial new concept. *Lancet Neurology*, *19*(3), 266–270. https://doi.org/10.1016/S1474-4422(19)30398-9.

Louis, E. D., Benito-León, J., Vega-Quiroga, S., & Bermejo-Pareja, F. (2010a). Faster rate of cognitive decline in essential tremor cases than controls: A prospective study. *European Journal of Neurology*, *17*(10), 1291–1297. https://doi.org/10.1111/j.1468-1331.2010.03122.x.

Louis, E. D., Benito-León, J., Vega-Quiroga, S., & Bermejo-Pareja, F. (2010b). Group TND in CS (NEDICES) S. Cognitive and motor functional activity in non-demented community-dwelling essential tremor cases. *Journal of Neurology, Neurosurgery & Psychiatry*, *81*(9), 997–1001. https://doi.org/10.1136/jnnp.2009.202838.

Louis, E. D., & Faust, P. L. (2020). Essential tremor: The most common form of cerebellar degeneration? *Cerebellum Ataxias*, 7(1), 12. https://doi.org/10.1186/s40673-020-00121-1.

Louis, E. D., Faust, P. L., Vonsattel, J. P. G., et al. (2007). Neuropathological changes in essential tremor: 33 cases compared with 21 controls. *Brain*, *130*(12), 3297–3307. https://doi.org/10.1093/brain/awm266.

Louis, E. D., & Ferreira, J. J. (2010). How common is the most common adult movement disorder? Update on the worldwide prevalence of essential tremor. *Movement Disorders, 25*(5), 534–541. https://doi.org/10.1002/mds.22838.

Louis, E. D., Ford, B., Frucht, S., Barnes, L. F., X-Tang, M., & Ottman, R. (2001). Risk of tremor and impairment from tremor in relatives of patients with essential tremor: A community-based family study. *Annals of Neurology, 49*(6), 761–769. https://doi.org/10.1002/ana.1022.

Louis, E. D., Gerbin, M., & Galecki, M. (2013). Essential tremor 10, 20, 30, 40: Clinical snapshots of the disease by decade of duration. *European Journal of Neurology, 20*(6), 949–954. https://doi.org/10.1111/ene.12123.

Louis, E. D., Huey, E. D., & Cosentino, S. (2021). Features of "ET plus" correlate with age and tremor duration: "ET plus" may be a disease stage rather than a subtype of essential tremor. *Parkinsonism & Related Disorders, 91*, 42–47. https://doi.org/10.1016/j.parkreldis.2021.08.017.

Louis, E. D., Joyce, J. L., & Cosentino, S. (2019). Mind the gaps: What we don't know about cognitive impairment in essential tremor. *Parkinsonism & Related Disorders, 63*, 10–19. https://doi.org/10.1016/j.parkreldis.2019.02.038.

Louis, E. D., Kerridge, C. A., Chatterjee, D., et al. (2019). Contextualizing the pathology in the essential tremor cerebellar cortex: A patholog-omics approach. *Acta Neuropathologica, 138*(5), 859–876. https://doi.org/10.1007/s00401-019-02043-7.

Louis, E. D., Lee, M., Babij, R., et al. (2014). Reduced Purkinje cell dendritic arborization and loss of dendritic spines in essential tremor. *Brain, 137*(12), 3142–3148. https://doi.org/10.1093/brain/awu314.

Louis, E. D., & McCreary, M. (2021). How common is essential tremor? Update on the worldwide prevalence of essential tremor. *Tremor and Other Hyperkinetic Movements, 11*(1), 28. https://doi.org/10.5334/tohm.632.

Louis, E. D., & Ottman, R. (2014). How many people in the USA have essential tremor? Deriving a population estimate based on epidemiological data. *Tremor and Other Hyperkinetic Movements (New York, NY), 4*. https://doi.org/10.7916/D8TT4P4B.

Louis, E. D., Radler, K. H., Huey, E. D., & Cosentino, S. (2021). Progression to dementia in patients with essential tremor. *Parkinsonism & Related Disorders, 83*, 123–124. https://doi.org/10.1016/j.parkreldis.2020.12.011.

Luo, R., Pan, P., Xu, Y., & Chen, L. (2019). No reliable gray matter changes in essential tremor. *Neurological Sciences, 40*(10), 2051–2063. https://doi.org/10.1007/s10072-019-03933-0.

Mameli, F., Tomasini, E., Scelzo, E., et al. (2013). Lies tell the truth about cognitive dysfunction in essential tremor: An experimental deception study with the guilty knowledge task. *Journal of Neurology, Neurosurgery, and Psychiatry, 84*(9), 1008–1013. https://doi.org/10.1136/jnnp-2012-304674.

Manca, R., Stabile, M. R., Bevilacqua, F., et al. (2019). Cognitive speed and white matter integrity in secondary progressive multiple sclerosis. *Multiple Sclerosis and Related Disorders, 30*, 198–207. https://doi.org/10.1016/j.msard.2019.02.021.

Marras, C., Beck, J. C., Bower, J. H., et al. (2018). Prevalence of Parkinson's disease across North America. *NPJ Parkinson's Disease, 4*. https://doi.org/10.1038/s41531-018-0058-0.

Miki, Y., Foti, S. C., Asi, Y. T., et al. (2019). Improving diagnostic accuracy of multiple system atrophy: A clinicopathological study. *Brain, 142*(9), 2813–2827. https://doi.org/10.1093/brain/awz189.

Miller, B. L., Seeley, W. W., Mychack, P., Rosen, H. J., Mena, I., & Boone, K. (2001). Neuroanatomy of the self: Evidence from patients with frontotemporal dementia. *Neurology, 57*(5), 817–821. https://doi.org/10.1212/wnl.57.5.817.

Montero-Odasso, M., Pieruccini-Faria, F., Bartha, R., et al. (2017). Motor phenotype in neurodegenerative disorders: Gait and balance platform study design protocol for the

Ontario neurodegenerative research initiative (ONDRI). *Journal of Alzheimer's Disease, 59*(2), 707–721. https://doi.org/10.3233/JAD-170149.

Novellino, F., Vasta, R., Saccà, V., et al. (2020). Hippocampal impairment in patients with essential tremor. *Parkinsonism & Related Disorders, 72*, 56–61. https://doi.org/10.1016/j.parkreldis.2020.02.006.

Pan, J. J., Lee, M., Honig, L. S., Vonsattel, J. P. G., Faust, P. L., & Louis, E. D. (2014). Alzheimer's-related changes in non-demented essential tremor patients vs. controls: Links between tau and tremor? *Parkinsonism & Related Disorders, 20*(6), 655–658. https://doi.org/10.1016/j.parkreldis.2014.03.003.

Park, I. S., Oh, Y. S., Lee, K. S., et al. (2015). Subtype of mild cognitive impairment in elderly patients with essential tremor. *Alzheimer Disease and Associated Disorders, 29*(2), 141–145. https://doi.org/10.1097/WAD.0000000000000054.

Passamonti, L., Novellino, F., Cerasa, A., et al. (2011). Altered cortical-cerebellar circuits during verbal working memory in essential tremor. *Brain, 134*(Pt 8), 2274–2286. https://doi.org/10.1093/brain/awr164.

Paul, M. S., & Limaiem, F. (2021). Histology, Purkinje cells. In *StatPearls* StatPearls Publishing. http://www.ncbi.nlm.nih.gov/books/NBK545154/. (Accessed 25 September 2021).

Pauletti, C., Mannarelli, D., De Lucia, M. C., et al. (2015). Selective attentional deficit in essential tremor: Evidence from the attention network test. *Parkinsonism & Related Disorders, 21*(11), 1306–1311. https://doi.org/10.1016/j.parkreldis.2015.08.035.

Pedrosa, D. J., Nelles, C., Maier, F., et al. (2016). Time reproduction deficits in essential tremor patients. *Movement Disorders, 31*(8), 1234–1240. https://doi.org/10.1002/mds.26630.

Peng, J., Wang, L., Li, N., Li, J., Duan, L., & Peng, R. (2020). Distinct non-motor features of essential tremor with head tremor patients. *Acta Neurologica Scandinavica, 142*(1), 74–82. https://doi.org/10.1111/ane.13242.

Peterson, A., Chapman, S., Radler, K., et al. (2021). Tremor features associated with cognition in cognitively normal subjects with essential tremor. *Movement Disorders, 36*, S594. https://doi.org/10.1002/mds.28794.

Pletnikova, O., Sloane, K. L., Renton, A. E., et al. (2014). Hippocampal sclerosis dementia with the C9ORF72 hexanucleotide repeat expansion. *Neurobiology of Aging, 35*(10). https://doi.org/10.1016/j.neurobiolaging.2014.04.009. 2419.e17-21.

Prasad, S., Shah, A., Bhalsing, K. S., Ingalhalikar, M., Saini, J., & Pal, P. K. (2019). Clinical correlates of abnormal subcortical volumes in essential tremor. *Journal of Neural Transmission, 126*(5), 569–576. https://doi.org/10.1007/s00702-019-02004-0.

Prasad, S., Shah, A., Bhalsing, K. S., et al. (2019). Abnormal hippocampal subfields are associated with cognitive impairment in essential tremor. *Journal of Neural Transmission, 126*(5), 597–606. https://doi.org/10.1007/s00702-019-01992-3.

Putzke, J. D., Whaley, N. R., Baba, Y., Wszolek, Z. K., & Uitti, R. J. (2006). Essential tremor: Predictors of disease progression in a clinical cohort. *Journal of Neurology, Neurosurgery, and Psychiatry*, 77(11), 1235–1237. https://doi.org/10.1136/jnnp.2006.086579.

Ratajska, A. M., Lopez, F. V., Kenney, L., et al. (2021). Cognitive subtypes in individuals with essential tremor seeking deep brain stimulation. *The Clinical Neuropsychologist*, 1–23. Published online February 10, 2021 https://doi.org/10.1080/13854046.2021.1882578.

Rodríguez-Aranda, C., Waterloo, K., Johnsen, S. H., et al. (2016). Neuroanatomical correlates of verbal fluency in early Alzheimer's disease and normal aging. *Brain and Language, 155-156*, 24–35. https://doi.org/10.1016/j.bandl.2016.03.001.

Şahin, H. A., Terzi, M., Uçak, S., Yapıcı, O., Başoğlu, T., & Onar, M. (2006). Frontal functions in young patients with essential tremor: A case comparison study. *The Journal of Neuropsychiatry and Clinical Neurosciences, 18*(1), 64–72. https://doi.org/10.1176/jnp.18.1.64.

Sánchez, S. M., Duarte-Abritta, B., Abulafia, C., et al. (2020). White matter fiber density abnormalities in cognitively normal adults at risk for late-onset Alzheimer's disease. *Journal of Psychiatric Research, 122*, 79–87. https://doi.org/10.1016/j.jpsychires.2019.12.019.

Sánchez-Ferro, Á., Benito-León, J., Louis, E. D., et al. (2017). Cognition in non-demented Parkinson's disease vs essential tremor: A population-based study. *Acta Neurologica Scandinavica, 136*(5), 393–400. https://doi.org/10.1111/ane.12752.

Schmahmann, J. D. (1998). Dysmetria of thought: Clinical consequences of cerebellar dysfunction on cognition and affect. *Trends in Cognitive Sciences, 2*(9), 362–371. https://doi.org/10.1016/s1364-6613(98)01218-2.

Schmahmann, J. D. (2000). The role of the cerebellum in affect and psychosis. *Journal of Neurolinguistics, 13*(2–3), 189–214. https://doi.org/10.1016/S0911-6044(00)00011-7.

Schmahmann, J. D. (2019). The cerebellum and cognition. *Neuroscience Letters, 688*, 62–75. https://doi.org/10.1016/j.neulet.2018.07.005.

Schmahmann, J. D., Guell, X., Stoodley, C. J., & Halko, M. A. (2019). The theory and neuroscience of cerebellar cognition. *Annual Review of Neuroscience, 42*(1), 337–364. https://doi.org/10.1146/annurev-neuro-070918-050258.

Schmahmann, J. D., & Sherman, J. C. (1998). The cerebellar cognitive affective syndrome. *Brain, 121*(Pt 4), 561–579. https://doi.org/10.1093/brain/121.4.561.

Schrag, A., Ben-Shlomo, Y., & Quinn, N. (1999). Prevalence of progressive supranuclear palsy and multiple system atrophy: A cross-sectional study. *The Lancet, 354*(9192), 1771–1775. https://doi.org/10.1016/S0140-6736(99)04137-9.

Schrag, A., Münchau, A., Bhatia, K. P., Quinn, N. P., & Marsden, C. D. (2000). Essential tremor: An overdiagnosed condition? *Journal of Neurology, 247*(12), 955–959. https://doi.org/10.1007/s004150070053.

Schwartz, M., Badarny, S., Gofman, S., & Hocherman, S. (1999). Visuomotor performance in patients with essential tremor. *Movement Disorders, 14*(6), 988–993. https://doi.org/10.1002/1531-8257(199911)14:6<988::AID-MDS1012>3.0.CO;2-M.

Selikhova, M., Kempster, P. A., Revesz, T., Holton, J. L., & Lees, A. J. (2013). Neuropathological findings in benign tremulous parkinsonism. *Movement Disorders, 28*(2), 145–152. https://doi.org/10.1002/mds.25220.

Sengul, Y., Sengul, H. S., Yucekaya, S. K., et al. (2015). Cognitive functions, fatigue, depression, anxiety, and sleep disturbances: Assessment of nonmotor features in young patients with essential tremor. *Acta Neurologica Belgica, 115*(3), 281–287. https://doi.org/10.1007/s13760-014-0396-6.

Shaked, D., Sunderaraman, P., Piscitello, J., et al. (2019). Modification of everyday activities and its association with self-awareness in cognitively diverse older adults. *PLoS One, 14*(11), e0222769. https://doi.org/10.1371/journal.pone.0222769.

Shalash, A. S., Mohamed, H., Mansour, A. H., et al. (2019). Clinical profile of non-motor symptoms in patients with essential tremor: Impact on quality of life and age-related differences. *Tremor and Other Hyperkinetic Movements (New York, NY), 9*. https://doi.org/10.7916/tohm.v0.736.

Shany-Ur, T., Lin, N., Rosen, H. J., Sollberger, M., Miller, B. L., & Rankin, K. P. (2014). Self-awareness in neurodegenerative disease relies on neural structures mediating reward-driven attention. *Brain, 137*(Pt. 8), 2368–2381. https://doi.org/10.1093/brain/awu161.

Shill, H. A., Adler, C. H., & Beach, T. G. (2012). Pathology in essential tremor. *Parkinsonism & Related Disorders, 18*(Suppl. 1), S135–S137. https://doi.org/10.1016/S1353-8020(11)70042-6.

Shill, H. A., Adler, C. H., Sabbagh, M. N., et al. (2008). Pathologic findings in prospectively ascertained essential tremor subjects. *Neurology, 70*(16 Pt. 2), 1452–1455. https://doi.org/10.1212/01.wnl.0000310425.76205.02.

Shill, H. A., Hentz, J. G., Jacobson, S. A., et al. (2014). Essential tremor in the elderly and risk for dementia. *Journal of Neurodegenerative Diseases, 2014*, 1–5. https://doi.org/10.1155/2014/328765.

Shill, H. A., Mehta, S., & Adler, C. H. (2021). Essential tremor and cognitive decline. *Parkinsonism & Related Disorders, 83*, 125. https://doi.org/10.1016/j.parkreldis.2020.12.012.

Sinoff, G., & Badarny, S. (2014). Mild cognitive impairment, dementia, and affective disorders in essential tremor: A prospective study. *Tremor and Other Hyperkinetic Movements (New York, NY), 4*, 227. https://doi.org/10.7916/D85B00KN.

Sollberger, M., Rosen, H. J., Shany-Ur, T., et al. (2014). Neural substrates of socioemotional self-awareness in neurodegenerative disease. *Brain and Behavior: A Cognitive Neuroscience Perspective, 4*(2), 201–214. https://doi.org/10.1002/brb3.211.

Starkstein, S. E. (2014). Anosognosia in Alzheimer's disease: Diagnosis, frequency, mechanism and clinical correlates. *Cortex, 61*, 64–73. https://doi.org/10.1016/j.cortex.2014.07.019.

Stoodley, C. J., & Schmahmann, J. D. (2009). Functional topography in the human cerebellum: A meta-analysis of neuroimaging studies. *NeuroImage, 44*(2), 489–501. https://doi.org/10.1016/j.neuroimage.2008.08.039.

Stoodley, C. J., Valera, E. M., & Schmahmann, J. D. (2010). An fMRI study of intra-individual functional topography in the human cerebellum. *Behavioural Neurology, 23*(1–2), 65–79. https://doi.org/10.3233/BEN-2010-0268.

Stoodley, C. J., Valera, E. M., & Schmahmann, J. D. (2012). Functional topography of the cerebellum for motor and cognitive tasks: An fMRI study. *NeuroImage, 59*(2), 1560–1570. https://doi.org/10.1016/j.neuroimage.2011.08.065.

Tarakad, A., & Jankovic, J. (2018). Essential tremor and Parkinson's disease: Exploring the relationship. *Tremor and Other Hyperkinetic Movements (New York, NY), 8*, 589. https://doi.org/10.7916/D8MD0GVR.

Thawani, S. P., Schupf, N., & Louis, E. D. (2009). Essential tremor is associated with dementia: Prospective population-based study in New York. *Neurology, 73*(8), 621–625. https://doi.org/10.1212/WNL.0b013e3181b389f1.

Tröster, A. I., Fields, J. A., Pahwa, R., et al. (1999). Neuropsychological and quality of life outcome after thalamic stimulation for essential tremor. *Neurology, 53*(8), 1774–1780. https://doi.org/10.1212/wnl.53.8.1774.

Tröster, A. I., Woods, S. P., Fields, J. A., et al. (2002). Neuropsychological deficits in essential tremor: An expression of cerebello-thalamo-cortical pathophysiology? *European Journal of Neurology, 9*(2), 143–151. https://doi.org/10.1046/j.1468-1331.2002.00341.x.

Vilariño-Güell, C., Soto-Ortolaza, A. I., Rajput, A., et al. (2011). MAPT H1 haplotype is a risk factor for essential tremor and multiple system atrophy. *Neurology, 76*(7), 670–672. https://doi.org/10.1212/WNL.0b013e31820c30c1.

Woodworth, D. C., Nguyen, H. L., Khan, Z., Kawas, C. H., Corrada, M. M., & Sajjadi, S. A. (2021). Utility of MRI in the identification of hippocampal sclerosis of aging. *Alzheimer's & Dementia, 17*(5), 847–855. https://doi.org/10.1002/alz.12241.

Yu, M., Ma, K., Faust, P. L., et al. (2012). Increased number of Purkinje cell dendritic swellings in essential tremor. *European Journal of Neurology, 19*(4), 625–630. https://doi.org/10.1111/j.1468-1331.2011.03598.x.

CHAPTER EIGHT

Does essential tremor increase the risk of dementia? No

Alex Rajput*
Division of Neurology, Saskatchewan Movement Disorders Program, University of Saskatchewan/Saskatchewan Health Authority, Saskatoon, SK, Canada
*Corresponding author: e-mail address: alex.rajput@usask.ca

Contents

Abstract

Essential tremor (ET) is the most common neurological cause of tremor affecting adult humans affecting about 6% of those over age 65 years. In the United States, dementia has a prevalence of 15% in those age 68 and older. Overlap of the two conditions is therefore not surprising.

Several studies report mild subclinical cognitive dysfunction in non-demented people with ET, likely related to overactivity of fronto-cerebellar circuitry involved in tremor pathophysiology. Frontal/executive dysfunction is often though not exclusively noted, and some studies have even shown areas of cognitive strengths.

Mild cognitive impairment (MCI) is impairment which a person is aware of but does not interfere significantly with daily activities. While MCI has been considered to presage dementia this is not necessarily the case, as some persons with MCI revert to normal cognition in follow-up. Dementia is a clinical syndrome with cognitive impairment interfering with daily activities. Population-based and clinic-based studies have shown mixed results regarding rates of MCI in ET. A handful of studies have looked at development of

International Review of Neurobiology, Volume 163
ISSN 0074-7742
https://doi.org/10.1016/bs.irn.2022.02.012

dementia in ET with differing results. Brain pathology studies in ET and dementia or investigating Alzheimer-type pathology have thus far been unrevealing. There is evidence by some investigators supporting a greater risk of dementia for those having older onset tremor, while those having ET at a younger age faring at least as well as controls regarding risk of cognitive impairment and dementia.

At present the evidence is inconclusive that ET as a group are at a greater risk of developing dementia.

1. Introduction

Essential tremor (ET) is the most common neurological cause of tremor affecting adult humans. In a recent meta-analysis, ET prevalence in all those over age 65 years approaches 6% (Louis & McCreary, 2021). Dementia affects an estimated 5.8 million people in the United States, with a prevalence of 15% in those older than age 68 (Arvanitakis, Shah, & Bennett, 2019). More than 50 million worldwide have dementia (Samarasekera, 2021), with that number projected to exceed 150 million persons by 2050 (Samarasekera, 2021). It is therefore not surprising that many people would have both conditions. To answer the question "does ET cause dementia," one has to establish that there is an association between the two conditions occurring more than simply by chance. The next step is to determine whether pre-existing ET leads to the subsequent development of dementia.

2. Question

The question posed: does ET increase the risk of dementia? My response is "No." In this chapter, I will provide supporting evidence that ET does *not* in fact increase the risk of dementia.

3. Background and definitions

3.1 Description and definition of essential tremor (ET)

Tremor in humans was noted centuries ago by various civilizations (DiDio & Koller, 2005). It was not, however, until 1887 when Dana authored his detailed report on multiple persons affected with a familial form of tremor (Dana, 1887). Over half a century later, Critchley used the term "essential tremor" in his seminal work including prior reports on tremor and his own personal observations (Critchley, 1949). Upper limb postural and/or action tremor is the hallmark of essential tremor (ET), though tremor may be noted elsewhere (most often the head or voice).

In the first consensus statement of the Movement Disorder Society (MDS) on tremor (Deuschl, Bain, Brin, & Ad Hoc Scientific Committee, 1998), ET was defined as a visible and persistent bilateral, largely symmetric postural or kinetic tremor involving the hands or forearms. Isolated head tremor or head tremor in addition to the upper limb tremor, but without abnormal posturing, was also included in the diagnosis of ET. There were several exclusionary criteria. Duration of tremor as specified by the TRIG (Tremor Investigation Group) of 3 years for probable ET and 5 years for definite ET were eliminated from the MDS criteria as duration was considered arbitrary (Deuschl, Bain, Brin, & Ad Hoc Scientific Committee, 1998). There was no mention of the terms "dementia" or "cognition" anywhere in the text.

Recognizing the heterogeneity among ET, Deuschl and Elble suggested a simple yet practical revision of diagnosing ET into one of three groups: (1) Hereditary ET—fulfilling the published consensus criteria of 1998 but also having unequivocal evidence of a first degree affected relative, with both the subject and the individual having onset before age 65; (2) Sporadic ET—onset before age 65 and no affected family member; (3) Senile tremor—tremor after age 65, with or without a family history of tremor (Deuschl & Elble, 2009). The words "cognitive" and "dementia" did appear in this publication.

The latest revision of the consensus statement for classification of tremors was published in 2018 (Bhatia et al., 2018). This was a fundamental shift in classification of tremor, moving to a two axis diagnosis similar to what has been used for dystonia: Axis 1—clinical characteristics, and Axis 2—etiology. If there is no axis 2 cause identified, the diagnosis remains as Axis 1 (which includes ET). The revised definition of ET is an isolated action tremor of both upper limbs of at least 3 years duration, with or without tremor in other sites (head, voice, lower limbs). Other neurological signs such as dystonia, parkinsonism or ataxia should be absent but they did not quantify to what degree. In recognition of "soft" signs that are not sufficient for an additional or different diagnosis but are not normal, the authors came up with the term "ET Plus." This is defined as "Tremor with the characteristics of ET and additional neurological signs of uncertain significance such as impaired tandem gait, questionable dystonic posturing, memory impairment, or other mild neurologic signs of unknown significance that do not suffice to make an additional syndrome classification or diagnosis. ET with tremor at rest should be classified here." (Bhatia et al., 2018).

While the terminology "ET Plus" arose from a consensus statement, it has not yet gained widespread acceptance and is a source of controversy (Fasano, Lang, & Espay, 2018; Louis et al., 2020). The recognition of these "soft signs" and allowing a diagnosis of "ET Plus" was not intended to attribute causality or significance, but rather to foster further investigation into ET (Deuschl, Bhatia, Elble, & Hallett, 2018). These cases remain practically as ET.

3.2 Definition of dementia

The diagnosis of dementia is "defined by chronic, acquired loss of two or more cognitive abilities caused by brain disease or injury." (Arvanitakis et al., 2019). In this review article on dementia, Arvanitakis and colleagues reinforce that "...dementia is a *clinical syndrome* with variable manifestations... while research studies have defined a 'preclinical' AD [Alzheimer's disease], in clinical care, AD is not diagnosed before symptom onset."

3.3 Definition of mild cognitive impairment (MCI)

Mild cognitive impairment (MCI) is defined as impairment in memory or other cognitive domains confirmed by objective testing with preserved activities of daily living. Higher order activities (instrumental activities of daily living), however, may be affected. The person with MCI is aware of the deficits (Lopez, 2013).

3.4 Significance of MCI

Overall MCI prevalence is reported at 8.4% for those age 65–69 and up to 25.2% for ages 80–84 (Petersen et al., 2018). For those with MCI older than age 65, cumulative dementia incidence was almost 15% for individuals followed for 2 years. Reversion from MCI to normal cognition has been reported in up to 38% in some class I studies, however, increased rates of conversion to dementia even after reversion to normal cognition may occur (Petersen et al., 2018).

A recent publication (Angevaare et al., 2022) in a diverse (multiethnic) community based study followed 2903 persons age >65 years old with normal cognition at baseline for the development of MCI. Of those who went on to be diagnosed with MCI, after an average of 2.4 years follow-up, 47.9% were considered to have normal cognition while only 13% were diagnosed with dementia.

3.5 Cognitive dysfunction

Detailed neuropsychological testing can reveal cognitive dysfunction—below normal performance—which the individual is not aware of (i.e., subclinical deficits). This is at a level below MCI (awareness of deficits), which itself is below the level of dementia (functional impairment).

4. Studies

Many publications reporting on ET and cognition comment on cognitive dysfunction and MCI. I will categorize these studies as either population-based or clinic-based studies, followed by a brief review of neuropathological studies related to ET and dementia.

4.1 Population-based studies

4.1.1 NEDICES studies

A large project called NEDICES (Neurological Disorders in Central Spain) examined a cohort of elderly persons age >65 years residing in three communities in central Spain. Baseline evaluation was done in 1994–95 and follow-up evaluation in 1997–98. There have been many publications based on this cohort. Various inclusion/exclusion criteria and methodology have been used to answer the numerous questions posed by investigators.

In 2006, Benito-Leon, Louis, Bermejo-Pareja, and On behalf of the Neurological Disorders in Central Spain (NEDICES) Study Group (2006) reported on all incident and prevalent cases of ET who completed their second evaluation (3 years after baseline evaluation) ($n = 232$ ET) and frequency matched controls ($n = 696$) in an elderly (>65 years old) population. Results from the second evaluation were compared. The primary goal of the study was to compare neuropsychological test performance with ET and controls and they did not publish data on dementia in the groups. Those with ET were more likely to answer "yes" to the screening answer for depression (40.9% vs 26.4%). After adjustment for age, education, gender, medication, premorbid intelligence, medications and depressive symptoms, ET subjects scored lower on the 37-MMSE (27.0 ± 6.7 vs 28.9 ± 5.9, $P < 0.001$) and frontal executive function (trail making test number of errors, 8.7 ± 11.0 in ET vs 3.8 ± 7.6 in controls, $P < 0.001$). Subjectively, forgetfulness was reported in a larger percentage of ET than controls (50.4% vs 43.1%, $P = 0.05$).

Later that year, the same group (Benito-León, Louis, Bermejo-Pareja, & On behalf of the Neurological Disorders in Central Spain (NEDICES)

Study Group, 2006) noted that while more ET cases than controls had dementia (31/273, 11.4% vs 204/3382, 6.0%), this was *non-significant* when adjusted for age, stroke and educational level (OR = 1.35, 95% CI = 0.87 to 2.18, $P = 0.17$). When the model was adjusted for age of ET onset, ET with tremor >age 65 years were at greater risk of having dementia than those with ET <65 years onset (OR = 1.70, 95% CI = 1.04 to 2.76, $P = 0.03$). Those with younger onset ET were non-significantly less likely to be demented (OR = 0.38, 95% CI = 0.09 to 1.73, $P = 0.21$). Of the 31 ET cases with dementia, 2 were young onset (2.2% of all young onset ET) while 29 ET cases were old onset (15.9% of all old onset ET) ($P = 0.001$). Tremor preceded dementia in 26 of the 31 ET + dementia cases (83.9%) while in 5 cases, dementia occurred before ET onset. This cross-sectional study was designed to answer the question of whether ET was associated with prevalent dementia. The majority (74.9%) had Alzheimer's disease.

In 2007, Bermejo-Pareja and colleagues reported on incident dementia in the NEDICES cohort (Bermejo-Pareja, Louis, Benito-León, & On behalf of the Neurological Disorders in Central Spain (NEDICES) Study Group, 2007). The 3891 participants were followed for mean of 3.2 years. All subjects were free of dementia at baseline. At follow-up (i.e., the second visit), 16/206 ET (7.8%) cases and 145/3685 controls (3.9%) developed dementia; the adjusted relative risk was bordering on significance (RR-1.66, 95% CI = 0.99–2.80, $P = 0.054$). When they adjusted the model for cases who had tremor onset after age 65, those with *older onset ET were twice as likely to develop dementia* vs *controls* (RR = 1.98, CI = 1.14–3.45, $P = 0.01$), while those with *ET onset age <65 years had similar rate of dementia to controls* (RR = 0.74, 95% CI = 0.19–3.20, $P = 0.79$). Those with incident dementia were older at baseline (79.5 ± 6.8 years vs 73.0 ± 6.2 years, $P < 0.001$), and those with ET onset after age 65 were older than those with ET onset <65 years (78.3 ± 5.9 years vs 70.4 ± 4.9 years, $P < 0.001$). Of the 16 incident dementia cases with ET, two had younger onset ET while 14 were older onset ET ($P = 0.004$). This study did not report on MCI. Elble and colleagues (Elble, Dubinsky, & Ala, 2007) commented on the results (Bermejo-Pareja, Louis, Benito-León, & On behalf of the Neurological Disorders in Central Spain (NEDICES) Study Group, 2007) in a thoughtful editorial.

Louis et al. (Louis, Benito-Leon, Vega-Quiroga, Bermejo-Pareja and The Neurological Disorders in Central Spain NEDICES Study Group, 2010a) reported on rate of cognitive decline in non-demented ET and non-demented controls in a population based study comparing baseline with follow-up assessment 3 years later. Baseline MMSE-37 (maximum score of 37) was lower in ET than controls (mean 28.8 ± 5.8 vs 30.2 ± 4.8, $P = 0.02$). They report that the rate of decline was 7 times faster in those with ET than controls based on the 37-MMSE score (mean decline of 0.70 vs 0.11, $P = 0.03$). While cases with dementia at baseline were excluded, they also excluded control participants who developed incident ET ($n = 56$) and participants who developed dementia ($n = 145$) during the study period, as their goal was to assess non-demented patients. They did not report details of the 145 participants who developed incident dementia. In the accompanying editorial, Tolosa and Molineuvo commented that it is conceivable that tremor performance could have interfered with performance on copying tasks (and the severity of tremor not explained) (Tolosa & Molinuevo, 2010). They questioned whether such cognitive changes are part of a condition in which progressive and widespread pathological changes lead to deteriorating cognition or reflect the functional involvement of anatomical loops responsible for both the progressive tremor and the cognitive decline. They favored the latter, although did not define what was meant by anatomical loops (Tolosa & Molinuevo, 2010).

Among non-demented ET, a population-based study reported worse scores on a functional activities questionnaire (FAQ) involving both motor (such as writing down an appointment) and cognitive tasks (such as remembering an appointment) (Louis, Benito-León, Vega-Quiroga, Bermejo-Pareja and The Neurological Disorders in Central Spain NEDICES Study Group, 2010b). Worse performance was correlated with a lower MMSE score, and persisted even accounting for mood and use of medication with CNS effects. ET patients ($n = 208$) had a worse mean MMSE-37 score (maximum score 37) than controls ($n = 3616$); (27.9 ± 5.8 vs 29.3 ± 5.4, $P = 0.002$) and ET patients were older than controls (75.1 ± 6.9 years vs 73.5 ± 6.4 years, $P = 0.001$). Tremor duration was mean 9.6 years in this study.

Benito-León et al. identified all persons with MCI and those with ET from the NEDICES cohort (Benito-León, Louis, Mitchell, & Bermejo-Pareja, 2011). Persons who developed dementia at the second screening were excluded as the authors wanted to assess whether ET was associated with MCI. No details on those who developed dementia ($n = 283$) were provided. Of the final sample of 2679 participants who participated in both baseline and the second (e.g., follow-up evaluation), 207 subjects had ET and 441 had MCI. They found 20.3% (42/207) of ET and 16.1% (399/2472) of controls had MCI (odds ratio 1.32, $P = 0.12$), and adjustment of models made no difference. When they stratified the ET cases according to age at onset (<65 years vs ≥65 years), in adjusted analyses those with older onset ET were more likely to have MCI than controls (OR = 1.57, 95% CI = 1.03–2.38, P = 0.03). Those with younger onset were at no greater risk to have MCI (OR = 0.73, 95% CI = 0.34–1.57, P = 0.43).

In another NEDICES study, Benito-León and colleagues (Benito-León, Louis, Sánchez-Ferro, & Bermejo-Pareja, 2013) reported on 2375 participants, with 135 prevalent ET (i.e., ET noted at baseline) and 56 incident ET cases (i.e., cases who were diagnosed as ET at follow-up visit but did not have ET at baseline). They noted over a mean follow-up of 3.4 years, the MMSE-37 score declined by 0.7 ± 3.3 points in prevalent ET cases, 1.1 ± 3.5 points in the incident ET, and only 0.1 ± 3.9 points in the controls. They noted that the differences were significant between incident ET cases and controls (P = 0.046) and between prevalent ET cases and controls (P = 0.027). Cases who developed incident dementia (i.e., dementia noted at second visit) were excluded from analysis. There was no mention of the term "MCI," "mild cognitive impairment" or "cognitive impairment" in this publication. Tremor severity was not captured. The authors used the term "premotor" for incident ET cases but did not comment on any particular characteristics of this group.

In a study of elderly subjects >65 years old with mean follow-up 3.2 years, Benito-Leon and colleagues (Benito-León, Contador, Louis, Cosentino, & Bermejo-Pareja, 2016) reported that ET patients with lower educational level were more likely to develop dementia than those with higher educational attainment (8/48 or 16.7% developed incident dementia vs 1/30 or 3.3% higher education level). This difference only occurred among the incident ET (no ET at baseline but were diagnosed with ET at follow-up visit). All subjects were non-demented at baseline. Those with incident dementia (n = 161) were older than those without incident dementia (n = 3717), age 79.5 ± 6.8 years vs 72.9 ± 6.2 years (P < 0.001). This study only focused on dementia and not on MCI (Benito-León et al., 2016). Of those diagnosed with incident dementia at follow-up, (161/3878 or 4.1%), the majority (71.4%) were diagnosed with Alzheimer's dementia. They reported that all 9 incident ET cases who developed dementia had AD, while only 11/16 (68.7%) of those with prevalent ET who developed dementia were diagnosed with AD.

4.1.2 Washington-Inwood heights

In a study of 2285 elderly individuals age >65 years old enrolled in the Washington/Hamilton Heights and Inwood Aging Project (WHICAP) II cohort, Thawani et al. (Thawani, Schupf, & Louis, 2009) reported higher cross-sectional prevalence of dementia in ET than controls (adjusted OR = 1.84, P = 0.01). Over a mean follow-up of 3.8 years, they reported that the adjusted risk of incident dementia in ET vs controls was 1.64 (95% CI 0.99–2.72, P = 0.055). Older age was noted in both those with ET and those with dementia. Interestingly, ET was not associated with an increased risk of incident MCI.

To assess tremor, a medical student was trained to rate tremor severity in six items (five shapes and a trail making test). There was high interrater agreement between the student and senior movement disorders neurologist (intraclass correlation coefficient 0.80). Each item was rated in 0.5 increments from 0 to 2, with a total maximum score of 12 (Bain and Findley rating scale score of 2.0 equivalent to a 1.0 in the scale used in this study) (Bain & Findley, 1993). An additional handwriting test was rated by the senior movement disorders neurologist. A diagnosis of ET was made when the total tremor score was greater than 5.5 or handwriting sample scored >2 (score >5 in Bain & Findley scale). The mean, median, and range of tremor scores was not reported. There was no mention of association of tremor scores to dementia or MCI, nor was data on age of ET onset available.

4.1.3 Danish twin study

In the Longitudinal Study of Aging Danish Twins (LSADT), over 2000 participants age 70+ years answered screening questions on ET. Tremor score as measured by spiral drawing increased with age, and symptomatic tremor was present in almost 10% of subjects older than 90 years. Deuschl and colleagues reported increased mortality, lower grip strength and worse ADL scores in those with tremor age 70 and older compared with controls (Deuschl, Petersen, Lorenz, & Christensen, 2015).

Those with definite ET (defined as tremor duration >5 years) had *significantly better* cognitive and ADL scores compared to controls, as well as a trend toward decreased mortality (non-significant). While a screening questionnaire, spiral drawing assessment by experienced raters, and a prior diagnosis of ET were used to determine whether someone was considered ET or control, a limitation of the study is that the vast majority of in the control group were not examined by an expert neurologist, raising the possibility of diagnostic misclassification among the controls. It was suggested that what had been called "sporadic ET" (onset after age 65, with or without family history) be changed to "age-related tremor" (Deuschl et al., 2015).

In an accompanying editorial, Bain suggested using the somewhat cumbersome but appropriately descriptive term "late-onset upper limb action tremor" (Bain, 2015).

4.2 Clinic-based studies

Lombardi et al. (Lombardi, Woolston, Roberts, & Gross, 2001) were among the first to report on cognitive deficits in ET. They performed neuropsychological testing on 18 moderate to severe ET and 18 PD patients as part of presurgical workup. Normative samples were used for comparison. The ET and PD were of comparable age (66.1 vs 66.8 years) and mean MMSE was identical between the groups (score 26.9). Duration of ET symptoms was at least 10 years (mean 36.3 years). When the groups were compared to each other, the ET showed greater impairment of verbal fluency and working memory than the PD, while PD had greater impairment in facial perception. Performance was not different comparing those on primidone or propranolol for tremor to those who were not. Cognitive impairment in ET was noted on tests primarily relying on intact prefrontal cortex function (e.g., verbal fluency, naming, recent memory, set-shifting). However, the ET group performed *above average* on tests of general verbal and intellectual abilities. Both ET and PD had high levels of depression noted.

In a study by Sahin and colleagues (Sahin et al., 2006) comparing young, healthy ET patients ($n = 16$, mean age 29.6 years) and controls ($n = 16$, mean age 28 years), the ET subjects performed worse on tests of visuospatial function. They also found a statistically significant inverse correlation between tremor severity and tests of executive function (i.e., patients with higher tremor scores performed worse). No difference in regional blood flow between ET and subjects was identified as measured by SPECT, but there were technical challenges so not all subjects were imaged. (Those with depression or other psychiatric conditions were excluded as were those on specific medication for tremor.)

Kim et al. (Kim et al., 2009) reported on 34 consecutive non-demented ET patients (mean age 67.6 years, mean tremor duration 7.3 years) seen in a specialty clinic and 33 age matched controls without symptoms of tremor or cognitive dysfunction. Impairments in frontal executive functions, attention and verbal memory were significantly worse in the ET group. Cognitive impairment in the ET group was significantly correlated with age and educational status, but not with duration of tremor or activities of daily living (Barthel index).

Sengul and colleagues (Sengul et al., 2015) reported on cognitive impairment in young otherwise healthy ET patients compared with controls (mean age 24.55 ± 7.16 years, mean duration of ET 4.36 ± 3.94 years). MoCA total score was statistically worse in the ET (25.80 ± 2.76) than controls (28.23 ± 1.69) ($P < 0.001$). The ET group scored worse on scales measuring sleep quality, depression, anxiety, and fatigue. Higher tremor score was non-significantly associated with worse score on the MoCA. There was no association between depression or anxiety and the total MoCA score. The authors did not report whether there was any correlation between fatigue or sleepiness scales and MoCA score.

In a recent paper comparing 60 dementia-free ET patients with 50 age-matched healthy controls with detailed neuropsychological testing and brain MRI, subclinical but significantly lower memory scores were noted in the ET group (Novellino et al., 2020). There was also a significantly increased medial diffusivity (MD) of both hippocampi on brain MRI in the ET group compared to controls, and a significant correlation between MD and memory scores. The ET group, however, compared within normal limits on cognitive testing. Attention, visuospatial skills, verbal fluency, language comprehension and global cognitive function, however, did not differ between ET and controls.

In a cross-sectional study, detailed neuropsychological assessment of 148 non-demented ET subjects age >55 years was performed. Subjects had no history of deep brain stimulation surgery and had enrolled in the brain donor program (Cersonsky et al., 2020). Cluster analysis resulted in subjects falling into one of three clusters: those with comparatively low memory scores ($n = 64$), ($P < 0.001$) those with relatively low attention and visuospatial scores ($n = 39$) ($P < 0.001$) and those ET who performed well across all domains ($n = 45$).

In another recent article, advanced ET cases ($n = 201$, mean age 68.9 ± 8.9 years) pursuing DBS (deep brain stimulation) surgery were evaluated by detailed neuropsychological testing (Ratajska et al., 2021). Nearly half of subjects were deemed to be cognitively normal. They arrived at a three cluster solution: low executive ($n = 64$), low memory multidomain ($n = 41$), and cognitively normal ($n = 96$). More than one quarter of subjects met criteria for MCI.

4.3 Studies measuring conversion to dementia

I have set aside three studies as they specifically targeted development (i.e., incidence) of dementia in an ET population.

Sinoff and Badarny reported on baseline prevalence of MCI in ET patients and evolution in a 2 year prospective study of 52 ET patients over the age of 50 (mean age 68.2 years, mean duration of ET 11.7 years, mean onset age 55.8 years) recruited from a movement disorders specialty clinic (Sinoff & Badarny, 2014). At baseline, no subjects had dementia (had to have MMSE $\geq$24 to participate in the study) but there was a very high baseline prevalence rate of MCI (more than 2/3 of subjects). Deficits noted included verbal memory, concentration, phonemic and semantic fluency. There was a high number lost to follow-up, with only 24 subjects (46.2%) available for evaluation at T2 (mean 27.5 months after baseline visit). Of those, 4/24 converted to dementia and all had MCI at baseline. They calculated an 8.4% annual conversion rate to dementia among those who could be followed for 2 years. There was a significant percentage of ET with anxiety (25%) and depression (17.6%). There was no control group in this observational study.

In a study of elderly ET, Shill and colleagues (Shill et al., 2014) prospectively followed ET subjects and controls over 5 years (median 5.4 years, range 0.9–12.1 years) enrolled in a longitudinal aging study (Arizona Study of Aging and Neurodegenerative Disorders (AZSAND)). All participants had an annual UPDRS and Fahn Tolosa Marin (FTM) tremor assessment by a movement disorders neurologist and all had annual neuropsychological testing. After excluding neurodegenerative conditions and others who did not meet criteria for ET, they had 83 persons with ET and 424 controls. The ET group was older at baseline than controls (80 ± 5.9 years vs 76.9 ± 8.5 years, $P = 0.002$). Mean age of tremor onset was 66 years (range near birth to age 91) and mean duration of tremor 14 years (range 0–72 years, median 5.2 years). Baseline cognitive status did not differ between the groups. They reported a dementia incidence within 5 years of 6% for ET and 8% for controls (hazard ratio between ET and controls of 0.79, 95% CI 0.33–1.83, $P = 0.58$). When subjects with baseline MCI were excluded (6/83 or 7% of ET; 39/424 or 9% of controls) from analysis, the hazard ratio of developing dementia was only 1.06 (95% CI 0.17–1.23). The hazard ratio for ET onset age $\geq$65 years vs onset $<$65 years was 2.1 (95% CI 0.24–18), and hazard ratio for association between tremor duration $>$5 years vs those with $\leq$5 years tremor duration was 0.46 (95% CI 0.08–2.6). When adjusted for age and sex, (HR 0.50, 95% CI 0.21–1.20, $P = 0.12$) or for age, sex and ApoE4 status, (HR 0.46, 95% CI 0.17–1.23, $P = 0.12$) there was a further decrease in the association between tremor and dementia though this was non-significant. An association between older onset tremor and dementia was suggested based on the above data, but this was not tested statistically as the authors stated the sample size was too small to assess a potential relationship. The conclusion was that the ET population in this closely followed cohort was not at increased risk of dementia.

Radler and colleagues reported on 167 ET cases followed in a prospective, longitudinal study (Radler et al., 2020). Methodology was similar to previously described (Cersonsky et al., 2020), including confirmation of ET diagnosis by a movement disorders neurologist using videotaped neurological exam. Subjects were evaluated at baseline (T1), 18 months after baseline (T2) and 36 months after baseline (T3). After cases were excluded (no second evaluation, alternative diagnoses), they had 26 ET cases (age 82.7 ± 7.7 years) who had MCI at baseline and evaluated 18 months later. Three of 26 (11.5%, 95% CI 0.0–23.8%) converted from MCI to dementia (7.7% annual conversion rate) at T2. There were 23 ET-MCI cases at the start of T2 (i.e., 18 months after baseline assessment) who were available for follow-up at T3 another 18 months later. Of those 23 cases, the authors note that 8 were incident MCI cases (newly diagnosed at T2), therefore, one infers that 8 of the initial 26 ET-MCI cases noted at baseline were not available for assessment at T3. At T3 (i.e., 36 months after baseline assessment), 6 of 23 ET-MCI cases (26.1%, 95% CI 8.1–44.0%) had converted to dementia. They reported an annual conversion rate from MCI to dementia of 12.5%. Strengths of the study included three assessments. Duration of follow-up was only 3 years. A limitation of the study was that there was no control group available for direct comparison, although the authors cited numerous studies of the general population in which conversion from MCI dementia was in the 2.6–6.3% range. Comparisons between ET cases who converted and ET cases who were non-converters were underpowered and presented as exploratory analyses.

Strengths and shortcomings of the two larger studies were debated by the authors (Louis, Radler, Huey, & Cosentino, 2021; Shill, Mehta, & Adler, 2021).

4.4 Pathological studies

Dementia is a clinical syndrome with multiple etiologies. After excluding reversible causes of cognitive impairment, diagnoses can be confirmed by brain pathology. The most common cause of dementia is Alzheimer's disease, and definitive diagnosis of dementia requires brain pathological evaluations (Elahi & Miller, 2017). Alzheimer's disease is a dual proteinopathy consisting of extracellular amyloid plaques and intracellular neurofibrillary tangles. There is, however, not a perfect correlation with brain pathology and clinical findings in dementia. The same brain pathology may be interpreted differently based on the clinical picture (Jack et al., 2018).

Kim et al. addressed brain tau pathology in ET. Fresh frozen samples from the temporal lobe were assessed with monoclonal antibody stains (Kim et al., 2021). They reported on 13 ET cases (3 normal, 4 MCI, and 6 with dementia) who died between ages 83–99 (median age 90). Diagnoses of MCI and dementia were by reviews of assessment data by a neuropsychologist. There was no report of tremor duration or severity of ET subjects. Total tau was significantly lower in ET ($n=13$) compared to PSP ($n=10$), Pick disease ($n=2$) and Alzheimer's disease ($n=7$) without N-terminal inserts (i.e., 0N) but was not lower in the ET with other N-terminal inserts. They noted "a varying burden of p [phosphorylated]-tau, with some ET cases completely devoid of p-tau pathology or with only scattered thread pathology." The authors reported: "... the neuropathological degenerative changes with respect to tau pathology in ET patients are variable and diverge considerably from those with AD." (Kim et al., 2021).

Alternative splicing of exon 10 of the MAPT (microtubule-associated protein tau) yields either 3R or 4R tau; ratio of 3R to 4R tau has been associated with various neurodegenerative conditions (Boyarko & Hook, 2021). Alternative splicing of exons 2 and 3 results in different N-terminus isoforms (0N,1N,2N) of which little is known. They reported the total 0N levels in ET were significantly lower compared to the AD group in this study, however, no control brains were studied for comparison (Kim et al., 2021).

Neurofilament light chain (NFL) is a non-specific marker of neurodegeneration, and elevated serum levels of NFL have been found in various conditions (Huang, Huang, Zhang, Shen, & Sun, 2022). A recent study showed elevated serum levels of NFL in those with Parkinson's compared to ET and controls, who had similar values to one another (Huang et al., 2022). Cerebellar cortex levels of NFL are no different comparing brains of ET with controls (Louis et al., 2012).

Pan et al. reported on Alzheimer-related changes in an autopsy study of 40 ET (mean age 82.6 ± 6.0 years) and 32 age matched elderly controls (Pan et al., 2014). Cases were excluded if they had additional diagnoses, including brain pathology of intermediate or high likelihood of Alzheimer's. Both groups were highly educated (mean 14.3 ± 2.7 in ET and 15.7 ± in controls), and ET cases were of long duration (mean 41.9 ± 21.4 years). Brain weight was similar. Braak stage for neurofibrillary tangles (NFT) was higher in ET than controls (mean 2.2 ± 1.2 and median 2, vs 1.2 ± 1.1, $P < 0.001$). (Braak stage 1—transentorhinal NFT lesions; Braak stage 2—NFT lesions extending into the entorhrinal region) (Braak, Alafuzoff, Arzberger, Kretzschmar, & Del Tredici, 2006). CERAD scoring for amyloid plaque score was similar between the groups. As none of the ET or controls had dementia or pathological criteria to diagnose Alzheimer's, one could argue that in this group of longstanding ET, a severe disturbance of cognition was not present.

In a comparison of ET ($n = 26$) and controls ($n = 73$), Farrell and colleagues noted no differences between ET and controls for β-amyloid plaques, cerebral amyloid angiopathy, Braak NFT stage, presence of Lewy bodies, or hippocampal sclerosis within each cognitive stratum (normal, MCI, and dementia) (Farrell et al., 2019). The finding of statistical significance was a higher number of NFT-positive neurons in the neocortex of the ET who were cognitively normal and in the ET + MCI group compared with controls of the same cognitive status. The authors concluded that this represents an increased pre-dementia tau burden in ET. There was no difference, however, in NFT-positive neurons in the demented ET and controls, and no difference between ET and controls among any of the three cognitive groups (normal, MCI, dementia) in the medial temporal lobe.

4.4.1 Nature of cognitive deficits in ET

Bermejo-Pareja & Puertas-Martin summarize historical data on ET and cognitive deficits associated with ET. They note the most consistent findings are deficits in working memory and attention-concentration, executive function, and (explicit) verbal memory (Bermejo-Pareja & Puertas-Martin, 2012). They compare the cognitive and affective disturbances in ET as similar to those with cerebellar cognitive affective syndrome but that the deficits in ET are "mild… and subclinical." Cerebellar cognitive affective syndrome was reported in patients with lesions isolated to the cerebellum (stroke, resection, post-infectious cerebellitis and cerebellar atrophy) (Schmahmann & Sherman, 1998). The full syndrome included impairment in executive function, spatial cognition, personality change and language difficulties causing obvious impairment to patients, with deficits evident on standard bedside testing. A recent detailed volumetric evaluation implicated the hippocampal formation with cognitive dysfunction in ET (Prasad et al., 2019), and amnestic MCI has been reported as more frequent than non-amnestic MCI in ET (Collins et al., 2017).

The nature of deficits reported in ET is not specific for ET or even cerebellar dysfunction. Tröster and colleagues noted clinically subtle deficits on pre-operative neuropsychological assessment of 101 patients with ET. They commented that the "observed pattern of cognitive deficits [attention and some executive functions] is consistent with cerebello-thalamo-cortical circuit dysfunction." (Tröster et al., 2002).

Supporting evidence for a circuitry problem as the cause of cognitive dysfunction in ET was presented by Tröster et al. in their report on unilateral thalamic DBS of 40 ET patients. Patients underwent neuropsychological testing pre- and postprocedure. In addition to benefits with unilateral (dominant hand) tremor, they noted clinically modest but statistically significant improvement in tasks of visual attention, visuoperceptual and constructional ability, delayed word list recognition, and prose recall (Tröster et al., 1999). Lexical verbal fluency, however, significantly declined after surgery. While only a single case of ET, Lucas et al. reported that bilateral DBS improved verbal fluency and recall (in addition to the expected motor improvement), and these worsened when the stimulator was turned off (Lucas, Rippeth, Uitti, Shuster, & Wharen, 2000). Other cognitive performance was unaffected.

For those ET who actually go on to develop dementia, the evidence thus far supports that the majority have Alzheimer's disease. This is not surprising given its frequency in the general population. Cases with concomitant pathology or clinical conditions are excluded from a number of studies, but there is no overwhelming tendency for a particular or rare pathology. There is not enough evidence supporting a greater burden of Alzheimer's type changes in ET based on pathological studies.

5. Discussion

ET has been described as "both a paradoxical and an elusive disorder" (Louis, 2000). It has not been resolved if ET is only one disease (Hopfner & Deuschl, 2018; Louis, 2018; Quinn, Schneider, Schwingenschuh, & Bhatia, 2011), and whether it is appropriate to call ET a disease (Lenka & Louis, 2020) or a syndrome (Elble, 2020).

The mild subclinical cognitive dysfunction in non-demented ET is not surprising with overactivity of fronto-cerebellar circuitry involved in tremor pathophysiology. While deficits tend to be related to frontal/executive dysfunction, the pattern of deficits reported is by no means universal in each study. Some studies have even shown areas of relative cognitive strengths in those with ET.

Assuming that cognitive dysfunction begets MCI, or MCI will lead to dementia is fraught with missteps. Brain pathology studies investigating ET and Alzheimer pathology have thus far not shed much light either. In a recent detailed review, Louis and colleagues acknowledge a paucity of information in ET regarding conversion rate to MCI, rate of conversion of ET patients with MCI to dementia, predictors of conversion to dementia (aside from older age), rate of decline, and underlying pathology/pathomechanisms (Louis, Joyce, & Cosentino, 2019).

Surveys of the general population will yield different rates of conditions than those who attend subspecialty clinics, and those who have primary cognitive issues will be more likely to go to a dementia clinic than to a movement disorders clinic. There are clinical diagnostic errors even among expert clinicians, though the longer someone is followed the less likely that occurs.

If implying that ET is a "degenerative" condition, it stands to reason that persons with longer duration of tremor and those with more severe/rapidly progressive tremor would have more advanced brain pathology than those without. That has not proven to be the case. Indeed, younger onset ET tend to fare slightly better if no worse than controls regarding cognition and mortality.

The most recent consensus definition of ET is that of an identifiable clinical entity, yet the definition is relatively broad and inclusive (Bhatia et al., 2018). This is not inappropriate, however, a 50 year old with tremor since childhood and many family members affected can be given the same diagnosis as an 80 year old with bilateral upper limb action tremor for 4 years and no family history of tremor. Lumping such clinically heterogeneous cases together for epidemiological, clinical and pathological studies can prove problematic, and likely is a significant contributor to the disparate results reported by investigators.

We have not yet determined clinical or biomarkers to stratify ET cases, however, there is evidence to support that what is operationally called ET is likely not the same underlying condition, despite having similar clinical expression (Hopfner et al., 2016; Muthuraman et al., 2015).

Population-based studies cross a wide swath and can detect cases who would not otherwise seek medical attention, e.g., milder cases not limiting function, or more obvious cases with barriers to medical care. Atypical and more severe cases can be overrepresented in specialty clinic studies, and interest and ability to participate in studies may influence the results. Clinic based studies, however, allow for detailed evaluation and follow-up, while population-based studies tend to use methodology that can be deployed for hundreds or thousands of subjects. The findings of population based studies, while one could argue as being more (or less) sensitive to detecting differences from controls depending on the methods used, may find results that are of limited clinical significance.

A number of studies employing detailed neuropsychological testing have detected abnormalities in those with ET, typically at the subclinical level. There are certainly some ET cases with MCI—noticeable difficulties but still functioning at a reasonable cognitive level in daily life—as well as ET cases with dementia, most commonly Alzheimer s disease.

6. Conclusion

As ET is currently defined (Bhatia et al., 2018), the evidence is mixed regarding ET and subsequent development of dementia. Further stratifying ET cases for studies, particularly by age of onset and family history, would allow for more appropriate comparisons between groups. This should lead us to a better understanding and management of essential tremor, or what we may determine, "the essential tremors" (Louis, 2018).

References

Angevaare, M. J., Vonk, J. M. J., Bertola, L., Zahodne, L., Watson, C. W.-M., Boehme, A., et al. (2022). Predictors of incident mild cognitive impairment and its course in a diverse community-based population. *Neurology*, *98*, e15–e26.

Arvanitakis, Z., Shah, R. C., & Bennett, D. A. (2019). Diagnosis and management of dementia: Review. *JAMA*, *322*(6), 1589–1599.

Bain, P. G. (2015). Essential tremor and senile varieties of action tremor an evolving ART. *Movement Disorders*, *30*, 1301–1303.

Bain, P. G., & Findley, L. J. (1993). *Assessing tremor severity*. London: Smith-Gordon and Compant, Ltd.

Benito-León, J., Contador, I., Louis, E. D., Cosentino, S., & Bermejo-Pareja, F. (2016). Education and risk of incident dementia during the premotor and motor phases of essential tremor (NEDICES). *Medicine (Baltimore)*, *95*(33), e4607.

Benito-Leon, J., Louis, E. D., Bermejo-Pareja, F., & On behalf of the Neurological Disorders in Central Spain (NEDICES) Study Group. (2006). Population-based case-control study of cognitive function in essential tremor. *Neurology*, *66*, 69–74.

Benito-León, J., Louis, E. D., Bermejo-Pareja, F., & On behalf of the Neurological Disorders in Central Spain (NEDICES) Study Group. (2006). Elderly-onset essential tremor is associated with dementia. *Neurology*, *66*, 1500–1505.

Benito-León, J., Louis, E. D., Mitchell, A. J., & Bermejo-Pareja, F. (2011). Elderly-onset essential tremor and mild cognitive impairment: A population-based study (NEDICES). *Journal of Alzheimer's Disease*, *23*, 727–735.

Benito-León, J., Louis, E. D., Sánchez-Ferro, A., & Bermejo-Pareja, F. (2013). Rate of cognitive decline during the premotor phase of essential tremor: A prospective study. *Neurology*, *81*, 60–66.

Bermejo-Pareja, F., Louis, E. D., Benito-León, J., & On behalf of the Neurological Disorders in Central Spain (NEDICES) Study Group. (2007). Risk of incident dementia in essential tremor: A population-based study. *Movement Disorders, 22*, 1573–1580.

Bermejo-Pareja, F., & Puertas-Martin, V. (2012). Cognitive features of essential tremor: A review of the clinical aspects and possible mechanistic underpinnings. *Tremor and Other Hyperkinetic Movements, 2* (02-74-541-1).

Bhatia, K. P., Bain, P., Bajaj, N., Elble, R. J., Hallett, M., Louis, E. D., et al. (2018). Consensus statement on the classification of tremors. From the task force on tremor of the International Parkinson and Movement Disorder Society. *Movement Disorders, 33*, 75–87.

Boyarko, B., & Hook, V. (2021). Human tau isoforms and proteolysis for production of toxic tau fragments in neurodegeneration. *Frontiers in Neuroscience, 15*, 702788.

Braak, H., Alafuzoff, I., Arzberger, T., Kretzschmar, H., & Del Tredici, K. (2006). Staging of Alzheimer disease-associated neurofibrillary pathology using paraffin sections and immunocytochemistry. *Acta Neuropathologica, 112*, 389–404.

Cersonsky, T. E. K., Kellner, S., Chapman, S., Huey, E. D., Louis, E. D., & Cosentino, S. (2020). Profiles of Normal cognition in essential tremor. *Journal of the International Neuropsychological Society, 26*, 197–209.

Collins, K., Rohl, B., Morgan, S., Huey, E. D., Louis, E. D., & Cosentino, S. (2017). Mild cognitive impairment subtypes in a cohort of elderly essential tremor cases. *Journal of the International Neuropsychological Society, 23*, 390–399.

Critchley, M. (1949). Observations on essential (heredo familial) tremor. *Brain, 72*, 113–139.

Dana, C. L. (1887). Hereditary tremor: A hitherto undescribed form of motor neurosis. *American Journal of the Medical Sciences, 94*, 386–393.

Deuschl, G., Bain, P., Brin, M., & Ad Hoc Scientific Committee. (1998). Consensus statement of the Movement Disorder Society on tremor. *Movement Disorders, 13*(Suppl 3), 2–23.

Deuschl, G., Bhatia, K. P., Elble, R., & Hallett, M. (2018). Understanding the new tremor classification. *Movement Disorders, 33*, 1267–1268.

Deuschl, G., & Elble, R. (2009). Essential tremor – Neurodegenerative or nondegenerative disease towards a working definition of ET. *Movement Disorders, 24*, 2033–2041.

Deuschl, G., Petersen, I., Lorenz, D., & Christensen, K. (2015). Tremor in the elderly: Essential and aging-related tremor. *Movement Disorders, 30*, 1327–1334.

DiDio, A. S., & Koller, W. C. (2005). The history of essential tremor. In K. E. Lyons, & R. Pahwa (Eds.), *Handbook of essential tremor and other tremor disorders* (pp. 3–11). Boca Raton, FL: Taylor & Francis Group.

Elahi, F. M., & Miller, B. L. (2017). A clinicopathological approach to the diagnosis of dementia. *Nature Reviews. Neurology, 13*, 457–476.

Elble, R. J. (2020). Do we belittle essential tremor by calling it a syndrome rather than a disease? No. *Frontiers in Neurology, 11*, 586606.

Elble, R. J., Dubinsky, R. M., & Ala, T. (2007). Alzheimer's disease and essential tremor finally meet (editorial). *Movement Disorders, 22*, 1525–1527.

Farrell, K., Cosentino, S., Iida, M. A., Chapman, S., Bennett, D. A., Faust, P. L., et al. (2019). Quantitative assessment of pathological tau burden in essential tremor: A postmortem study. *Journal of Neuropathology and Experimental Neurology, 78*, 31–37.

Fasano, A., Lang, A. E., & Espay, A. J. (2018). What is "essential" about essential tremor? A diagnostic placeholder. *Movement Disorders, 33*, 58–61.

Hopfner, F., Ahlf, A., Lorenz, D., Klebe, S., Zeuner, K. E., Kuhlenbaumer, G., et al. (2016). Early- and late-onset essential tremor patients represent clinically distinct subgroups. *Movement Disorders, 31*, 1560–1566.

Hopfner, F., & Deuschl, G. (2018). Is essential tremor a single entity? *European Journal of Neurology, 25*, 71–82.

Huang, Y., Huang, C., Zhang, Q., Shen, T., & Sun, J. (2022). Serum NFL discriminates Parkinson disease from essential tremor and reflect motor and cognition severity. *BMC Neurology, 22*(1), 39.

Jack, C. R., Jr., Bennett, D. A., Blennow, K., Carrillo, M. C., Dunn, B., Haeberlein, S. B., et al. (2018). NIAA-AA research framework: Toward a biological definition of Alzheimer's disease. *Alzheimer's & Dementia, 14*, 535–562.

Kim, J.-S., Song, I.-U., Shim, Y.-S., Park, J.-W., Yoo, J.-Y., Kim, Y.-I., et al. (2009). Cognitive impairment in essential tremor without dementia. *Journal of Clinical Neurology, 5*, 81–84.

Kim, S. H., Farrell, K., Cosentino, S., Vonsattel, J.-P. G., Faust, P. L., Cortes, E. P., et al. (2021). Tau isoform profile in essential tremor diverges from other tauopathies. *Journal of Neuropathology and Experimental Neurology, 80*, 835–843.

Lenka, A., & Louis, E. D. (2020). Do we belittle essential tremor by calling it a syndrome rather than a disease? Yes. *Frontiers in Neurology, 11*, 522687.

Lombardi, W. J., Woolston, D. J., Roberts, J. W., & Gross, R. E. (2001). Cognitive deficits in patients with essential tremor. *Neurology, 57*, 785–790.

Lopez, O. L. (2013). Mild cognitive impairment. *Continuum, 19*, 411–424.

Louis, E. (2000). Essential tremor. *Archives of Neurology, 57*, 1522–1524.

Louis, E. D. (2018). Essential tremor: "Plus" or "minus". Perhaps now is the time to adopt the term "the essential tremors". *Parkinsonism & Related Disorders, 56*, 111–112.

Louis, E. D., Benito-Leon, J., Fahn, S., Frucht, S. J., Jankovic, J., Ondo, W. G., et al. (2020). Essential tremor-plus: A controversial new concept. *Lancet Neurology, 19*, 266–270.

Louis, E. D., Benito-Leon, J., Vega-Quiroga, S., Bermejo-Pareja, F., & The Neurological Disorders in Central Spain (NEDICES) Study Group. (2010a). Cognitive and motor functional activity in non-demented community-dwelling essential tremor cases. *Journal of Neurology, Neurosurgery, and Psychiatry, 81*, 997–1001.

Louis, E. D., Benito-León, J., Vega-Quiroga, S., Bermejo-Pareja, F., & The Neurological Disorders in Central Spain (NEDICES) Study Group. (2010b). Faster rate of cognitive decline in essential tremor cases than controls: A prospective study. *European Journal of Neurology, 17*, 1291–1297.

Louis, E. D., Joyce, J. L., & Cosentino, S. (2019). Mind the gaps: What we don't know about cognitive impairment in essential tremor. *Parkinsonism & Related Disorders, 63*, 10–19.

Louis, E. D., Ma, K., Babij, R., Cortés, E., Liemb, R. K., Vonsattel, J.-P. G., et al. (2012). Neurofilament protein levels: Quantitative analysis in essential tremor cerebellar cortex. *Neuroscience Letters, 518*, 49–54.

Louis, E. D., & McCreary, M. (2021). How common is essential tremor? Update on the worldwide prevalence of essential tremor. *Tremor and Other Hyperkinetic Movements, 11*, 28.

Louis, E. D., Radler, K. H., Huey, E. D., & Cosentino, S. (2021). Progression to dementia in patients with essential tremor. *Parkinsonism & Related Disorders, 83*, 123–124.

Lucas, J. A., Rippeth, J. D., Uitti, R. J., Shuster, E. A., & Wharen, R. E. (2000). Neuropsychological functioning in a patient with essential tremor with and without bilateral VIM stimulation. *Brain and Cognition, 42*, 253–267.

Muthuraman, M., Deuschl, G., Anwar, A. R., Mideksa, K. G., von Helmolt, F., & Schneider, S. A. (2015). Essential and aging-related tremor: Differences of central control. *Movement Disorders, 30*, 1673–1680.

Novellino, F., Vastaa, R., Saccà, V., Nisticòa, R., Morelli, M., Arabia, G., et al. (2020). Hippocampal impairment in patients with essential tremor. *Parkinsonism & Related Disorders, 72*, 56–61.

Pan, J. J., Lee, M., Honig, L. S., Vonsattel, J.-P. G., Faust, P. L., & Louis, E. D. (2014). Alzheimer's-related changes in non-demented essential tremor patients vs. controls: Links between tau and tremor? *Parkinsonism & Related Disorders, 20*, 655–658.

Petersen, R. C., Lopez, O., Armstrong, M. J., Getchius, T. S. D., Ganguli, M., Gloss, D., et al. (2018). Practice guideline update summary: Mild cognitive impairment. *Neurology, 90*, 126–135.

Prasad, S., Shah, A., Bhalsing, K. S., Kumar, K. J., Saini, J., Ingalhalikar, M., et al. (2019). Abnormal hippocampal subfields are associated with cognitive impairment in essential tremor. *Journal of Neural Transmission, 126*, 597–606.

Quinn, N. P., Schneider, S. A., Schwingenschuh, P., & Bhatia, K. P. (2011). Tremor—Some controversial aspects. *Movement Disorders, 26*, 18–23.

Radler, K. H., Zdrodowska, M. A., Dowd, H., Cersonsky, T. E. K., Huey, E. D., Cosentino, S., et al. (2020). Rate of progression from mild cognitive impairment to dementia in an essential tremor cohort: A prospective, longitudinal study. *Parkinsonism & Related Disorders, 74*, 38–42.

Ratajska, A. M., Lopez, F. V., Kenney, L., Jacobson, C., Foote, K. D., Okun, M. S., et al. (2021). Cognitive subtypes in individuals with essential tremor seeking deep brain stimulation. *The Clinical Neuropsychologist*, 1–23. *Published online Feb 10, 2021.*

Sahin, H. A., Terzi, M., Ucak, S., Yapici, O., Basoglu, T., & Onar, M. (2006). Frontal functions in young patients with essential tremor: A case comparison study. *The Journal of Neuropsychiatry and Clinical Neurosciences, 18*, 64–72.

Samarasekera, U. (2021). Modelling dementia. *The lancet. Neurology, 20*(8), 599–600.

Schmahmann, J. D., & Sherman, J. C. (1998). The cerebellar cognitive affective syndrome. *Brain, 121*, 561–579.

Sengul, Y., Sengul, H. S., Yucekaya, S. K., Yucel, S., Bakim, B., Pazarci, N. K., et al. (2015). Cognitive functions, fatigue, depression, anxiety, and sleep disturbances: Assessment of nonmotor features in young patients with essential tremor. *Acta Neurologica Belgica, 115*, 281–287.

Shill, H. A., Hentz, J. G., Jacobson, S. A., Belden, C., Sabbagh, M. N., Beach, T. G., et al. (2014). Essential tremor in the elderly and risk for dementia. *Journal of Neurodegenerative Diseases, 2014*, 328765.

Shill, H. A., Mehta, S., & Adler, C. H. (2021). Essential tremor and cognitive decline. *Parkinsonism & Related Disorders, 83*, 125.

Sinoff, G., & Badarny, S. (2014). Mild cognitive impairment, dementia, and affective disorders in essential tremor: A prospective study. *Tremor and Other Hyperkinetic Movements, 4*, 1–7.

Thawani, S. P., Schupf, N., & Louis, E. D. (2009). Essential tremor is associated with dementia: Prospective population-based study in New York. *Neurology, 73*, 621–625.

Tolosa, E., & Molinuevo, J. L. (2010). Cognition in essential tremor: Should we worry about progressive cognitive decline? (editorial). *European Journal of Neurology, 17*, 1227–1228.

Tröster, A. I., Fields, J. A., Pahwa, R., Wilkinson, S. B., Strait-Tröster, K. A., Lyons, K., et al. (1999). Neuropsychological and quality of life outcome after thalamic stimulation for essential tremor. *Neurology, 53*, 1774–1780.

Tröster, A. I., Woods, S. P., Fields, J. A., Lyons, K. E., Pahwa, R., Higginson, C. I., et al. (2002). Neuropsychological deficits in essential tremor: An expression of cerebello-thalamo-cortical pathophysiology? *European Journal of Neurology, 9*, 143–151.

SECTION V

Is essential tremor a disorder primarily due to GABA dysfunction?

Sheng-Han Kuo[a,b] and Elan D. Louis[c]

[a]Department of Neurology, College of Physicians and Surgeons, Columbia University, New York, NY, United States
[b]Initiative for Columbia Ataxia and Tremor, Columbia University, New York, NY, United States
[c]Department of Neurology, University of Texas Southwestern Medical Center, Dallas, TX, United States

As part of the current series, which highlights current controversies in the essential tremor (ET) field, we commissioned authors to submit papers with opposing viewpoints on the pathophysiology of ET, one using available data to put forth the argument that ET is primarily a disorder resulting from gamma aminobutyric acid (GABA)ergic dysfunction (Gironell, 2022), and the other, using available data to put forth an argument against such a theory (Bellows and Jimenez-Shahed, 2022). We are grateful to both sets of authors for thoughtfully making their cases and here we provide additional commentary.

The GABA hypothesis of ET derives from several sources, including the observation that in a subset of ET patients tremor is responsive to ethanol and primidone (Lou & Jankovic, 1991; Zesiewicz & Kuo, 2015), which presumably act via the GABAergic system. In addition, a recent study demonstrated that a novel $GABA_A$ receptor allosteric modulator can reduce tremor in ET patients (Paskavitz, Bullock, Nguyen, et al., 2019), further fueling the GABA theory of ET. Indeed, the vast majority of medications that are used to treat ET modulate the GABAergic system. Arising from these types of observations, GABA deficiency has been hypothesized to be at the center of ET pathophysiology. Where is the location of GABA deficiency in ET? The location has been postulated to be at the Purkinje cell-deep cerebellar nuclear synaptic levels. Purkinje cells release GABA to modulate the firing patterns of deep cerebellar nuclear neurons (Tang, Blenkinsop, & Lang, 2019). In the examination of the postmortem ET cerebellum, there is Purkinje cell loss and Purkinje cell axonal pathology (Babij, Lee, Cortés, Vonsattel, et al., 2013; Louis, Faust, Vonsattel, et al., 2007), which suggests insufficient GABA neurotransmission from Purkinje cells onto deep cerebellar nuclear neurons.

If tremor in ET comes from insufficient GABA neurotransmission at the Purkinje cell-deep cerebellar nuclear synapses, reduction in or elimination of such synaptic transmission might be expected to recreate ET-like tremor. However, studies in mouse models of tremor do not demonstrate this. Elimination of neurotransmission between Purkinje cells and deep cerebellar neurons, using genetic approaches, does not produce tremor (Brown, White, van der Heijden, et al., 2020), and acute inhibition of Purkinje cell inputs to the deep cerebellar nuclear neurons, using optogenetics, in mouse models of tremor lead to *inhibition* of tremor, rather than *enhancement* of tremor (Pan et al., 2020). Of course, these are animal models of tremor and one must be careful in extrapolating observations to humans with the specific disease ET, in which the pathophysiology is distinctive. While the above studies in a mouse model do not support that GABA deficiency at the Purkinje cell-deep cerebellar nuclear synapses causes tremor, other factors must be taken into further consideration. ET is a chronic, progressive disease, and tremor thus may derive from a chronic, compensatory change in the GABAergic system of the cerebellum, in response to partial Purkinje cell loss. To date, although such compensatory changes of the GABA system in ET have not been extensively studied, there is some emerging evidence for Purkinje cell compensatory changes (Louis & Faust, 2020; Louis, Hernandez, & Dyke, 2018).

Genetic studies in ET have not identified any targets related to GABAergic neurotransmission (Welton et al., 2021), and RNAseq analysis of the postmortem ET brain also has not revealed any transcript level alterations of the GABAergic system (Martuscello et al., 2020). Although further studies are needed, these initial studies do not provide support for the theory that ET is primarily a disorder of GABAergic dysfunction. Interestingly, the recent identification of excitatory synaptic pathology in ET may provide an alternative explanation. Specifically, abnormal climbing fiber-Purkinje cell synapses have been identified in the postmortem ET cerebellum (Lin et al., 2014). Interestingly, a mouse model, *hotfoot17J*, with ET-like climbing fiber-Purkinje cell synaptic pathology develops an action tremor that shares features with ET and that can be suppressed by both ethanol and primidone (Pan et al., 2020). These data suggest that action tremor can be generated primarily from abnormal excitatory synapses in the cerebellum but still can secondarily be responsive to GABAergic medications. The detailed pathophysiology will need to be further explored.

These complex issues are discussed by Drs. Gironell (Gironell, 2022), Bellows, and Jimenez-Shahad (Bellows and Jimenez-Shahed, 2022), and with important implications for future therapy development of ET. Drs. Lang, Handforth (Lang and Handforth, 2022), and Manto (Manto, 2022) in this series also have provided in-depth discussion of ET pathophysiology. There is no doubt that GABA neurotransmission is involved and is important in the pathophysiology of ET, but whether GABA dysfunction is the primary cause of tremor has been brought to the center of the debate. It is likely that the therapeutic efforts may be more effective when targeting the primary cause, rather than the secondary cause.

In addition to the argument whether GABA deficiency is the primary cause of tremor in ET, there are related questions in the field. First, what types of GABA dysfunction are important in ET? While a decrease in $GABA_A$ and $GABA_B$ receptor levels has been identified in the postmortem ET dentate nucleus (Paris-Robidas, Brochu, Sintes, et al., 2012), an extra-synaptic up-regulation of other types of GABA receptors has been hypothesized to explain ethanol responsiveness in the face of reduced GABA receptor levels (Handforth & Lang, 2021). These extra-synaptic GABA receptors have not been studied in the postmortem ET brain. Second, where is the exact location of GABAergic dysfunction in ET? GABAergic neurotransmission exists across several different synapses in the olivocerebellum: (1) Purkinje cells onto deep cerebellar nuclear neurons, (2) interneurons onto Purkinje cells, (3) deep cerebellar nuclear neurons onto inferior olivary neurons. Perturbed GABAergic synaptic transmission in each place requires further exploration. Third, only about two thirds of ET patients report some responsiveness to ethanol (Lou & Jankovic, 1991). Does this mean that ET patients who do not respond to ethanol do not have GABAergic dysfunction? Could ethanol responsiveness or other biomarkers be used to predict pharmacological response of tremor? Answers to these questions will be important to further our understanding of GABA dysfunction in ET.

References

Babij, R., Lee, M., Cortés, E., Vonsattel, J. P., et al. (2013). Purkinje cell axonal anatomy: Quantifying morphometric changes in essential tremor versus control brains. *Brain, 136*, 3051–3061.

Bellows, S., & Jimenez-Shahed, J. (2022). Is essential tremor a disorder of GABA dysfunction? No. *International Review of Neurobiology, 163*, 285–310.

Brown, A. M., White, J. J., van der Heijden, M. E., et al. (2020). Purkinje cell misfiring generates high-amplitude action tremors that are corrected by cerebellar deep brain stimulation. *eLife, 9*, e51928.

Gironell, A. (2022). Essential tremor is a disorder of primary GABA dysfunction? Yes. *"Where there is smoke, there is fire". International Review of Neurobiology, 163*, 259–284.

Handforth, A., & Lang, E. J. (2021). Increased Purkinje cell complex spike and deep cerebellar nucleus synchrony as a potential basis for syndromic essential tremor. A review and synthesis of the literature. *Cerebellum, 20*, 266–281.

Lang, E. J., & Handforth, A. (2022). Is the inferior olive central to essential tremor? Yes. *International Review of Neurobiology, 163*, 133–165.

Lin, C. Y., Louis, E. D., Faust, P. L., Koeppen, A. H., Vonsattel, J. P., & Kuo, S. H. (2014). Abnormal climbing fibre-Purkinje cell synaptic connections in the essential tremor cerebellum. *Brain, 137*, 3149–3159.

Lou, J. S., & Jankovic, J. (1991). Essential tremor: Clinical correlates in 350 patients. *Neurology, 41*, 234–238.

Louis, E. D., & Faust, P. L. (2020). Essential tremor pathology: Neurodegeneration and reorganization of neuronal connections. *Nature Reviews. Neurology, 16*, 69–83.

Louis, E. D., Faust, P. L., Vonsattel, J. P., et al. (2007). Neuropathological changes in essential tremor: 33 cases compared with 21 controls. *Brain, 130*, 3297–3307.

Louis, E. D., Hernandez, N., & Dyke, J. P. (2018). In vivo dentate nucleus gamma-aminobutyric acid concentration in essential tremor vs. controls. *Cerebellum., 17*, 165–172.

Manto, M. (2022). Essential tremor is a genuine cerebellar disorder and the cerebellar cortex is the culprit. *International Review of Neurobiology, 163*, 167–187.

Martuscello, R. T., Kerridge, C. A., Chatterjee, D., Hartstone, W. G., Kuo, S.-H., Sims, P. A., et al. (2020). Gene expression analysis of the cerebellar cortex in essential tremor. *Neuroscience Letters, 721*, 134540.

Pan, M. K., Li, Y. S., Wong, S. B., Ni, C. L., Wang, Y. M., Liu, W. C., et al. (2020). Cerebellar oscillations driven by synaptic pruning deficits of cerebellar climbing fibers contribute to tremor pathophysiology. *Science Translational Medicine*, 12, eaay1769.

Paris-Robidas, S., Brochu, E., Sintes, M., et al. (2012). Defective dentate nucleus GABA receptors in essential tremor. *Brain, 135*, 105–116.

Paskavitz, J., Bullock, A., Nguyen, D., et al. (2019). GABA-A receptor positive allosteric modulators: Phase 2 proof of concept studies of brexanolone injection and SAGE-217 in essential tremor. *MDS Abstract*. https://www.mdsabstracts.org/abstract/gaba-a-receptor-positive-allosteric-modulators-phase-2-proof-of-concept-studies-of-brexanolone-injection-and-sage-217-in-essential-tremor/.

Tang, T., Blenkinsop, T. A., & Lang, E. J. (2019). Complex spike synchrony dependent modulation of rat deep cerebellar nuclear activity. *eLife, 8*, e40101.

Welton, T., Cardoso, F., Carr, J. A., Chan, L. L., Deuschl, G., Jankovic, J., et al. (2021). Essential tremor. *Nature Reviews. Disease Primers*, 7, 83.

Zesiewicz, T. A., & Kuo, S. H. (2015). Essential tremor. BMJ. *Clinical Evidence, 2015*, 1206.

CHAPTER NINE

Is essential tremor a disorder of primary GABA dysfunction? Yes

Alexandre Gironell*
Movement Disorders Unit, Department of Neurology, Sant Pau Hospital, Autonomous University of Barcelona, Catalonia, Spain
*Corresponding author: e-mail address: agironell@santpau.cat

Contents

Abstract

Dysfunction in gamma-aminobutyric acid (GABA) neurotransmission has emerged as a prime suspect for the underlying neurochemical dysfunction in essential tremor (ET). This dysfunction has been termed the GABA hypothesis. We review findings to date supporting the 4 steps in this hypothesis in studies of cerebrospinal fluid, pathology, genetics, animal models, imaging, computational models, and human drugs, while not overlooking the evidence of negative studies and controversies. It remains to be elucidated whether reduced GABAergic tone is a primary contributing factor to ET pathophysiology, a consequence of altered Purkinje cell function, or even a result of Purkinje

International Review of Neurobiology, Volume 163
ISSN 0074-7742
https://doi.org/10.1016/bs.irn.2022.02.005

cell death. More studies are clearly needed to confirm both the neurodegenerative nature of ET and the reduction in GABA activity in the cerebellum. Also necessary is to test further therapies to enhance GABA transmission specifically focused on the cerebellar area.

The etiology, pathophysiology, and precise anatomy of essential tremor (ET) remain to be fully elucidated. Current knowledge of the neurochemical abnormalities underlying ET is incomplete, although there is undeniable evidence to support the role of the dysfunctional gamma-aminobutyric acid (GABA) system (Gironell & Marin-Lahoz, 2016; Jiménez-Jiménez, Alonso-Navarro, García-Martín, & Agúndez, 2020).

Dysfunction in GABA neurotransmission has emerged as a prime suspect for the underlying neurochemical dysfunction in ET. This dysfunction has been termed the GABA hypothesis (Gironell, 2014). Supporting evidence in each step of this hypothesis point to the pathophysiological origin of ET. The aim of this review is to provide a clear overview of the relationship between GABA and ET.

1. Background

1.1 The GABA hypothesis

The GABA hypothesis—which postulates a role for underlying disturbances in the GABAergic system, in particular in the cerebellum (Marin-Lahoz & Gironell, 2016)—has, for the past 3 decades, been the main theory to explain the pathophysiological basis of ET (Gironell, 2014; Louis, 1999). This hypothesis implies a disturbance of the GABAergic system in ET, specially involving the cerebellum (Marin-Lahoz & Gironell, 2016). The hypothesis comprises 4 main steps (Fig. 1): (1) cerebellar degeneration with Purkinje cell loss, (2) decreased GABA activity in deep cerebellar neurons, (3) disinhibition of pacemaker-like activity in deep cerebellar neurons, and (4) increased rhythmic thalamus and thalamo-cortical circuit activity causing tremor.

1.2 Cerebellar degeneration model

Proposed to explain the pathophysiology of ET are 2 main models: (1) the traditional olivary model, which posits that ET is the result of pathological pacemaker-like activity in the inferior olivary nucleus, and (2) the more recent cerebellar degeneration model, which posits that ET is a neurodegenerative disease that originates in the cerebellum and is possibly related to

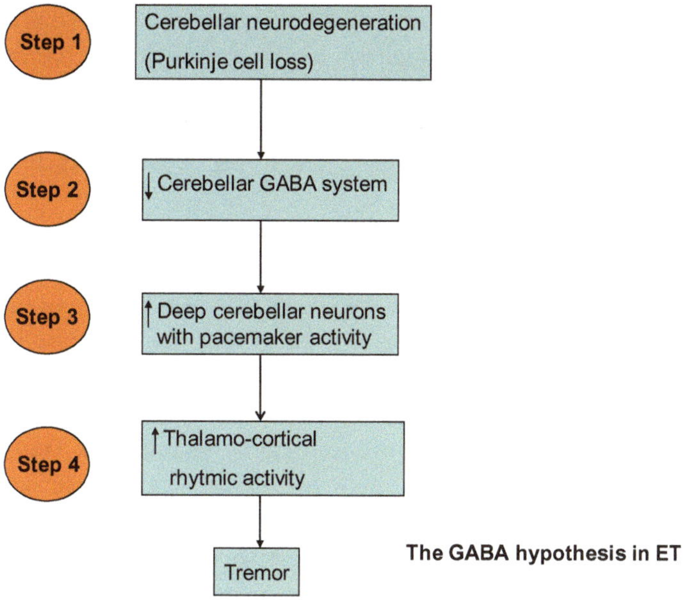

Fig. 1 Schematic representation of the GABA hypothesis in essential tremor (Gironell, 2014).

deficient or abnormal Purkinje cell outputs (step 1 in the GABA hypothesis). An underlying biochemical dysfunction in GABA neurotransmission is the possible link in the second model (Louis, 2014a, 2014b; Shaefer, Rodriguez, & Louis, 2017).

Motor control is the most important and most widely studied function of the cerebellum, which can accurately control movement through real-time integration of motor afferents. The computations performed by the cerebellum, however, are not fully understood. For simulation purposes, one of the main models used is the proportional integral derivative controller (of several possible manufacturing control systems), which, when a setting fails, tends to produce a rhythmic oscillating error at the target. When the final output of the system is position, this oscillation becomes tremor (Jo & Massaquoi, 2004).

Leaving aside this modeling logic, several studies have suggested that the pathology of ET arises in the cerebellum. Verified cerebellar disorders (spinocerebellar ataxias) are known to cause tremor (Gan et al., 2017), but in some cases, ET has improved following a cerebellar stroke (Dupuis, Delwaide, Boucquey, & Gonsette, 1989). While this may seem contradictory, it is less so if we consider cerebellar structure and functioning. Cerebellar afferents (parallel fibers and climbing fibers) synapse with Purkinje cells in the cerebellar cortex and, through inhibition, these cells link the cerebellar cortex with the deep nuclei. Since these deep nuclei activity project to the rest of the central nervous system (CNS), a lesion in this zone would diminish that output, improving various kinds of tremor including ET (Dupuis et al., 1989).

Cerebellar Purkinje cells are large neurons with multiple branching extensions. Found in the cortex of the cerebellum, they play a fundamental role in motor control (Hallett & Dubinsky, 1993). Characterized by cell bodies with numerous branching dendrites, they are flask-like in shape, and have a single long axon. Most Purkinje cells release the neurotransmitter GABA, which inhibits the dentate nucleus and certain deep cerebellar output neurons, reducing the transmission of nerve impulses. Those inhibitory functions enable Purkinje cells to regulate and coordinate motor movements.

Postmortem studies have pointed to neuropathological changes in the cerebellum of patients with ET that are not present in healthy controls (Axelrad et al., 2008; Louis et al., 2007; Louis, Babij, Lee, Cortés, & Vonsattel, 2013; Louis, Faust, & Vonsattel, 2012). These studies identified abnormalities in the cerebellum, which primarily affected Purkinje cells, basket cells and climbing fibers. It has been suggested that some of these pathological changes (for example, focal Purkinje cells axonal swellings, swellings in and regression of the Purkinje cells dendritic arbor and Purkinje cells death) are likely to be primary and degenerative. By contrast, other changes, such as an increase in Purkinje cells recurrent axonal collateral formation and hypertrophy of GABAergic basket cell axonal processes, could be compensatory responses to restore cerebellar GABAergic tone and cerebellar cortical inhibitory efficacy. Such compensatory responses are likely to be insufficient, enabling the disease to progress (Louis & Faust, 2020). Other studies reported no Purkinje cell loss in autopsied patients with ET (Rajput, Robinson, Rajput, Robinson, & Rajput, 2012; Shill et al., 2008; Symanski et al., 2014). However, methological flaws have been described for these negative studies (Louis & Faust, 2020).

Neuroimaging studies, studying structural and functional brain changes in ET patients reveal a possible gray matter damage in the cerebellum, although some conflicting results have been obtained, which may be explained by differences in the clinical features of the ET patients studied (Boecker et al., 1996; Colebatch, Findley, Frackowiak, Marsden, & Brooks, 1990; Daniels et al., 2006; Hallett & Dubinsky, 1993; Louis et al., 2002; Louis, Shungu,

Mao, Chan, & Jurewicz, 2004; Wills, Jenkins, Thompson, Findley, & Brooks, 1994). Furthermore, widespread white matter damage and cerebello-thalamo-cortical functional alterations were consistent findings in many reports. Overall, both structural and functional magnetic resonance imaging (MRI) studies suggested cerebellar and cerebello-thalamo-cortical circuit damage in ET patients (Pitracupa, Bolonga, Tommasin, Berardelli, & Pantano, 2021).

There is electrophysiological evidence of an association between the cerebellum and ET. Short-term (Gironell et al., 2002) and long-term (Popa et al., 2013) tremorlytic effects have been reported for repetitive transcranial magnetic stimulation (rTMS) targeting the cerebellum, possibly the result of the activation of Purkinje cells through parallel fibers inhibiting deep cerebellar nuclei (Popa, Russoc, & Meunierc, 2010).

The progressive nature of ET clearly supports the cerebellar degeneration model, and consequently, the GABA hypothesis. As shown by both prospective and retrospective studies, ET is a slowly progressive disorder. Elble and coworkers (Elble, 2000), using accelerometry, followed 44 patients with ET over 4 years, finding a 7.4% annual worsening of tremor from baseline. Putzke and coworkers (Putzke, Whaley, Baba, Wizolek, & Uiti, 2006), in a prospective study of 45 patients with a mean follow-up of 3.6 years, estimated an annual worsening of 12% from baseline. For 83 patients followed for a mean of 5 years, mean annual worsening rates from baseline were between 3.1% and 5.3% (Louis, Agnew, Gillman, Gerbin, & Viner, 2011), and for a longitudinal spirograph analysis of 116 patients followed for a mean of 5.8 years, spiral scores worsened by a mean of 0.12 points annually (Louis, Michalec, & Gillman, 2014).

Louis and coworkers (Louis, Gerbin, & Galecki, 2013) performed a cross-sectional study to determine the clinical situation of ET in 335 patients at 10-year milestones. At each milestone, patients underwent a single evaluation of tremor severity. It was found that between 20% and 60% of patients had a large tremor amplitude after disease duration of over 40 years, and fewer than 10% were incapacitated. The authors concluded that ET progressively worsens over decades and also noted that patients with long-standing ET did not necessarily converge at the same severe end-stage of functionally incapacitating tremor.

In a very long-term retrospective analysis of patients with ET aimed at assessing tremor severity, our group used the Glass Scale, which is similar to the Hoehn and Yahr scale used for Parkinson's disease (Gironell, Martinez-Corral, Pagonabarraga, & Kulisevsky, 2010). The Glass Scale score is obtained by asking the patient a single question: "When you are seated at the table, how do you drink water from a glass with your dominant hand?" By means of extensive interviews with the patients and their relatives, Glass Scale scores were retrospectively obtained throughout the patient's life in 50 consecutive patients with severe ET. Authors asked when they first noted tremor in arms (Glass Scale I), when tremor increased, making it difficult to drink from a glass using one hand (Glass Scale II), when tremor provoked the need to use both hands to drink (Glass Scale III), and when patients needed a straw to drink (Glass Scale IV). Common milestone events were used to help to remember tremor severity changes, such as: school leaving age, age of military service in the case of men, marriage, birth or wedding of a child, and retirement. Authors confirmed that ET is a slowly progressive disorder. Mean $\pm$ SD years to attain the different Glass scale scores were 17 ± 10 years for score II, 25 ± 13 years for score III, and 36 ± 15 years for score IV (Gironell, Ribosa-Nogué, Gich, Marin-Lahoz, & Pascual-Sedano, 2015).

2. GABA

GABA, the main inhibitory neurotransmitter in mammals (Bowery et al., 2002), is involved in all CNS functions and structures, including the cerebellum. As such, the GABA system is targeted by a wide range of CNS agents, such as anxiolytics, sedative-hypnotics, general anesthetics, and anticonvulsants (Shaefer et al., 2017). Fig. 2 shows a schematic view of GABA synapse.

GABA is synthesized in a single step from l-glutamate by l-glutamic acid decarboxylase. Glutamate must therefore be available from certain nerve endings for GABA biosynthesis. Much of the glutamate and GABA used for neurotransmission is derived from glial storage pools of glutamine, with GABA released from electrically stimulated inhibitory nerve cells. The application of GABA and structural analogues to cells innervated by GABAergic neurons produces effects on the target cells identical to those produced by stimulating inhibitory innervation. GABA is removed from the synaptic cleft by GABA transporters and metabolized. GABA can also be converted into glutamate.

GABA acts through two types of receptors: $GABA_A$ and $GABA_B$. $GABA_A$ receptors are linked to chloride (Cl^-) channels, which, when open, lead to fast hyperpolarization, while $GABA_A$ receptors are pentamers that usually consist of 2 alpha, 2 beta, and either 1 gamma or 1 delta subunit.

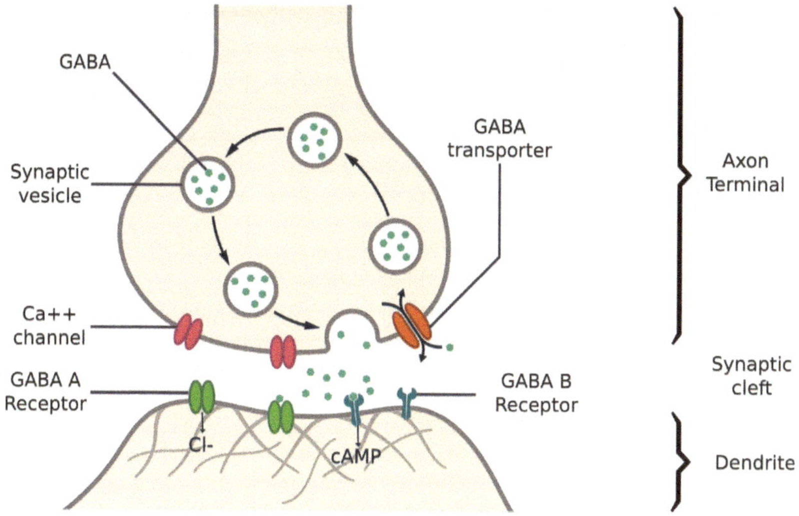

Fig. 2 GABA synapsis scheme (Marin-Lahoz & Gironell, 2016).

To add to this complexity, several distinct forms of each subunit exist (e.g., alpha1 to alpha 6). The representation of these subunits in $GABA_A$ receptors across different brain regions is detailed in Table 1, along with a list of known receptor agonists (Shaefer et al., 2017).

The pharmacology of the $GABA_A$ receptor is interesting. While drugs for most other neurotransmitters act by binding to the original receptor site, those acting on the GABAergic system usually bind to a different site, thereby modulating the receptor's properties. Although ethanol has multiple CNS effects, it is mainly considered to be a positive allosteric modulator of $GABA_A$ receptors. Importantly, this could be considered to be the only effect at low doses (De Haas et al., 2012; Zoethout et al., 2012). Gabapentin does not bind to the $GABA_A$ receptor; rather, it increases GABA synthesis (Bowery et al., 2002; De Haas et al., 2012). Primidone is metabolized into phenobarbital, which also acts as an allosteric modulator enhancing GABA transmission, and, at higher doses, acts as a $GABA_A$ agonist.

$GABA_B$ is a metabotropic receptor coupled to G proteins that mediates inhibitory responses (mainly via potassium channels), reducing the activity of adenylyl cyclase and neuronal conductance to Ca++ (Chen, Li, Ye, Zhang, & Wang, 2005; Mouginot & Gahwiler, 1996). $GABA_B$ receptors mediate slow responses, whereas $GABA_A$ receptors mediate fast responses. The main known agonists of $GABA_B$ receptors apart from GABA are

Table 1 GABA receptors brain location and known receptor agonists.

$GABA_A$ receptor	Locations	Agonists
Alpha 1	Widespread; most in cerebellum and thalamus	Benzodiazepines, zolpidem
Alpha 2	Limbic structures, hippocampus, striatum, hypothalamus	Benzodiazepines, zolpidem
Alpha 3	Cerebellar dentate nucleus, hippocampus, ION spinal cord	Benzodiazepines, zolpidem
Alpha 4	Thalamus, hippocampus, striatum	Ethanol, topiramate
Alpha 5	Hippocampus	Benzodiazepines
Alpha 6	Cerebellar granule cells	Ethanol, topiramate
Gamma	Widespread	Benzodiazepines, zolpidem
Delta	Cerebellar granule cells, hippocampus, thalamus	Ethanol
$GABA_B$ receptor	Widespread, most in thalamus, cerebellar molecular layer, interpeduncular nucleus	Baclofen

baclofen and gamma-hydroxybutyrate. Baclofen, studied in a harmaline model of tremor but not of ET, was found to improve tremor in a rat model (Tariq, Arshaduddin, Biary, Al Moutaery, & Al Deeb, 2001) but not in a mouse model (Paterson, Malekiani, Foreman, Olivier, & Hanania, 2009). More interestingly, gamma-hydroxybutyrate has shown some benefits for ET in 2 open-label clinical trials (Frucht, Bordelon, Houghton, & Reardan, 2005; Frucht, Houghton, Bordelon, Greene, & Louis, 2005).

3. GABA studies in essential tremor

3.1 Cerebrospinal fluid studies

In 1996, Mally and coworkers (Mally, Baranyi, & Vizi, 1996) studied cerebrospinal fluid and serum amino acid concentrations in 20 patients with ET compared to 10 controls, reporting a slight increase in glutamate concentrations, reduced GABA levels, and decreased glycine and serine levels in the patients with ET. Their finding suggesting GABAergic dysfunction in ET was among the first to lend support to the GABA hypothesis. The pathophysiological basis of ET seemed simple: a decrease in GABA, the main inhibitory neurotransmitter in the brain. However, this is the only study to have analyzed GABA levels in cerebrospinal fluid and the data have not been replicated.

3.2 Pathology studies

A 1996 neuropathology study revealed that $GABA_B$ receptors are present on the presynaptic terminals of Purkinje cells in the deep cerebellar nuclei (Mouginot & Gahwiler, 1996). These anatomic structures have since been attributed a probable pathogenic role in ET. The results of a more recent postmortem neuropathological study of the GABA hypothesis further support this role; Paris-Robidas and coworkers (Paris-Robidas et al., 2012) reported a 35% reduction in $GABA_A$ and a 22–31% reduction in $GABA_B$ receptors in the dentate nucleus of the cerebellum in 10 individuals with ET compared to 10 individuals with Parkinson's disease and to 16 controls (all evaluated by receptor-binding autoradiography); furthermore, concentrations of $GABA_B$ receptors in the dentate nucleus were inversely correlated with ET symptom duration, suggesting that $GABA_B$ receptors are lost as the disease progresses. Those authors suggested that a reduction in GABA receptors in the dentate nucleus leads to disinhibition of cerebellar pacemaker output activity, propagated along the cerebello-thalamo-cortical pathways and thus generating tremors (Paris-Robidas et al., 2012). Also lending support to the GABA dysfunction in ET are findings of decreased concentrations of parvalbumin (a marker of GABAergic neurons) in the locus coeruleus of patients with ET (Shill et al., 2008, 2012). However, a recent post-mortem study did not find any consistent patterns of $GABA_B$ receptor 1 (GBR1) changes in the cerebellar cortex of patients with ET (Luo, Rajput, Robinson, & Rajput, 2012).

3.3 Electrophysiological studies

Basic electrophysiology has yielded invaluable data for the GABA hypothesis in ET. In 2008, Uusisaari and Knopfel (2008), using electrophysiological recordings of deep cerebellar nuclei neurons, demonstrated that all dentate nucleus cells are sensitive to GABAergic input from Purkinje cells, and that both $GABA_A$ and $GABA_B$ receptors mediate the inhibitory input of these cells. Findings by other authors indicate that deep cerebellar nuclei neurons belong to the family of single cell oscillators and have pacemaker-like activity (Jahnsen, 1986; Llinas, 1988; Llinas & Muhlethaler, 1988). It has also been shown that this activity is under tonic control by GABAergic input from Purkinje cells (Mouginot & Gahwiler, 1996). Even more compelling is evidence that deep cerebellar nuclei neurons can impose rhythmicity in the form of highly regular patterns of activity on targeted thalamic neurons (Hua, Lenz, Zirh, Reich, & Dougherty, 1998; Pinault & Deschenes, 1992).

A recent study (Chuang, Huang, Lu, & Chen, 2014) of 13 patients with ET and 18 healthy controls involving continuous theta-burst stimulation (an efficient form of rTMS) of the primary motor cortex and premotor cortex showed a suppressive effect on motor cortical excitability in the patients with ET, together with a reduction or absence of changes in measures of intracortical inhibition, and a decrease in tremor amplitude. Those findings are consistent with an abnormal GABA modulation in ET outside of the cerebellum.

3.4 Genetic studies

Genetic studies in ET are characterized by methodological shortcomings, including small sample size, a lack of stringent diagnostic criteria and of biomarker use, a high phenocopy rate, evidence of non-Mendelian inheritance, and high locus heterogeneity in presumably monogenic ET (Kuhlenbäumer, Hopfner, & Deuschl, 2014).

The first genetic study of GABA and ET, conducted in 2006 by Deng and coworkers (Deng, Xie, Le, Huang, & Jankovic, 2006) in 76 patients with familial ET, reported a nonsignificant association between the $GABA_A$ receptor alpha-1 subunit (GABRA1) and ET, with the authors concluding that missense, nonsense, or splice site mutations in the coding regions of the $GABA_{A1}$ gene were not a major genetic cause of ET. More recent studies have also shown negative results. In 2011, Thier and coworkers (Thier et al., 2011), in a study of 503 patients with ET and 818 controls, found no association between ET and single nucleotide polymorphisms (SNPs) in 15 $GABA_A$ and 4 GABA transporter genes. García-Martin and coworkers published two studies with negative results in 2011. In the first (García-Martin et al., 2011a), they investigated a possible role of the GABA receptor subtypes rho1, rho2, and rho3 and allelic variants of the SNPs GABRR1-M26V, GABRR1-H27R, GABRR2-T455M, and GABRR3-Y205X in the development of ET, finding no significant differences in the frequencies of the GABBR1 genotypes and allelic variants between patients with ET and controls, and also finding no link to age at tremor onset, gender, tremor location, or tremor response to ethanol. In the second study (García-Martin et al., 2011b) using TaqMan genotyping, of 200 patients with familial ET and 250 healthy controls, the authors again found no differences in allelic and genotype frequencies for the GABRA4-L26M, GABRE-S102A, and GABRQ-I478F SNPs. Their results suggest that the studied GABRA4, GABRE, and GABRQ SNPs are not associated with risk of familial ET.

Although the above studies did not find any associations between most GABA receptor genes and ET risk, more recent studies have pointed to a possible role for other GABA-related genes. Liu and coworkers (Liu, Hernadez, Kisselev, et al., 2016) identified variants, in 3 different families with ET, that they predicted would affect function in *KCNS2* (KV9.2), *HAPLN4* (BRAL2), and *USP46* that co-segregated with the disease. These genes are highly expressed in the cerebellum and Purkinje cells and influence function of the GABA system. Tio and coworkers (Tio et al., 2016) developed transgenic *Drosophila* expressing the Q290X mutation in *FUS*, a gene that had been previously identified in a large family with ET, whose members improved in both motor dysfunction and impairment of the GABAergic pathway in response to gabapentin.

3.5 Experimental animal models

In 2005, Kralic and coworkers (Kralic, Criswell, Osterman, et al., 2005) investigated the mechanisms underlying ET using $GABA_A$ receptor alpha1 subunit knockout mice exhibiting ET-like postural and kinetic tremor and motor incoordination. Tests with current ET drugs (primidone, propranolol, and gabapentin) reduced the amplitude of the tremor, while non-sedative doses of ethanol eliminated the tremor. While the cerebellar Purkinje cells lost all responses to synaptic or exogenous GABA, they showed no differences in abundance, gross morphology, or spontaneous synaptic activity. The authors concluded that their genetic mouse model elucidated a GABAergic dysfunction mechanism in the major motor pathway, as well as potential targets for pharmacotherapy of ET.

A more recent study, used transgenic mice to selectively knock out the $GABA_A$ receptor alpha1 subunit from cerebellar Purkinje cells (Nietz et al., 2020). They found that these animals showed a tremor phenotype reminiscent of ET. They concluded that selective knockout of $GABA_A$ receptor alpha1 from Purkinje cells is sufficient to produce a tremor phenotype, although this tremor is less severe than seen in animals with a global knockout. Those findings illustrate that the cerebellum can play a key role in the genesis of the observed tremor phenotype.

3.6 Imaging studies

Neuroimaging findings lend relatively little support to the hypothesis that the inferior olivary nucleus is the primary oscillator of abnormal neuronal activity, and reveal robust evidence indicating that the cerebellum plays a major role in ET pathophysiology (Holtbern & Shah, 2021).

Volumetric MRI study results differ, and voxel-based morphometry studies have not clearly shown correlations with histopathological findings of cerebellar neurodegeneration. Disputing a uniform pattern of cerebellar cell loss is the fact that the topography of cerebellar regions displaying atrophy is inconsistent across studies. A recent meta-analysis that included 16 voxel-based morphometry studies and some 350 patients with ET highlighted differences in structural brain alterations reported for ET (Luo, Pan, Xu, & Chen, 2019). Diffusion tensor imaging studies have more consistently revealed microstructural alterations of the cerebellum, while functional MRI studies have clearly demonstrated abnormal cerebellar function and altered connectivity in cerebello-thalamo-cortical circuitry (Revuelta, McGill, Jensen, & Bonilha, 2019; Shin, Han, Kim, & Lee, 2008; Tikoo et al., 2020).

Radiotracer studies further underpin the crucial role of the cerebellum in tremor genesis, as they have shown increased regional cerebral blood flow in patients with ET. Meanwhile, findings from ^{18}F-fluorodeoxyglucose (^{18}F-FDG) positron emission tomography (PET) studies are mixed, with some such studies reporting extensive cortical cerebellar hypometabolism and evidence for increased thalamic activity (Song, Ha, Yang, & Chung, 2015), but with other studies reporting opposite findings—in particular, conflicting results for cerebellar metabolic activity (Ha et al., 2015; Hallett & Dubinsky, 1993; Sun et al., 2019).

While data from magnetic resonance spectroscopy (MRS) studies are insufficient to draw firm conclusions, the few such studies that do exist provide some evidence for thalamic and cerebellar neuronal damage (Louis, Hernandez, Dyke, Ma, & Dydak, 2018; Tapper, Göransson, Lundberg, Tisell, & Zsigmond, 2020). One MRS study reported a positive correlation between the cerebellar GABA/glutamate + glutamine ratio and clinical tremor scores for a small ET cohort (Pagan, Butman, Dambrosia, & Hallett, 2003), while a study employing proton MRS (^{1}H-MRS) failed to demonstrate significant differences in GABA concentrations in the dentate nuclei between 45 patients with ET and 28 controls (Louis et al., 2018).

Few in vivo imaging studies have analyzed the role of GABA in ET. For 8 patients with bilateral ET and 11 healthy controls, Boecker and coworkers (Boecker et al., 2010) performed a comparative PET study using $[^{11}C]$flumazenil, a tracer that specifically binds to the central benzodiazepine receptor sites of the $GABA_A$ receptor complex. For focally altered $[^{11}C]$flumazenil binding at the sites of tremor genesis—mainly for the cerebellum and interconnected thalamo-cortical pathways—they calculated parametric distribution volume images, finding that the patients with ET showed significant increases in $[^{11}C]$flumazenil binding at the benzodiazepine receptor site of the $GABA_A$ receptor in the cerebellum, the ventrolateral thalamus, and the lateral premotor cortex. In a correlation analysis of clinical scale scores and parametric images of binding potential for $[^{11}C]$flumazenil PET in 10 ET patients at different stages of clinical severity, tremor severity was found to be statistically correlated with GABA receptor binding abnormalities observed in the cerebellar vermis, bilateral posterior lobes, and right anterior lobe (Gironell et al., 2012).

It can therefore be concluded that findings from MRS and radiotracer studies lend support to the hypothesis that GABAergic system dysfunction is involved in ET pathophysiology.

3.7 Computational models

Zhang and Santaniello (2019), in a study that used sophisticated computational modeling to investigate tremor generation in ET, showed that tremor-related activity, originating in the olivocerebellar loop in response to dysfunction and compensatory upregulation of GABA receptors in the dentate nucleus of cerebellum, projects to the thalamus and reaches the cortico-thalamic motor loop. In their computational model of the cortico-cerebello-thalamo-cortical loop, they simulated the effects of a progressive loss of $GABA_A$ receptor alpha 1 subunit and upregulation of receptor alpha 2 and alpha 3 subunits in the dentate nucleus, studying the evolution of the firing patterns along the loop, and closely reproducing experimental evidence for each structure in the loop. They showed that altered amplitudes and decay times in GABAergic currents to the dentate nucleus can facilitate sustained oscillatory activity at tremor frequency, as well as a robust burst of activity in the thalamus. Tremor-related oscillations thus initiated in small neuronal populations spread to a larger network as synaptic dysfunction increases (Zhang & Santaniello, 2019). Those results would suggest a cerebellar dysfunction mechanism for tremor generation that may explain the origin of ET.

3.8 Human drugs

Weakening the olivary hypothesis is the fact that a significant effect in human patients with ET has not been demonstrated for T-type calcium channel blockers (Shaefer et al., 2017). One of the main arguments in favor of the GABA hypothesis is that anti-tremor effects are observed for certain GABAergic drugs, e.g., gabapentin. Overall, however, only about 30% to 60% patients respond to GABAergic drugs with declared anti-tremor effects, and the lack of response to several GABAergic drugs in controlled studies is also a concern. It may be that enhancing GABA neurotransmission is too broad a solution, as loss of this transmission in ET may simply be a marker of the underlying problem, namely, Purkinje cell loss.

3.9 Drugs that have shown some efficacy

3.9.1 Theophylline

A 1991 double-blind cross-over study of the effects of theophylline in 20 patients with ET reported significant improvements in tremor after 4 weeks of treatment (Mally & Stone, 1991). The hypothesis was that theophylline-enhanced GABA might explain the anti-tremor effect. In fact, previous studies in mice showed that after theophylline treatment GABA significantly potentiated *N*-methyl-D-aspartate depolarization. The enhancement of GABA sensitivity might be an important common factor in decreasing ET after theophylline treatment (Mally, Connick, & Stone, 1991).

3.9.2 Muscimol

In 1999, Pahapill and coworkers (Pahapill et al., 1999) performed an interesting experiment, in which they injected the $GABA_A$ agonist muscimol into the ventralis intermedius nucleus of the thalamus of 6 patients undergoing stereotactic procedures for ET. The drug was administered in areas where tremor-synchronous cells were electrophysiologically identified with microelectrode recordings, and where tremor reduction occurred with electrical microstimulation. Injections of muscimol, but not of saline solution, consistently reduced tremor in the patients. The authors concluded that GABA-mediated thalamic neuronal inhibition may represent a mechanism underlying the effectiveness of surgery for tremor, and also that GABA analogues could potentially be used therapeutically.

3.9.3 Gabapentin

Gabapentin, which is structurally related to GABA, penetrates the blood-brain barrier but does not modify $GABA_A$ or $GABA_B$ radioligand binding. It is not converted metabolically into GABA or a GABA agonist and does not inhibit GABA uptake or degradation. While its mechanism of action is not fully understood, recent data indicate that it can increase human brain GABA levels and reduce intracortical excitability. In 1999, Gironell et al. (1999) performed a randomized placebo-controlled comparative trial of gabapentin and propranolol in 16 ET patients not receiving anti-tremor medication. By day 15, both gabapentin and propranolol demonstrated significant and comparable efficacy in reducing baseline tremor according to all clinical and accelerometry tremor measures. Results of that study are in accordance with the central mechanism hypothesis that implicates the GABAergic system in the pathogenesis of ET, with the authors concluding that gabapentin merits a role in the pharmacological treatment of ET. Also in 1999, Pahwa and coworkers (Pahwa, Lyons, Hubble, Busenbark, & Rienerth, 1999) performed a controlled study of gabapentin as treatment for ET in 20 patients. Although some patients showed benefits for tremor, the results were not statistically significant for total tremor scores. In 2000, Ondo and coworkers (Ondo, Hunter, Dat Wuong, Scwartz, & Jankovic, 2000) performed a multiple-dose double-blind placebo-controlled trial with gabapentin in 25 patients with ET. While some endpoints of the study were positive (patient global assessment, water pouring scores, observed tremor scores, and activities of daily living), accelerometry, spirograph and investigator global impression scores did not improve. The results were similar for high and low doses, and the authors' conclusion was that gabapentin may be effective in some cases of ET.

The evidence-based American Academy of Neurology (AAN) guidelines for the treatment of ET consider that gabapentin is probably effective in reducing upper limb tremor (Zesiewicz et al., 2011). By contrast, the Movement Disorder Society Task Force on Tremor (MDS-TFT) considers that there is insufficient evidence on the benefits of gabapentin (Ferreira et al., 2019).

3.9.4 Topiramate

Topiramate is an anticonvulsant that blocks sodium channels and potentiates GABA activity. In view of the GABA hypothesis, several trials have tested this drug for ET. In 2002, in a controlled study in 24 patients with ET, Connor and coworkers (Connor, 2002) reported significant anti-tremor properties of topiramate (it improved all tremor clinical scores). In a multicenter controlled study of 208 patients with ET, Ondo and coworkers (Ondo et al., 2006) reported significant mean percentage improvement in overall scores of 29%. In 2006, Frima and Grünewald (2006) performed a controlled crossover study of topiramate in 16 ET patients, reporting no evidence of a therapeutic benefit for ET. In 2008, Connor and coworkers (Connor, Edwuards, & Tarsy, 2008) performed a double-blind placebo-controlled crossover study of topiramate in 62 patients with ET, finding a significant improvement (18–23%) in all tremor measures; however, the drop-out rate in that study was approximately 40% due to appetite suppression, weight loss, paresthesia, anorexia, and concentration difficulties.

The AAN guidelines consider that topiramate is probably effective in reducing upper limb tremor (Zesiewicz et al., 2011). The MDS-TFT goes a step further in classifying topiramate as efficacious (Ferreira et al., 2019). However, a systematic review on the efficacy and safety of topiramate for ET concluded that data were limited and only very low- to low-quality evidence supports topiramate efficacy (Bruno, Nicoletti, Filippini, et al., 2017; Bruno, Nicoletti, Quatrocchi, et al., 2017).

3.9.5 Pregabalin

Pregabalin is a structural derivative of the inhibitory neurotransmitter GABA. Like gabapentin, it does not bind directly to $GABA_A$, $GABA_B$, or benzodiazepine receptors, does not increase GABA responses in cultured neurons, and does not alter rat brain GABA concentrations or have acute effects on GABA uptake or degradation. However, in cultured neurons, prolonged application of pregabalin increases GABA transporter protein density and the rate of functional GABA transport.

In 2007, Zesievick and coworkers (Zesiewicz et al., 2007) published a double-blind study of pregabalin in 22 patients with ET, finding significant improvement in accelerometry and tremor clinical scores. However, in a 2009 controlled study of pregabalin in 20 patients with ET during a titration period of 6 weeks, no improvement was found for any tremor measure (Ferrara et al., 2009), while findings were likewise negative (no improvements in any tremor measure) for a multisite double-blind 2012 study of pregabalin in 29 patients with ET (Zesiewicz et al., 2012).

The AAN guidelines consider that there is insufficient evidence to either support or refute the usefulness of pregabalin in treating ET (Zesiewicz et al., 2011), while the MDS-TFT considers it to be non-efficacious (Ferreira et al., 2019).

3.10 Other GABAergic transmission-enhancing drugs

Several classes of drugs that enhance GABAergic transmission by binding to sites on the $GABA_A$ receptors are reported to be effective in the treatment of ET. These include barbiturates (phenobarbital and primidone) and benzodiazepines (alprazolam and clonazepam) (Chakrabarti & Pearce, 1981; Koller & Vetere-Overfield, 1989).

Primidone is a first-line anti-tremor agent, similar in efficacy to beta blockers, and considered efficacious by both the AAN (Zesiewicz et al., 2011) and the MDS-TFT (Ferreira et al., 2019). Alprazolam is also probably efficacious in the treatment of ET (Huber & Paulson, 1988), and is classified as such by the MDS-TFT. In contrast, the AAN and a systematic review published in 2015 (Bruno et al., 2015) concluded that there was insufficient evidence to support or refute alprazolam use for ET.

Ethanol, which is often effective in reducing the severity of tremor in ET, potentiates the activity of the $GABA_A$ receptor (Manto & Laute, 2008). Alcohol, in fact, is an indirect GABA agonist. It is believed to mimic the effect of GABA on the brain, binding to GABA receptors and inhibiting neuronal signaling.

While propranolol is believed to have a therapeutic effect on ET by antagonistically binding to non-specific peripheral beta-adrenergic receptors, there is some evidence that it may also act by enhancing GABA receptivity in the CNS (Cleeves & Findley, 1988).

3.10.1 Zonisamide

Zonisamide is a sulfonamide with anti-epileptic action. While its mechanism of action is not fully understood, it involves T-type calcium channels, glutamatergic neurotransmission, and GABAergic modulation (it binds allosterically to GABA receptors and inhibits GABA reuptake). In a $GABA_A$ receptor alpha 1 subunit-null mouse model, zonisamide significantly improved tremor (Handforth, Homanics, Covey, et al., 2010). A review of neuropharmacological treatments for ET reported some degree of clinical improvement with zonisamide in some studies, but no improvement in others (Alonso-Navarro, García-Martín, Agúndez, & Jiménez-Jiménez, 2020).

According to the AAN (Zesiewicz et al., 2011), the MDS-TFT (Ferreira et al., 2019), and a systematic review (Bruno, Nicoletti, Filippini, et al., 2017; Bruno, Nicoletti, Quatrocchi, et al., 2017), there is insufficient evidence to support or to refute the usefulness of zonisamide for ET.

3.11 Drugs that have not demonstrated efficacy

3.11.1 Progabide

In the first controlled ET trial of the GABA agonist progabide, of 18 patients with ET, Mondrup and coworkers (Mondrup, Dupont, & Pederson, 1983) found no significant differences between progabide and placebo in tremor scores. In another controlled trial with progabide in 10 ET patients 4 years later, Koller and coworkers (Koller, Rubino, & Gupta, 1987) likewise found no differences in tremor scores between progabide and placebo, concluding that altered GABA neurotransmission does not appear to be involved in the pathogenesis of ET.

3.11.2 Tiagabine

Tiagabine is a centrally-acting GABA uptake inhibitor that makes more GABA available for post-synaptic receptor binding, and so, in accordance with the GABA hypothesis, might have anti-tremor properties of benefit to patients with ET. For an open-label trial with 7 ET patients, however, Gironell and coworkers (Gironell, Martínez-Corral, Pagonabarraga, & Kulisevsky, 2008) reported negative findings, and, at the dosage tested, found no significant improvement in tremor amplitudes in ET.

4. GABA hypothesis: Supporting evidence

The findings and observations reported above (see Table 2) support GABA's role in the pathophysiology of ET. Below we summarize evidence in favor of the GABA hypothesis in each of its 4 steps.

Step 1: Cerebellar degeneration with Purkinje cell loss.

- Several post-mortem studies have pointed to Purkinje cell loss in ET, while the neurodegenerative nature of ET is supported by its progressive course.

Step 2: Decreased GABA activity in deep cerebellar neurons.

- Cerebrospinal fluid studies point to decreased GABA levels in patients with ET.
- Pathology studies indicate a loss of GABA receptors in the dentate nucleus.
- In vivo imaging studies point to increased flumazenil binding in GABA receptors in the cerebellar area that suggest a deficit of GABA transmitters.
- Animal models of GABA dysfunction (knockout mice) have shown symptoms similar to those observed in patients with ET.

Table 2 Studies published concerning the GABA hypothesis in ET.

	CSF	Pukinje cell loss	Pathology GABA	Electrophysiology	Genetic	Animal models	Imaging GABA	Computation model	Drug therapies
Pros	Mally et al. (1996)	Louis et al. (2007) Axelrad et al. (2008) Louis et al. (2012) Louis et al. (2013)	Mouginot and Gahwiler (1996) Paris-Robidas et al. (2012) Shill et al. (2012)	Uusisaari and Knopfel (2008) Pinault and Deschenes (1992) Hua et al. (1998) Chuang et al. (2014)	Liu et al. (2016) Tio et al. (2016)	Kralic et al. (2005) Nietz et al. (2020)	Boecker et al. (2010) Gironell et al. (2012)	Zhang and Santaniello (2019)	Ethanol, 1988 Primidone, 1981 Alprazolam, 1989 Theophylline, 1991 Muscimol, 1991 Gabapentin, 1999, 2000 Topiramate, 2002, 2006, 2008 Pregabaline, 2007 Zonisamide, 2018
Cons		Rajput et al. (2012) Shill et al. (2008) Symanski et al. (2014)	Luo et al. (2012)		Deng et al. (2006) Thier et al. (2011) García-Martín et al. (2011a, 2011b)				Progabide, 1983, 1987 Gabapentin, 1999 Topiramate, 2006 Tiagabine, 2008 Pregabaline, 2009, 2012 Zonisamide, 2019

CSF: cerebrospinal fluid.

- Computational models have shown that tremor activity can originate from the olivocerebellar loop in response to GABA receptor dysfunction in the dentate nucleus of cerebellum.
- Drugs that enhance GABA activity have an anti-tremor effect.

Step 3: Disinhibition of pacemaker-like activity in deep cerebellar neurons.

- Electrophysiological studies indicate that dentate nucleus cells are sensitive to GABAergic input from Purkinje cells and that deep cerebellar nuclei neurons exhibit pacemaker-like activity.
- PET studies with ^{18}F-FDG point to increased cerebellar metabolic output.
- The anti-tremor effects of inhibitory rTMS applied to the cerebellum supports an increase in hypermetabolism and activity in this area.

Step 4: Increased rhythmic thalamus and thalamo-cortical circuit activity causing tremor.

- Thalamic rhythmic or tremoric neurons, specifically in the ventral intermediate nucleus, have been observed in electrophysiological recordings in deep brain stimulation surgery.

5. Final remarks

The GABA hypothesis, to date, remains the most robust pathophysiological theory in explaining ET. There is supporting evidence in each step of the GABA hypothesis, although negative studies and controversies cannot be overlooked. It remains to be elucidated whether the reduced GABAergic tone observed in ET is a primary contributing factor to the pathophysiology of this disease, a consequence of altered Purkinje cell function, or even the result of cell death.

More studies are needed to confirm the neurodegenerative nature of ET and decreased GABA activity in the cerebellum. Research is also needed into treatments that enhance GABA transmission in the cerebellar area.

References

Alonso-Navarro, H., García-Martín, E., Agúndez, J. A. G., & Jiménez-Jiménez, F. J. (2020). Current and future neuropharmacological options for the treatment of essential tremor. *Current Neuropharmacology*, *18*, 518–537.

Axelrad, J. E., Louis, E. D., Honig, L. S., Flores, I., Ross, G. W., Pahwa, R., et al. (2008). Reduced Purkinje cell number in essential tremor: A postmortem study. *Archives of Neurology*, *65*, 101–107.

Boecker, H., Weindl, A., Brooks, D. J., Ceballos-Baumann, A. O., Liedtke, C., Miederer, M., et al. (2010). GABAergic dysfunction in essential tremor: An 11C-flumazenil PET study. *Journal of Nuclear Medicine*, *51*, 1030–1035.

Boecker, H., Wills, A. J., Ceballos-Baumann, A., Samuel, M., Thompson, P. D., Findley, L. J., et al. (1996). The effect of ethanol on alcohol-responsive essential tremor: A positron emission tomography study. *Annals of Neurology*, *39*, 650–658.
Bowery, N. G., Bettler, B., Froestl, W., Gallagher, J. P., Marshall, F., Raiteri, M., et al. (2002). International Union of Pharmacology. XXXIII. Mammalian gamma-aminobutyric acid (B) receptors: Structure and function. *Pharmacological Reviews*, *54*, 247–264.
Bruno, E., Nicoletti, A., Filippini, G., Quattrocchi, G., Colosimo, C., & Zappia, M. (2017). Zonisamide for essential tremor. *Cochrane Database of Systematic Reviews*, 8CD009684.
Bruno, E., Nicoletti, A., Quatrocchi, G., Allegra, R., Filippini, G., Colosimo, C., et al. (2017). Topiramate for essential tremor. *Cochrane Database of Systematic Reviews*, 4CD009683.
Bruno, E., Nicoletti, A., Quatrocchi, G., Filippini, G., Zappia, M., & Colosimo, C. (2015). Alprazolam for essential tremor. *Cochrane Database of Systematic Reviews*, 12CD009681.
Chakrabarti, A., & Pearce, J. M. S. (1981). Essential tremor: Response to primidone. *Journal of Neurology, Neurosurgery, and Psychiatry*, *44*, 650.
Chen, K., Li, H. Z., Ye, N., Zhang, J., & Wang, J. J. (2005). Role of GABAB receptors in GABA and baclofen-induced inhibition of adult rat cerebellar interpositus nucleus neurons in vitro. *Brain Research Bulletin*, *67*, 310–318.
Chuang, W. L., Huang, Y. Z., Lu, C. S., & Chen, R. S. (2014). Reduced cortical plasticity and GABAergic modulation in essential tremor. *Movement Disorders*, *29*, 501–507.
Cleeves, L., & Findley, L. J. (1988). Propranolol and propranolol-LA in essential tremor: A double blind comparative study. *Journal of Neurology, Neurosurgery, and Psychiatry*, *51*, 379–384.
Colebatch, J. G., Findley, L. J., Frackowiak, R. S., Marsden, C. D., & Brooks, D. J. (1990). Preliminary report: Activation of the cerebellum in essential tremor. *Lancet*, *336*, 1028–1030.
Connor, G. S. (2002). A double-blind placebo-controlled trial of topiramate treatment for essential tremor. *Neurology*, *59*, 132–134.
Connor, G. S., Edwuards, K., & Tarsy, D. (2008). Topiramate in essential termor: Findings from double-blind, placebo-controlled, crossover trials. *Clinical Neuropharmacology*, *31*, 97–103.
Daniels, C., Peller, M., Wolff, S., Alfke, K., Witt, K., Gaser, C., et al. (2006). Voxel-based morphometry shows no decreases in cerebellar gray matter volume in essential tremor. *Neurology*, *67*, 1452–1456.
De Haas, S. L., Zoethout, R. W. M., Van Dyck, K., De Smet, M., Rosen, L. B., Murphy, M. G., et al. (2012). The effects of TPA023, a GABAAα2,3 subtype-selective partial agonist, on essential tremor in comparison to alcohol. *Journal of Psychopharmacology (Oxford England)*, *26*, 282–291.
Deng, H., Xie, W. J., Le, W. D., Huang, M. S., & Jankovic, J. (2006). Genetic analysis of the GABRA1 gene in patients with essential tremor. *Neuroscience Letters*, *401*, 16–19.
Dupuis, M. J., Delwaide, P. J., Boucquey, D., & Gonsette, R. E. (1989). Homolateral disappearance of essential tremor after cerebellar stroke. *Movement Disorders*, *4*, 183–187.
Elble, R. J. (2000). Essential tremor frequency decreases with time. *Neurology*, *55*, 1547–1551.
Ferrara, J. M., Kenney, C., Davidson, A. L., Shinawi, L., Kissel, A. M., & Jankovic, J. (2009). Efficacy and tolerability of pregabalin in essential tremor: A randomized, double-blind, placebo-controlled, crossover trial. *Journal of the Neurological Sciences*, *285*, 195–197.
Ferreira, J. J., Mestre, T. A., Lyons, K. E., Benito-Leon, J., Tan, E. K., Abbruzzese, G., et al. (2019). MDS evidence-based review of treatments for essential tremor. *Movement Disorders*, *34*, 950–958.

Frima, N., & Grünewald, R. A. (2006). A double-blind, placebo-controlled, crossover trial of topiramate in essential tremor. *Clinical Neuropharmacology*, *2*, 94–96.

Frucht, S. J., Bordelon, Y., Houghton, W. H., & Reardan, D. (2005). A pilot tolerability and efficacy trial of sodium oxybate in ethanol-responsive movement disorders. *Movement Disorders*, *20*, 1330–1337.

Frucht, S. J., Houghton, W. C., Bordelon, Y., Greene, P. E., & Louis, E. D. (2005). A single-blind, open-label trial of sodium oxybate for myoclonus and essential tremor. *Neurology*, *65*, 1967–1969.

Gan, S. R., Wang, J., Figueroa, K. P., Pulst, S. M., Tomishon, D., Lee, D., et al. (2017). Postural tremor and ataxia progression in spinocrebellar ataxias. *Tremor and Other Hyperkinetic Movements*, 7. https://doi.org/10.7916/D9GM8KRH.

García-Martin, E., Martínez, C., Alonso-Navarro, H., Benito-Leon, J., Lorenzo-Betancor, O., Pastor, P., et al. (2011a). Gamma-aminobutyric acid (GABA) receptor rho (GABRR) polymorphisms and risk for essential tremor. *Journal of Neurology*, *2*, 203–211.

García-Martin, E., Martínez, C., Alonso-Navarro, H., Benito-Leon, J., Lorenzo-Betancor, O., Pastor, P., et al. (2011b). Gamma-aminobuityric acid GABRA4, GABRE, and GABRQ receptor plymorphisms and risk for essential tremor. *Pharmacogenetics and Genomics*, 7, 436–439.

Gironell, A. (2014). The GABA hypothesis in essential tremor: Lights and shadows. *Tremor and Other Hyperkinetic Movements*, *4*. https://doi.org/10.7916/D8SF2T9C.

Gironell, A., Figueiras, F. P., Pagonabarraga, J., Herance, J. R., Pascual-Sedano, B., Trampal, C., et al. (2012). Gaba and serotonin molecular neuroimaging in essential tremor: A clinical correlation study. *Parkinsonism & Related Disorders*, *18*, 876–880.

Gironell, A., Kulisevsky, J., Barbanoj, M., López-Villegas, D., Hernández, G., & Pascual-Sedano, B. (1999). A randomized placebo-controlled comparative trial of gabapentin and propranolol in essential tremor. *Archives of Neurology*, *56*, 475–480.

Gironell, A., Kulisevsky, J., Lorenzo, J., Barbanoj, M., Pascual-Sedano, B., & Otermin, P. (2002). Transcranial magnetic stimulation of the cerebellum in essential tremor: A controlled study. *Archives of Neurology*, *59*, 413–417.

Gironell, A., & Marin-Lahoz, J. (2016). La esencia del temblor esencial: Bases neuroquímicas. *Revista de Neurologia*, *62*, 507–515.

Gironell, A., Martínez-Corral, M., Pagonabarraga, X., & Kulisevsky, J. (2008). Tiagabine for essential tremor: An open-label trial. *Movement Disorders*, *23*, 1955–1956.

Gironell, A., Martinez-Corral, M., Pagonabarraga, J., & Kulisevsky, J. (2010). The glass scale: A simple tool to determine severity in essential tremor. *Parkinsonism & Related Disorders*, *16*, 412–414.

Gironell, A., Ribosa-Nogué, R., Gich, I., Marin-Lahoz, J., & Pascual-Sedano, B. (2015). Severity stages in essential tremor: A long-term retrospective study using the glass scale. *Tremor and Other Hyperkinetic Movements*. https://doi.org/10.7916/D8DV1HQC.

Ha, S. W., Yang, Y. S., Song, I. U., Chung, Y. A., Oh, J. K., & Chung, S. W. (2015). Changes in regional brain glucose 976 metabolism measured with F-18-FDG-PET in essential tremor. *Acta Radiologica (Stockholm, Sweden)*, *56*, 482–486.

Hallett, M., & Dubinsky, R. M. (1993). Glucose metabolism in the brain of patients with essential tremor. *Journal of the Neurological Sciences*, *114*, 45–48.

Handforth, A., Homanics, G. E., Covey, D. F., et al. (2010). T-type calcium channel antagonists suppress tremor in two mouse models of essential tremor. *Neuropharmacology*, *59*, 380–387.

Holtbern, F., & Shah, N. J. (2021). Imaging the pathophysiology of essential tremor—A systematic review. *Frontiers in Neurology*, *12*, 680254.

Hua, S. E., Lenz, F. A., Zirh, T. A., Reich, S. G., & Dougherty, P. M. (1998). Thalamic neuronal activity correlated with essential tremor. *Journal of Neurology, Neurosurgery, and Psychiatry*, *64*, 273–276.

Huber, S. J., & Paulson, G. W. (1988). Efficacy aof alprazolam ofor essential tremor. *Neurology*, *38*, 241–243.
Jahnsen, H. (1986). Extracellular activation and membrane conductances of neurones in the guinea-pig deep cerebellar nuclei in vitro. *The Journal of Physiology*, *372*, 149–168.
Jiménez-Jiménez, F. J., Alonso-Navarro, H., García-Martín, E., & Agúndez, J. A. G. (2020). An update on the neurochemistry of essential tremor. *Current Medicinal Chemistry*, *27*, 1690–1710.
Jo, S., & Massaquoi, S. G. (2004). A model of cerebellum stabilized and scheduled hybrid long-loop control of upright balance. *Biological Cybernetics*, *91*, 188–202.
Koller, W. C., Rubino, F., & Gupta, S. (1987). Pharmacologic probe with progabide of GABA mechanisms in essential tremor. *Archives of Neurology*, *44*, 905–906.
Koller, W. C., & Vetere-Overfield, B. (1989). Acute and chronic effects of propranolol and primidone in essential tremor. *Neurology*, *39*, 1587–1588.
Kralic, J. E., Criswell, H. E., Osterman, J. L., et al. (2005). Genetic essential tremor in gamma-aminobutyric acid A receptor alpha1 subunit knockout mice. *The Journal of Clinical Investigation*, *115*, 774–779.
Kuhlenbäumer, G., Hopfner, F., & Deuschl, G. (2014). Genetics of essential tremor: Meta-analysis and review. *Neurology*, *82*, 1000–1007.
Liu, X., Hernadez, N., Kisselev, S., et al. (2016). Identification of candidate genes for familiar early-onset essential tremor. *European Journal of Human Genetics*, *24*, 1009–1015.
Llinas, R. (1988). The intrinsic electrophysiological properties of mammalian neurons: Insights into central nervous system function. *Science*, *242*, 1654–1664.
Llinas, R., & Muhlethaler, M. (1988). Electrophysiology of guinea-pig cerebellar nuclear cells in the in vitro brain stem-cerebellar preparation. *The Journal of Physiology*, *404*, 241–258.
Louis, E. D. (1999). A new twist for stopping the shakes? Revisiting GABAergic therapy for essential tremor. *Archives of Neurology*, *56*, 807–808.
Louis, E. D. (2014a). Re-thinking the biology of essential tremor: From models to morphology. *Parkinsonism & Related Disorders*, *20*(Suppl. 1), S88–S93.
Louis, E. D. (2014b). Understanding essential tremor: Progress on the biological front. *Current Neurology and Neuroscience Reports*, *14*, 450.
Louis, E. D., Agnew, A., Gillman, A., Gerbin, M., & Viner, A. S. (2011). Estimating annual rate of decline: Prospective, longitudinal data on arm tremor severity in two groups of essential tremor cases. *Journal of Neurology, Neurosurgery, and Psychiatry*, *82*, 761–765.
Louis, E. D., Babij, R., Lee, M., Cortés, E., & Vonsattel, J. P. G. (2013). Quantification of cerebellar hemispheric purkinje cell linear density: 32 ET cases versus 16 controls. *Movement Disorders*, *28*, 1854–1859.
Louis, E. D., & Faust, P. L. (2020). Essential tremor pathology: Neurodegeneration and reorganization of neuronal connections. *Nature Reviews. Neurology*, *16*, 69–83.
Louis, E. D., Faust, P. L., & Vonsattel, J. P. G. (2012). Purkinje cell loss is a characteristic of essential tremor: Towards a more mature understanding of pathogenesis. *Parkinsonism & Related Disorders*, *18*, 1003–1004.
Louis, E. D., Faust, P. L., Vonsattel, J. P. G., Honig, L. S., Rajput, A., Robinson, C. A., et al. (2007). Neuropathological changes in essential tremor: 33 cases compared with 21 controls. *Brain: A Journal of Neurology*, *130*, 3297–3307.
Louis, E. D., Gerbin, M., & Galecki, M. (2013). Essential tremor 10,20,30,40: Clinical snapshots of the disease by decade of duration. *European Journal of Neurology*, *20*, 949–954.
Louis, E. D., Hernandez, N., Dyke, J. P., Ma, R. E., & Dydak, U. (2018). In vivo dentate nucleus gamma-aminobutyric acid concentration in essential tremor vs controls. *Cerebellum*, *17*, 165–172.
Louis, E. D., Michalec, M., & Gillman, A. (2014). Shaky drawing: What is the rate of decline during prospective follow-up of essential tremor? *BMJ Open*, *4*, e004626.

Louis, E. D., Shungu, D. C., Chan, S., Mao, X., Jurewicz, E. C., & Watner, D. (2002). Metabolic abnormality in the cerebellum in patients with essential tremor: A proton magnetic resonance spectroscopic imaging study. *Neuroscience Letters*, *333*, 17–20.

Louis, E. D., Shungu, D. C., Mao, X., Chan, S., & Jurewicz, E. C. (2004). Cerebellar metabolic symmetry in essential tremor studied with ^{1}H magnetic resonance spectroscopic imaging: Implications for disease pathology. *Movement Disorders*, *19*, 672–677.

Luo, R., Pan, P., Xu, Y., & Chen, L. (2019). No reliable gray matter changes in essential tremor. *Neurological Sciences*, *40*, 2051–2063.

Luo, C., Rajput, A. H., Robinson, C. A., & Rajput, A. (2012). Gammaaminobutyric acid (GABA)-B receptor 1 in cerebellar cortex of essential tremor. *Journal of Clinical Neuroscience*, *19*, 920–921.

Mally, J., Baranyi, M., & Vizi, E. S. (1996). Change in the concentrations of amino acids in CSF and serum of patients with essential tremor. *Journal of Neural Transmission*, *103*, 555–560.

Mally, J., & Stone, T. W. (1991). The effect of theophylline on essential tremor: The possible role of GABA. *Pharmacology, Biochemistry, and Behavior*, *39*, 345–349.

Mally, J., Connick, J. H., & Stone, T. W. (1991). Changes in neurotransmitter sensitivity in the mouse neocortical slice following propranolol and theophylline administration. *British Journal of Pharmacology*, *102*, 711–717.

Manto, M., & Laute, M. A. (2008). A possible mechanism for the beneficial effect of ethanol in essential tremor. *European Journal of Neurology*, *15*, 697–705.

Marin-Lahoz, J., & Gironell, A. (2016). Linking essential tremor to the cerebellum: Neurochemical evidence. *Cerebellum*, *15*, 243–252.

Mondrup, K., Dupont, E., & Pederson, E. (1983). The effect of the GABA-agonist, progabide, on benign essential tremor. *Acta Neurologica Scandinavica*, *68*, 248–252.

Mouginot, D., & Gahwiler, B. H. (1996). Presynaptic GABAB receptors modulate IPSPs evoked in neurons of deep cerebellar nuclei in vitro. *Journal of Neurophysiology*, *75*, 894–901.

Nietz, A., Krook-Magnuson, C., Gutierrez, H., Klein, J., Sauve, C., Hoff, I., et al. (2020). Selective loss of the GABA(A)alpha1 subunit form Purkinje cells is sufficient to induce a tremor phenotype. *Journal of Neurophysiology*, *124*, 1183–1197.

Ondo, W., Hunter, C., Dat Wuong, K., Scwartz, K., & Jankovic, J. (2000). Gabapentin for essential tremor: A multiple-dose, double-blind placebo-controlled trial. *Movement Disorders*, *15*, 678–682.

Ondo, W. G., Jankovic, J., Connor, G. S., Pahwa, R., Elble, R., Satcy, M. A., et al. (2006). Topiramate in essential tremor: A double-blind, placebo-controlled trial. *Neurology*, *66*, 672–677.

Pagan, F. L., Butman, J. A., Dambrosia, J. M., & Hallett, M. (2003). Evaluation of essential tremor with multi_795 voxel magnetic resonance spectroscopy. *Neurology*, *60*, 1344–1347.

Pahapill, P. A., Levy, R., Dostrovsky, J. O., Davis, K. D., Rezai, R., Tasker, R. R., et al. (1999). Tremor arrest with thalamic microinjections of muscimol in patients with essential tremor. *Annals of Neurology*, *46*, 249–252.

Pahwa, R., Lyons, K., Hubble, J. P., Busenbark, K., & Rienerth, J. D. (1999). Double-blind controlled trial of gabapentin in essential tremor. *Movement Disorders*, *13*, 465–467.

Paris-Robidas, A., Brochu, E., Sintes, M., Emond, V., Bousquet, M., Vandal, M., et al. (2012). Defective dentate nucleus GABA receptors in essential tremor. *Brain*, *135*, 105–116.

Paterson, N. E., Malekiani, S. A., Foreman, M. M., Olivier, B., & Hanania, T. (2009). Pharmacological characterization of harmaline-induced tremor activity in mice. *European Journal of Pharmacology*, *616*, 73–80.

Pinault, D., & Deschenes, M. (1992). The origin of rhythmic fast subthreshold depolarizations in thalamic relay cells of rats under urethane anaesthesia. *Brain Research, 595*, 295–300.

Pitracupa, S., Bolonga, M., Tommasin, S., Berardelli, A., & Pantano, P. (2021). The contribution of neuroimaging to the understanding of essential tremor pathophysiology: A systematic review. *Cerebellum*. https://doi.org/10.1007/s12311-021-01335-7.

Popa, T., Russo, M., Vidailhet, M., Roze, E., Lehéricy, S., Bonnet, C., et al. (2013). Cerebellar rTMS stimulation may induce prolonged clinical benefits in essential tremor, and subjacent changes in functional connectivity: An open label trial. *Brain Stimulation, 6*, 175–179.

Popa, T., Russoc, M., & Meunierc, S. (2010). Long-lasting inhibition of cerebellar output. *Brain Stimulation, 3*, 161–169.

Putzke, J. D., Whaley, N. R., Baba, Y., Wizolek, Z. K., & Uiti, R. J. (2006). Essential tremor: Predictors of disease progression in a clinical cohort. *Journal of Neurology, Neurosurgery, and Psychiatry, 77*, 1236–12137.

Rajput, A. H., Robinson, C. A., Rajput, M. L., Robinson, S. L., & Rajput, A. (2012). Essential tremor is not dependent upon cerebellar Purkinje cell loss. *Parkinsonism & Related Disorders, 18*, 626–628.

Revuelta, G., McGill, C., Jensen, J. H., & Bonilha, L. (2019). Characterizing Thalamo-cortical structural 703 connectivity in essential tremor with diffusional kurtosis imaging tractography. *Tremor and Other Hyperkinetic Movements (New York, NY)*. https://doi.org/10.7916/tohm.v0.690.

Shaefer, S. M., Rodriguez, A. V., & Louis, E. D. (2017). Brain circuits and neurochemical systems in essential tremo: Insights into current and future pharmacotherpeutic approaches. *Expert Review of Neurotherapeutics*. https://doi.org/10.1080/147375.2018.1413353.

Shill, H. A., Adler, C. H., Beach, T. G., Lue, L. F., Caviness, J. N., Sabbagh, M. N., et al. (2012). Brain biochemistry in autopsied patients with essential tremor. *Movement Disorders, 27*, 113–117.

Shill, H. A., Adler, C. H., Sabbagh, M. N., Connor, D. J., Caviness, J. N., Hentz, J. G., et al. (2008). Pathologic findings in prospectively ascertained essential tremor subjects. *Neurology, 70*, 1452–1455.

Shin, D. H., Han, B. S., Kim, H. S., & Lee, P. H. (2008). Diffusion tensor imaging in patients with essential tremor. *American Journal of Neuroradiology, 29*, 151–153.

Song, I. U., Ha, S. W., Yang, Y. S., & Chung, Y. A. (2015). Differences in regional glucose metabolism of the brain 980 measured with F-18-FDG-PET in patients with essential tremor according to their response to Beta_981 blockers. *Korean Journal of Radiology, 16*, 967–972.

Sun, X., Liu, F., Liu, Q., Gai, Y., Ruan, W., Wimalarathne, D. N., et al. (2019). Quantitative research of (11)C_915 CFT and (18)F-FDG PET in Parkinson's disease: A pilot study with NeuroQ software. *Frontiers in Neuroscience, 13*, 299. https://doi.org/10.3389/fnins.2019.00299.

Symanski, C., Shill, H. A., Dugger, B., Hentz, J. G., Adler, C. H., Jacobson, S. A., et al. (2014). Essential tremor is not associated with cerebellar Purkinje cell los. *Movement Disorders, 29*, 496–500.

Tapper, S., Göransson, N., Lundberg, P., Tisell, A., & Zsigmond, P. (2020). A pilot study of essential tremor: 836 cerebellar GABA+/Glx ratio is correlated with tremor severity. *Cerebellum & Ataxias*, 7, 8. https://doi.org/10.1186/s40673-020-00116-y.

Tariq, M., Arshaduddin, M., Biary, N., Al Moutaery, K., & Al Deeb, S. (2001). Baclofen attenuates harmaline induced tremors in rats. *Neuroscience Letters, 312*, 79–82.

Thier, S., Kuhlenbäumer, G., Lorenz, D., Nothnagel, M., Nebel, A., Christensen, K., et al. (2011). GABA(A) receptor and GABA transporter polymorphisms and risk for essential tremor. *European Journal of Neurology, 8*, 1098–1100.

Tikoo, S., Pietracupa, S., Tommasin, S., Bologna, M., Petsas, N., Bharti, K., et al. (2020). Functional 686 disconnection of the dentate nucleus in essential tremor. *Journal of Neurology, 267*, 1358–1367.

Tio, M., Wen, R., Lim, Y. L., Wang, H., Ling, S. C., Zhao, Y., et al. (2016). FUS-linked essential tremor associated with motor dysfunction in Drosophila. *Human Genetics, 135*, 1223–1232.

Uusisaari, M., & Knopfel, T. (2008). GABAergic synaptic communication in the GABAergic and non-GABAergic cells in the deep cerebellar nuclei. *Neuroscience, 156*, 537–549.

Wills, A. J., Jenkins, I. H., Thompson, P. D., Findley, L. J., & Brooks, D. J. (1994). Red nuclear and cerebellar but no olivary activation associated with essential tremor: A positron emission tomographic study. *Annals of Neurology, 36*, 636–642.

Zesiewicz, T. A., Elble, R. J., Louis, E. D., Gronsethc, G. S., Ondo, W. G., Dewey, R. B., Jr., et al. (2011). Evidence-based guideline update: Treatment of essential tremor: Report of the quality standards subcommittee of the American Academy of Neurology. *Neurology, 17*, 1752–1755.

Zesiewicz, T. A., Sullivan, K. L., Hinson, V., Stover, N. P., Fang, J., Jahan, I., et al. (2012). Multisite, double-blind, randomized, controlled study of pregabalin for essential tremor. *Movement Disorders, 28*, 249–250.

Zesiewicz, T. A., Ward, C. L., Hauser, R. A., Salemi, J. L., Siraj, S., Wilson, M. C., et al. (2007). A pilot, double-blind, placebo-controlled trial of pregabalin in the treatment of essential tremor. *Movement Disorders, 22*, 1660–1663.

Zhang, X., & Santaniello, S. (2019). Role of cerebellar GABAergic dysfunctions in the origins of essential 825 tremor. *Proceedings of the National Academy of Sciences of the United States of America, 116*, 13592–13601.

Zoethout, R. W. M., Iannone, R., Bloem, B. R., Palcza, J., Murphy, G., Chodakewitz, J., et al. (2012). The effects of a novel histamine-3 receptor inverse agonist on essential tremor in comparison to stable levels of alcohol. *Journal of Psychopharmacology (Oxford England), 26*, 292–302.

CHAPTER TEN

Is essential tremor a disorder of GABA dysfunction? No

Steven Bellows[a] and Joohi Jimenez-Shahed[b,*]
[a]Baylor College of Medicine, Houston, TX, United States
[b]Icahn School of Medicine at Mount Sinai, New York City, NY, United States
*Corresponding author: e-mail address: joohi.jimenez-shahed@mountsinai.org

Contents

Abstract

Although essential tremor is common, its underlying pathophysiology remains uncertain, and several hypotheses seek to explain the tremor mechanism. The GABA hypothesis states that disinhibition of deep cerebellar neurons due to reduced GABAergic input from Purkinje cells results in increased pacemaker activity, leading to rhythmic output to the thalamo-cortical circuit and resulting in tremor. However, some neuroimaging, spectroscopy, and pathology studies have not shown a clear or consistent GABA deficiency in essential tremor, and animal models have indicated that large reductions of Purkinje cell inhibition may improve tremor. Instead, tremor is increasingly attributable to dysfunction in oscillating networks, where altered (but not necessarily reduced) inhibitory signaling can result in tremor. Hypersynchrony of Purkinje cell activity may account for excessive oscillatory cerebellar output, with potential contributions along multiple sites of the olivocerebellar loop. Although older animal tremor models, such as harmaline tremor, have explored contributions from the inferior olivary body, increasing evidence has pointed to the role of aberrant climbing fiber synaptic organization in oscillatory cerebellar activity and tremor generation. New animal models such as *hotfoot17j* mice, which exhibit abnormal climbing fiber organization due to mutations in *Grid2*, have recapitulated many features of ET. Similar abnormal climbing fiber architecture and excessive cerebellar oscillations as measured by EEG have been found in

International Review of Neurobiology, Volume 163
ISSN 0074-7742
https://doi.org/10.1016/bs.irn.2022.02.006

humans with essential tremor. Further understanding of hypersynchrony and excessive oscillatory activity in ET phenotypes may lead to more targeted and effective treatment options.

1. Introduction

Despite its long recognition and common occurrence in the population, the underlying pathophysiology of essential tremor (ET) remains unclear. Attempts at clarifying the mechanisms or genetic underpinnings of ET are confounded by its phenotypic heterogeneity, which reflects that ET likely encompasses a family of diseases rather than a single entity (Louis, 2021). Nevertheless, increasing evidence has pointed to cerebellar involvement in ET, which has led to the development of the GABA hypothesis (Gironell, 2014). In this model (Fig. 1), a decrease in Purkinje cell (PC) output leads to decreased GABAergic inhibition of the dentate nucleus and other deep cerebellar neurons (DCN). This disinhibition results in increased output of dentate nucleus pacemaker activity, which is propagated through the dentato-rubro-thalamic tract (DRTT) to the thalamus and then motor cortex, leading to generation of tremor.

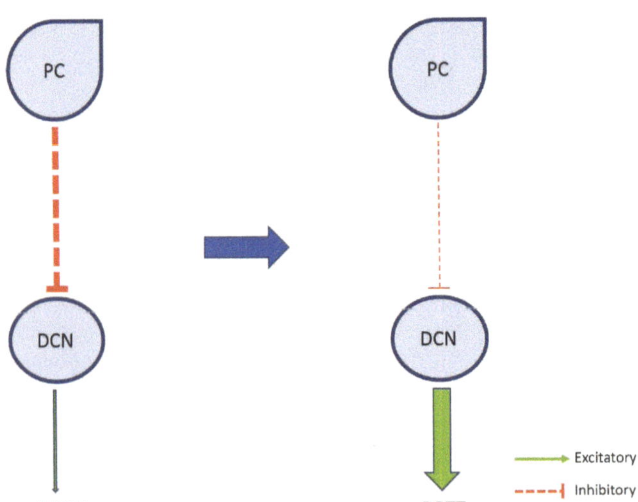

Fig. 1 GABA hypothesis. Reduced GABAergic inhibition of the deep cerebellar nuclei by Purkinje cells results in disinhibited pacemaker output to the dentato-rubro-thalamic tract, resulting in tremor. PC = Purkinje cell, DCN = deep cerebellar nuclei, DRTT = dentato-rubro-thalamic tract.

Multiple lines of evidence have supported this model. A mouse model with $GABA_A$ receptor alpha1 knockout was found to demonstrate postural and kinetic tremor responsive to many typical ET therapies, including primidone, propranolol, and ethanol (Kralic et al., 2005), although unexpected worsening was noted with diazepam, a GABAergic agonist. However, $GABA_A$ receptor alpha1 subunits are expressed on both PC and DCN (Kralic et al., 2005), and a knockout could result in decreased PC inhibition of DCN via loss of GABAergic inhibition onto PCs themselves. A PET study utilizing ^{11}C-flumazenil, which binds to the $GABA_A$ receptor complex, found increased binding of the radioligand in the cerebellums of 8 ET subjects versus 11 healthy controls (HC), which could suggest possible receptor upregulation in the setting of GABAergic deficiency (Boecker et al., 2010). Several pathological studies of ET brains have found reduced PC counts and PC densities as compared to HC (Axelrad et al., 2008; Choe et al., 2016; Louis et al., 2007; Wu et al., 2021), raising the possibility of a neurodegenerative process resulting in lower GABAergic inhibition of the DCN. However, key inconsistencies exist in the literature with regards to evidence for reduced GABA signaling and ET symptoms. In this article, we will explore the accumulating evidence against GABA-mediated tremor expression, and for the presence of oscillating networks in ET phenotypes, the role of PC hypersynchrony in the generation of tremor, as well as PC hypersynchrony along several points within oscillatory networks.

2. Accumulating evidence against GABA-mediated tremor expression

Studies of varying modalities have results that question the GABA hypothesis. Multiple MR spectroscopy (MRS) studies have not found evidence of decreased GABA levels in ET patients. In one study, there was no difference in dentate GABA concentration between 45 ET cases and 35 controls (Louis, Hernandez, Dyke, Ma, & Dydak, 2018). Another study utilizing MRS found no difference in cerebellar GABA or glutamate concentrations between 10 ET subjects and 6 HC, and instead found a positive correlation between tremor severity and the GABA/glutamate ratio in the ET subjects, although this finding may be impacted by MRS limitations as discussed below (Tapper, Göransson, Lundberg, Tisell, & Zsigmond, 2020). No differences in dentate GABA or glutamate levels or GABA/glutamate ratio by MRS were found in a separate cohort of 15 ET and 8 controls, although the overall small sample size and lack of complete subject matching limit the ability to draw definitive conclusions (Buijink et al., 2021).

While these findings call into question a direct link between low GABAergic inhibition and tremor generation, MRS studies do carry important limitations. As noted by Louis et al. in their study, a degenerative loss of GABA at the PC-DCN synapse could potentially be masked by compensatory terminal sprouting, a phenomenon found in other animal models with PC degeneration (Louis, Hernandez, et al., 2018). MRS analysis may also not have sufficient resolution to detect the loss of GABAergic PC-DCN synapses. Analyzed voxels are typically large and not isolated to the dentate nucleus (Louis, Hernandez, et al., 2018; Tapper et al., 2020), which may consequently mask specific loss of GABAergic function in the DCN. In the DCN itself, the presence of GABAergic interneurons may make it more difficult to detect the loss of GABA in neurons associated with PC terminals (Louis, Hernandez, et al., 2018).

Volumetric imaging studies provide conflicting evidence for a neurodegenerative process affecting the cerebellum that could cause a GABAergic deficit (Pietracupa, Bologna, Tommasin, Berardelli, & Pantano, 2021). Several studies have noted a reduction in cerebellar gray matter volumes in ET patients versus HC (Bagepally et al., 2012; Benito-León et al., 2009). Another study found that while whole cerebellar volumes were not different between ET patients and controls, several contiguous vermis segments were significantly reduced in ET patients (Shin et al., 2016). Yet other studies have found cerebellar atrophy only in subgroups of ET, such as those with head tremor (Cerasa et al., 2009; Choi et al., 2015; Dyke, Cameron, Hernandez, Dydak, & Louis, 2017), voice tremor (Dyke et al., 2017), or cognitive impairment (Bhalsing et al., 2014). In contrast, some investigations have not found clear evidence of cerebellar atrophy in ET patients (Buijink et al., 2016; Daniels et al., 2006; Pietracupa et al., 2019).

Several pathology studies have not found a difference in PC counts between ET and control brains (Rajput, Robinson, Rajput, Robinson, & Rajput, 2012; Symanski et al., 2014), or an association between PC counts and tremor duration (Symanski et al., 2014). One biochemical study compared parvalbumin concentrations, a marker of GABAergic neurons, between ET and HC brains (Shill et al., 2012). In the cerebellum, parvalbumin has been found to be expressed in PC, basket, and stellate cells (Caillard et al., 2000). While there was significantly less parvalbumin in the locus coeruleus of ET patients, there was no difference in cerebellar parvalbumin concentrations. While neurodegeneration has been posited as a possible cause for GABA deficiency, evidence for its presence in ET is inconsistent.

Several animal models have further suggested that the degree of GABA depletion is not proportional to tremor expression. In one mouse model where PC GABAergic output was silenced by deletion of PC vesicular GABA transporter, no tremor was noted in the mutant mice (Brown et al., 2020). Furthermore, unlike control mice, these mutant mice did not have a significantly increased tremor after administration of harmaline, a tremorgenic agent discussed in further detail below. This implies the requirement that some PC GABAergic function must be maintained to generate tremor. This is recapitulated by the demonstration of tremor in multiple animal models exhibiting PC loss (Handforth, 2016), but the absence of tremor in animal models with more severe PC loss, such as the *Purkinje cell generation* mouse model where most PCs are gone (Handforth, 2016). Similarly, cerebellar lesions have improved tremor in mouse models with faulty PC output. In one study using the *Weaver* mouse model of cerebellar ataxia, which demonstrates disorganized PC output, vermectomy of these mice abolished their tremor (Grüsser-Cornehls, Grüsser, & Bäurle, 1999). In another tremor mouse model where the mice also expressed inhibitory opsins in their PCs, inhibition of the PCs by green light illumination resulted in immediate suppression of the mouse tremor (Pan et al., 2020). These studies show that while decreased PC GABAergic output results in tremor in some models, a larger reduction appears to result in improvement or even abolition of tremor.

3. Tremor as a result of oscillating networks

While initial research into essential tremor looked for areas that could serve as single oscillatory generators of tremor (Louis & Lenka, 2017), increasing evidence points to essential tremor as a symptom of oscillating network dysfunction (Raethjen & Deuschl, 2012). A network of primary interest has been the cerebellothalamocortical loop (Fig. 2). A study utilizing magnetoencephalography (MEG) in 8 ET patients established that EMG-recorded tremor frequencies and their first harmonics were coherent with oscillations in the primary motor cortex contralateral from the recorded limb (Schnitzler, Münks, Butz, Timmermann, & Gross, 2009). The contralateral motor cortex, in turn, was found to be coherent from the premotor cortex,

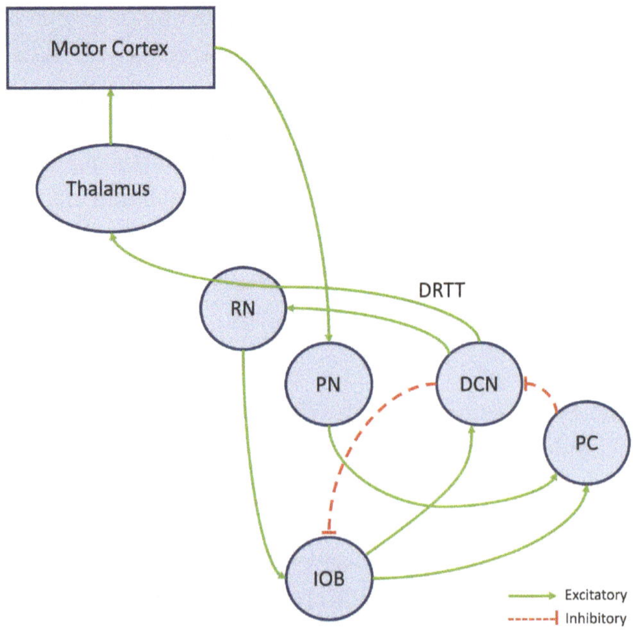

Fig. 2 Cerebello-thalamo-cortical loop. Purkinje cells modulate the excitatory output of deep cerebellar nuclei, which is directed to the red nucleus and thalamus through the dentato-rubro-thalamic tract. RN = red nucleus, PN = pontine nuclei, DRTT = dentato-rubro-thalamic tract, DCN = deep cerebellar nuclei, PC = Purkinje cell, IOB = inferior olivary body.

thalamus, and brainstem contralateral and the cerebellum ipsilateral to the recorded limb, supporting tremor generation from an oscillatory network.

This cerebellothalamocortical loop serves as the primary target of deep brain stimulation (DBS) in treatment of ET. The ventralis intermedius (VIM) nucleus of the thalamus is the most established DBS target in ET (Zesiewicz et al., 2011), and the improvement in tremor appears to be related to its involvement in the DRTT (Coenen et al., 2020). Other areas near the DRTT, such as the posterior subthalamic area (PSA) which includes the caudal zona incerta, have also shown efficacy in improving tremor when stimulated (Dembek et al., 2020). Specific targeting of the DRTT with tractography for DBS has resulted in significant improvement in contralateral arm tremor in ET (Fenoy & Schiess, 2017).

Oscillations in the network are not fixed, however. Cerebral cortical oscillations have been found to be only intermittently coherent with EMG recordings of limb tremor (Raethjen, Govindan, Kopper, Muthuraman, & Deuschl, 2007). Thalamic oscillation coherence with limb tremor has been demonstrated in some studies (Marsden, Ashby, Limousin-Dowsey, Rothwell, & Brown, 2000), but coherence has also been found to follow rather than precede tremor onset (Pedrosa et al., 2014). One study demonstrated cerebral and cerebellar coherence with ET, but found that improvement in tremor amplitude with alcohol only resulted in changes in cerebellar coherence (Pedrosa et al., 2017). Many other findings have increasingly drawn attention to the role of cerebellar dysfunction in development and manifestations of ET.

4. Cerebellar dysfunction and tremor manifestations

It is widely accepted that evidence for cerebellar dysfunction can be seen in the clinical examination of ET patients (Louis & Faust, 2020b). Ataxic gait features, such as tandem gait or balance impairment, have been noted in multiple studies (Louis, Rios, & Rao, 2010; Rao & Louis, 2019; Stolze, 2001). Intention tremor, associated with cerebellar disease, can be found in a sizable proportion of ET patients (Deuschl, 2000; Louis et al., 2017; Louis, Frucht, & Rios, 2009). Cerebellar oculomotor deficits (Helmchen et al., 2003), impaired hand-eye coordination (Trillenberg et al., 2006), impaired visuo-motor adaptation (Hanajima et al., 2016), and variability in rapid alternating movement tasks (Farkas, Szirmai, & Kamondi, 2006) have also been described. The cerebellum contains both feedback and feedforward loops involved in the modulation of action (Tanaka, Ishikawa, & Kakei, 2019), lesioning of which may explain oscillatory clinical phenomena such as intention tremor (Ibrahim, Beevis, & Empson, 2020).

Electrophysiological and functional imaging studies have provided further evidence of cerebellar involvement. Tremor frequency and its harmonics as captured by EMG have been found to be coherent with cerebellothalamic loop neural activity as measured by MEG, including the ipsilateral cerebellum (Schnitzler et al., 2009). Similarly, EMG-recorded tremor has been correlated with ipsilateral cerebellar activity as measured by fMRI (Contarino et al., 2012). Another study similarly found increased bilateral cerebellar activation by fMRI on induction of right arm postural tremor in ET patients (Broersma et al., 2016). PET studies in ET patients have demonstrated increased cerebellar blood flow at rest and with action (Colebatch, Findley, Frackowiak, Marsden, & Brooks, 1990; Jenkins et al., 1993; Wills, Jenkins, Thompson, Findley, & Brooks, 1995). Some MRI studies (Bagepally et al., 2012; Cerasa et al., 2009; Quattrone et al., 2008), but not all as noted above

(Daniels et al., 2006), have shown reduced cerebellar volumes in ET patients (Louis, Huang, Dyke, Long, & Dydak, 2014).

More recently, cerebellar electroencephalograms (EEGs) utilizing a unique montage to record from the cerebellar cortex have shown evidence of increased power in frequency ranges associated with ET. One study compared cerebellar EEG results between 10 ET and 10 controls and found significantly increased power in the 4 to 12 Hz frequency range in patients with ET. The group was expanded to 20 ET and 20 control patients, and showed correlation between the tremor severity of ET patients with the area under-the-curve of the 2 Hz band centered on the measured peak frequency (Pan et al., 2020). A follow-up study of 40 ET and 20 control patients similarly showed an increase in the cerebellar oscillatory index in ET patients, defined as the sum of normalized power spectrum densities at the 2 Hz frequency band centered at the peak frequency measured (Wong et al., 2021). This index measure also correlated with tremor severity in familial ET patients, but not sporadic ET patients. Therapeutic modalities targeting oscillations have been studied as well. Transcranial alternating stimulation to the cerebellum has been found to result in significant improvement in tremor amplitude, but only when phase-locked to the tremor, suggesting the particular importance of abnormal oscillatory activity in the pathologic generation tremor (Schreglmann et al., 2021).

Although pathologic studies have shown mixed results of degenerative cerebellar changes in ET patients, animal models demonstrate that a simple decrease in PC inhibition of the DCN alone does not appear to result in tremor, as significant reduction in PC output can improve tremor. Instead, the pattern rather than level of PC inhibition may potentially result in pathologic DCN activity (Brown et al., 2020), and several lines of evidence suggest that excessive PC synchronization in particular may generate an abnormal DCN output.

5. Purkinje cell hypersynchrony as a correlate of tremor

Features and potential causes of abnormal PC synchrony can be best explored through examination of the olivo-cerebellar loop (Fig. 3). In this loop, the inferior olivary body (IOB), a generator of 8–10 Hz rhythmic discharges, connects to PC via climbing fibers (CF). PC then send inhibitory GABAergic afferents to the DCN, which in turn send inhibitory GABAergic afferents back to the IOB (Helmich, Toni, Deuschl, & Bloem, 2013).

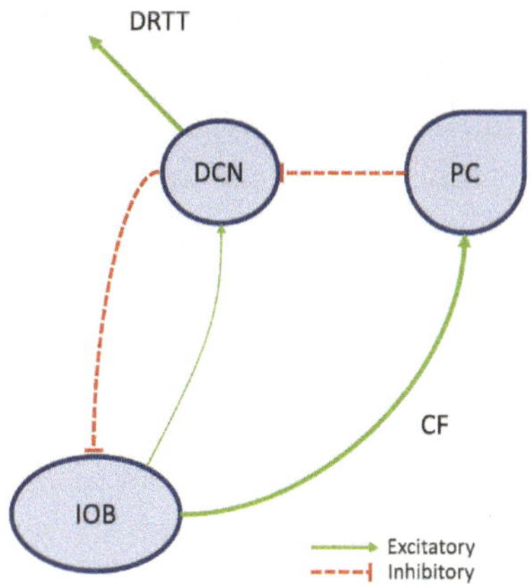

Fig. 3 Olivo-cerebellar loop. The inferior olivary body sends strong excitatory fibers to promote complex spike firing in Purkinje cells, which in turn inhibits the excitatory output of the loop through the dentato-rubro-thalamic tract. The inferior olivary body has weaker direct excitatory stimulation of the deep cerebellar nuclei as well. DRTT = dentato-rubro-thalamic tract, DCN = deep cerebellar nuclei, PC = Purkinje cell, IOB = inferior olivary body, CF = climbing fiber.

Purkinje cells express their output via high-frequency simple spike (SS) and low-frequency complex spike (CS) discharges (Streng, Popa, & Ebner, 2018). High-frequency SS discharges are derived from large amounts of parallel fiber outputs, which are driven by mossy fiber (MF) inputs from multiple locations in the CNS. In contrast, CS discharges are driven by CF input. Each PC in the adult cerebellum is innervated by one CF via a powerful excitatory synapse through a developmental process of elimination leading from multiple innervation in the early postnatal period to mono-innervation in the later postnatal period, though each CF retains multiple synaptic contacts with its PC (Kano, Watanabe, Uesaka, & Watanabe, 2018). Additionally, each CF innervates 5–10 Purkinje cells that tend to be located in a parasagittal plane leading to organization into "zonal compartments" that relate to a circumscribed region of the ION (Streng et al., 2018). Ultimately, CF activation depolarizes the whole PC, leading to a large CS (Streng et al., 2018) and gap junction coupling in the ION may lead to synchronized CF activation, resulting in a synchronized discharge of a wide band of PCs (Handforth & Lang, 2021). Synchronization of PC CS firing is required for accurate multi-joint movement control (Hoogland, De Gruijl, Witter, Canto, & De Zeeuw, 2015). One hypothesis is that these CS generated by CF represent error signals, used to modulate subsequent SS activity for behavior correction (Streng et al., 2018). The PC subsequently provides high-frequency, smaller amplitude SS and low-frequency, larger amplitude CS GABAergic inhibition of the DCN. The DCN provides excitatory input to the red nucleus and the VIM nucleus of the thalamus through the DRTT, which subsequently activates the motor cortex. The DCN also sends direct inhibitory fibers back to the IOB (Helmich et al., 2013).

In vivo whole cell patch clamp recordings of DCN cells have shown that PC cells provide a tonic inhibitory signal to DCN, such that individual PC cells have little effect on the DCN (Bengtsson, Ekerot, & Jörntell, 2011). However, synchronized PC activation via CF results in a synchronized DCN inhibition, which can subsequently lead to significant rebound activation of the DCN. Another study in rodents showed that train stimuli to the cerebellar cortex and inferior olive resulted in transient rebound increases in DCN neuron firing frequencies (Hoebeek, Witter, Ruigrok, & De Zeeuw, 2010). However, spike train stimulation alone may not be sufficient to generate a DCN burst, but may also require a sufficient pause in inhibition to allow for a rebound response (Dykstra, Engbers, Bartoletti, & Turner, 2016).

An alternative hypothesis suggests that DCN entrainment is not necessarily due to a rebound effect, but rather emerges as a phase-locked response to synchronized gaps in inhibitory input. One study showed that in vitro, DCN neurons were able to be entrained by synchronized inhibitory input due to the high firing rate, fast inhibitory postsynaptic current kinetics, and relatively lower PC to DCN convergence ratios (Person & Raman, 2012). When inhibitory postsynaptic currents (IPSCs) were asynchronous, DCN neuron firing remained suppressed. Synchronized IPSCs, however, allows for a gap between inhibitory signals, during which the DCN is disinhibited enough to fire. This was exhibited by phase locking of the DCN output to the stimulus frequency applied to PCs. Moreover, only a fraction of PCs synapsed to a DCN neuron need to be synchronized before generating some degree of phase-locked output.

Indeed, some studies have been able to generate tremor via rhythmic stimulation of PCs (Brown et al., 2020; Pan et al., 2020). In one study, optogenetically modified PCs were rhythmically stimulated at frequencies ranging from 1 to 20 Hz with sinusoidal pulses of light, which induced DCN bursting and tremor which matched the frequency of PC stimulation (Brown et al., 2020). Maximum tremor power occurred at 10 and 20 Hz, which may reflect the reflex loop and limb mechanical resonance frequency and harmonic (McAuley, 2000).

Perturbations at multiple points in the olivo-cerebellar loop have the potential to increase cerebellar oscillatory activity through increased PC synchronization. Each point has varying levels of evidence for potential involvement in ET. In the subsequent sections, we will explore the components of the olivo-cerebellar loop and their potential roles in tremor generation.

5.1 Inferior olivary body

The IOB has long been studied as a potential site of pathology in ET (Louis & Lenka, 2017), with its rhythmic oscillatory activity in the range of ET frequencies observed as early as 1971 (Lamarre & Mercier, 1971). Neurons in the IOB express Ca(V)3.1 calcium channels, which are involved in the generation of neuronal oscillations through low-threshold calcium spikes (Park et al., 2010). These rhythmic discharges then travel to neighboring neurons through gap junctions, resulting in synchronization (Raethjen & Deuschl, 2012). This synchronized IO cell firing, transmitted via CF, may then result in synchronized CS firing of PCs (Handforth & Lang, 2021).

Harmaline-induced tremor, which appears to act through the IOB, is the most studied ET animal model, as the harmaline-induced tremor has a similar 8 to 16Hz frequency range and responds to similar medications as ET (Handforth, 2016). Harmaline, a β-carboline alkaloid, is administered to animals subcutaneously, and within 10–15min generates a tremor which persists for hours. Harmaline transiently increases the IOB firing frequency to 8–12Hz burst-firing, which then travels via CF to induce PC complex spikes (Handforth, 2016). Both the frequency and regularity of the complex spike firing are increased (Brown et al., 2020). Cutting cerebellar peduncles (Handforth, 2016) or severing CF (Handforth & Lang, 2021) results in isolated IOB rhythmic activity in harmaline models. Ultrasound lesioning of the IOB in harmaline rats results in improved tremor (Sharabi et al., 2019). Suppression of IOB activity by lidocaine microinfusion in other mouse tremor models abolishes tremor as well (Pan et al., 2020). This implies that the IOB is involved in at least some models of tremor generation.

As a caveat, harmaline-induced tremor has several limitations as a model for ET. The tremor is only transient, and animals develop resistance to development of tremor with subsequent administrations (Louis & Lenka, 2017). While many agents that suppress ET also suppress harmaline tremor, such as primidone, gabapentin, and alcohol, there are other agents that appear effective in harmaline tremor but not ET (Handforth, 2016).

Levodopa and apomorphine, a D1/D2 agonist, both improve harmaline tremor in mice, but are not effective in ET (Handforth, 2012). Similarly, certain anti-epileptic medications such as brivaracetam, valproate, lacosamide, and carisbamate have demonstrated efficacy in harmaline tremor models but not in ET (Handforth, 2012).

In keeping with the mechanisms of rhythmic discharge generation in the IOB, knockout of Ca(V)3.1 calcium channels also reduces harmaline tremor (Park et al., 2010). However, gap junctions are not necessary for propagation of harmaline tremor, as mice without the connexin36 gap junction protein continue to demonstrate tremor when administered harmaline, despite reduced synchrony in IO cells (Long, Deans, Paul, & Connors, 2002). However, gap junction blockers such as carbenoxolone and mefloquine do appear to suppress harmaline tremor (Martin & Handforth, 2006). In contrast, direct GABA antagonism via picrotoxin injection into the IOB increases synchrony of neighboring neurons (Leznik, Makarenko, & Llinás, 2002), and results in increased synchronization of CS firing from PCs (Lang, Sugihara, & Llinas, 1996). This suggests that increased synchrony of IOB neurons plays a role in tremor generation in harmaline models, though how this relates to the disease state in ET remains less certain.

In regards to imaging and electrophysiological studies, increased metabolism has been noted by PET in the IOB in ET patients versus controls in one study (Hallett & Dubinsky, 1993). Impaired eyeblink conditioning, which involves the IOB in the error-signaling pathway, has been found in ET patients versus controls (Kronenbuerger et al., 2008; Kronenbuerger, Gerwig, Brol, Block, & Timmann, 2007). Abnormal vestibular-evoked myogenic potentials in ET patients have also been suggestive of brainstem pathology (Bayramoglu et al., 2021), although some have theorized that problems lie with the cerebellum instead (Uyaroglu, Ucar, Acarer, & Celebisoy, 2021).

There is mixed evidence for improvement in tremor using medications targeting the IOB. T-type calcium channel antagonists have been efficacious in harmaline animal models (Handforth et al., 2010). Zonisamide, an antagonist of T-type calcium channels, had efficacy in one evaluator blinded study (Handforth, Martin, Kang, & Vanek, 2009). While drugs targeting T-type calcium channels may improve tremor in animal models, a trial of ethosuxomide (Gironell & Marin-Lahoz, 2016) did not clearly improve ET in humans. In a phase 2 randomized controlled trial, CX-8998, a selective T-type calcium channel modulator, was shown to improve ET by clinical rating scale measures, but not by kinematic measurements

(Papapetropoulos et al., 2021). A separate study examining zonisamide in ET patients showed improvement by accelerometric measures but not by rating scale measures (Zesiewicz et al., 2007), in contrast to the evaluator-blinded study mentioned above (Handforth et al., 2009). These inconsistent findings suggest further caution be applied in the application of harmaline models of tremor to the ET disease state.

However, other studies have suggested that the IOB is not the main site of pathology in ET (Louis & Lenka, 2017). PET studies which showed increased cerebellar and red nucleus activity in ET patients versus controls found no difference in IOB activity (Wills, Jenkins, Thompson, Findley, & Brooks, 1994). An fMRI study noted activity in only 2 of 12 ET patients close to the olivary nucleus (Bucher, Seelos, Dodel, Reiser, & Oertel, 1997). There has been no clear difference in inferior olivary nucleus morphology on postmortem studies of ET and control brains (Louis, Babij, Cortés, Vonsattel, & Faust, 2013). In one case report, a patient with long-standing ET developed progressive ataxia with palatal tremor and associated IOB hypertrophy on MRI (Elkouzi, Kattah, & Elble, 2016). Despite the radiographic olivary body changes, his upper extremity tremor remained stable in both amplitude and frequency by electrophysiologic measures, suggesting that the IOB changes did not contribute to the generation of his tremor. Another case report described a patient with childhood onset ET with subsequent worsening in severity after age 85 who was found to have IOB degeneration on postmortem examination, presumably of late onset (Louis et al., 2018). The worsening of his tremor throughout his lifetime prior to presumed onset of pathologic evidence of IOB degeneration also suggested that the IOB was not involved in the generation of his tremor.

5.2 Climbing fibers

Despite conflicting evidence about their pathogenic role in ET, the IOB still provides a powerful excitatory signal to PC via CF. As discussed above, CF stimulation results in synchronized CS firing, and abnormal CF connections to the PC may cause excessive synchronicity of CS.

Abnormal CF pathology has been found in ET patients. In a comparison of post-mortem cerebellums from ET and control patients, CF in ET patients aberrantly extended into the territory of parallel fibers (PF) in the molecular layer (Lin et al., 2014). While CF typically have a higher synaptic density on thick PC dendrites, the CF in ET patients had more synapses on the thin spiny branchlets of the PC in the molecular layer, as measured by vesicular glutamate transporter type 2 immunohistochemistry. An expanded cohort of the original study continued to show CF extension into the molecular layer, and correlated tremor severity with the degree of CF-PC synaptic pathology (Louis, Lin, Faust, Koeppen, & Kuo, 2015). Similar decreased CF density and extension into the molecular layer was noted in subsequent pathological studies of ET brains, including across multiple ET subtypes (Lee, Gan, Faust, Louis, & Kuo, 2018; Pan et al., 2020). CFs typically synapse with PCs in a 1:1 ratio. However, CFs have also been noted to cross a higher number of PC somas and dendrites in ET patients, with the number of crossings also correlating with tremor severity (Wu et al., 2021). The increased number of PC somas and dendrites traversed by the CFs suggests potential for excess synapse formation. Excessive innervation of multiple PCs by a single CF could lead to increased PC synchrony (Wu et al., 2021). However, the study was unable to determine if these crossings also resulted in synapse formation. Other conditions involving the cerebellum, such as multiple system atrophy or spinocerebellar ataxia type 1, show reduced CF density but not aberrant CF extension into parallel fiber territory (Kuo et al., 2017). This suggests that aberrant CF extension may be unique to ET and not a general marker of cerebellar dysfunction. The causes for this abnormal CF organization, however, are uncertain. One group hypothesized that these findings may be secondary to PC damage and may reflect compensatory collateral sprouting from CFs (Wu et al., 2021). In rats who have undergone subtotal lesions of the inferior olive, surviving climbing fibers were shown to sprout collaterals to denervated PCs, increasing the amount of PCs innervated by a single CF by more than six-fold (Rossi, Wiklund, van der Want, & Strata, 1991).

Glutamate receptor delta 2 (GluRδ2) is a protein involved in the proper organization of CF and PF synapsing to PC dendrites (Pan et al., 2020). GluRδ2, exclusively expressed in PCs (Miyazaki et al., 2010), promotes PF-PC synapse formation, and prevents distal extension of CF (Watanabe, 2008). Ablation of GluRδ2 results in CF extension into PF territory and CF innervation of neighboring PCs (Miyazaki et al., 2010). Ataxia and tremor phenotypes have been described in patients with antibodies to GluRδ2 (Shiihara, Kato, Konno, Takahashi, & Hayasaka, 2007). Prior mouse models deficient in GluRδ2 have exhibited 10 Hz oscillations of CS firing in PCs, inducible by CF activation (Yoshida, Katoh, Ohtsuki, Mishina, & Hirano, 2004). GluRδ2 knockout mice exhibit increased CF collateral synapsing onto neighboring PCs, with resultant increased PC synchrony as compared to wild-type mice (Hashizume et al., 2013).

The *hotfoot* mice models have loss-of-function mutations in the *Grid2* gene encoding GluRδ2, often resulting in an ataxia phenotype (Lalouette, Guénet, & Vriz, 1998). *Grid2* knockout mice also exhibit severe ataxia, as well as innervation of PCs by multiple climbing fibers (Lalouette et al., 1998). The *hotfoot17j* mouse model was developed to explore the contribution of pathologic CF organization to tremor (Pan et al., 2020). In this model, the mice have a duplication in exon 3 of *Grid2*, which results in longer variants of GluRδ2. Rather than an absence of GluRδ2, these mice have GluRδ2 levels around 10% of normal, possibly due to degradation of the large protein with a smaller proportion of normal protein created via alternative splicing (Pan et al., 2020). As expected, *hotfoot17j* mice exhibit aberrant CF synapses on distal PC dendrites. Phenotypically, the mice develop a 20 Hz predominantly action tremor early in life which progressively worsens over time, and which improves with propranolol, primidone, and alcohol (Pan et al., 2020). Rescue of GluRδ2 levels by using a viral vector containing wild-type *Grid2* mRNA results in transient improvement in tremor (Pan et al., 2020). Optogenetic inhibition of the CF-PC synapse also improved tremor. Similarly to prior studies showing that tremor generation required some degree of PC output, optogenetic silencing of PC outputs resulted in immediate suppression of tremor (Pan et al., 2020).

In exploring these models of CF pathology on the generation of PC hypersynchrony, excessive oscillations in cerebellar LFPs measured by cerebellar surface electrodes on lobules V and VI were found in *hotfoot17j* mice as compared to wild-type mice. Furthermore, the phases of these oscillations were coupled to IOB spikes, and the oscillations (and tremor) were abolished following lidocaine infusion into the IOB, supporting that the oscillations relied on intact IOB input via CF. Artificially-induced oscillations could be generated by rhythmic optogenetic stimulation of PCs in wild-type mice, which resulted in coherent cerebellar LFP oscillations and tremor, which was measured by a weight-based vibration detection platform (Pan et al., 2020).

Taken together, the features of the *hotfoot17j* mouse model and concordant CF pathology and cerebellar oscillations found in ET patients provide intriguing evidence that, at least for some ET patients, aberrantly extensive CF connections to PCs may contribute to hypersynchronous CS firing and resultant tremor. While CF pathology was due to loss of GluRδ2 in the *hotfoot17j* mouse model, other mechanisms may underly abnormal CF organization in human ET patients. Abnormal CF synapses have correlated inversely with the number of PCs and positively with the number of torpedoes, a marker of PC degeneration, in one study (Lee et al., 2018). While suggestive that the CF pathology could be due to PC loss (Louis & Faust, 2020a), an alternative possibility is that excessive CF synapsing, which utilize glutamate neurotransmitters (Grandes, Ortega, & Streit, 1994; Wadiche & Jahr, 2001), may lead to excitotoxicity (Slemmer, De Zeeuw, & Weber, 2005) and result in PC loss (Kuo et al., 2016).

5.3 Purkinje cells

Besides abnormal CF connections to PCs, there are other potential intrinsic mechanisms that could contribute to excessive PC synchronization. As noted above, there have been several pathology studies showing evidence of PC loss and degeneration in ET patients (Choe et al., 2016; Louis et al., 2007), although other studies have conflicting results (Rajput et al., 2012; Symanski et al., 2014). While decreased GABA inhibition of the DCN alone may not be sufficient to induce tremor, altered PC configurations may play a role. In one study examining PC morphology, PCs in ET brains had significantly more axonal recurrent collaterals, axonal branching, and terminal axonal sprouting as compared to those in control brains (Babij et al., 2013). Terminal axonal sprouting could result in a single PC innervating a larger number of DCN neurons, resulting in increased DCN synchrony (Handforth & Lang, 2021). Terminal sprouting was speculated as a possible reason as to why GABA levels appear normal in the dentate nuclei of ET patients, as additional synapse formation by surviving PCs could lead to higher levels of GABA as measured by MR spectroscopy (Louis, Hernandez, et al., 2018). Recurrent PC inhibitory collaterals have been suggested to contribute to high-frequency synchrony among neighboring PCs (de Solages et al., 2008). Increased recurrent PC collaterals in ET patients may increase PC synchrony as compensation for loss of PCs (Louis & Faust, 2020a). Ephaptic current alone among neighboring PCs may contribute to synchronization as well, as changes in the local electrical field from PC depolarization may induce an action potential in a nearby PC (Han et al., 2018).

5.4 Deep cerebellar nuclei

While the DCN serve as the output from the cerebellum to the thalamus, they may be able to promote PC synchrony through their inhibitory feedback connections to the IOB (Fig. 3). GABAergic connections serve as a general inhibitory effect, but also can reduce the strength of coupling between IOB neurons (Hoang et al., 2020). Reduced IOB neuron coupling would reduce synchronized CF activation and subsequently synchronized PC firing. The nucleo-olivary neurons display a lower firing frequency and longer decay time of IPSC, effectively serving as a temporal summation of PC input to the DCN (Najac & Raman, 2015). Overall increased inhibitory input to the DCN, such as through increased SS firing from MF stimulation, could increase inhibition of the nucleo-olivary neurons, and subsequently reduce inhibition of the IOB and promote CS firing via CF excitation (Handforth & Lang, 2021). This double-inhibitory pathway could be a mechanism for increased SS firing from a variety of pathways to lead to increased CS firing. Indeed, a firing pattern has been demonstrated in PCs showing an increase in SS activity, followed by CS, and late inhibition of SS activity (Tang et al., 2017).

6. Conclusions and future directions

Despite its prevalence in the population, much more research is needed in clarifying the underlying pathophysiology of ET. However, accurate identification and classification of ET remains a continuing problem. Diagnostic errors are common, leading some to describe ET as a "waste basket" diagnosis (Amlang, Trujillo Diaz, & Louis, 2020). Criteria for ET have been repeatedly updated, including the more recent adoption of "ET plus" to describe patients with ET and other "soft" neurologic signs such as mild ataxia or questionable dystonia (Bhatia et al., 2018). Furthermore, significant heterogeneity in ET, such as differences in age at onset, tremor distribution, presence of family history, associated neurologic features, and response to treatment, suggest that ET likely represents a collection of various subtypes (Bellows & Jankovic, 2021; Lenka & Louis, 2020). Some of these subtypes may reflect pathology at different sites in ET-relevant networks, including the olivocerebellar loop.

We have presented evidence supporting the notion that alterations in cerebellar inhibitory and excitatory pathways are likely involved in ET, and that tremor generation does not appear attributable solely to a decrease in GABAergic function. MR spectroscopy has not demonstrated a clear deficiency of GABA in the cerebellum. Imaging and pathology studies have provided mixed evidence for the presence of PC degeneration as a source of GABAergic deficiencies. Animal models with severe or total reduction in GABAergic output from PCs shows reduced rather than worsened tremor as compared to models with only partial reduction.

Thus, tremor increasingly can be seen as a result of network dysfunction, with the potential for disturbance(s) at multiple sites contributing to the generation and propagation of tremor. This network dysfunction may result in tremor through altered patterns of GABAergic signaling, instead of simply reduced GABAergic tone. PC hypersynchrony shows promise as a pathophysiological explanation for some ET patients. While hypersynchrony may occur due to pathology at multiple sites along the olivocerebellar loop, recent studies have highlighted the potential role of aberrant CF organization in the generation of cerebellar oscillations. In further support of this mechanism, ViM DBS may reduce tremor by interrupting the oscillatory activity transmitted from the cerebellum to the thalamus (Elble, 2014). More recent studies have examined targeting of the DRTT directly (Fenoy & Schiess, 2017). Further research will hopefully yield data for more targeted and effective treatments directed at these pathways. For example, closed-loop stimulation of the DCN in a harmaline mouse model was found to reliably reduce tremor (Brown et al., 2020).

Better understanding of the nature of site-specific dysfunction and their effects on tremor generation could lead to identification of patients with specific ET pathology. For instance, cerebellar oscillations in ET were found to correlate with tremor severity in patients with familial ET but not sporadic ET (Wong et al., 2021). Improved classification and identification of ET subtypes along with consideration of the non-GABA mediated causes of tremor, could serve to both complete our understanding of ET pathophysiology and lead to improved patient-specific therapies.

References

Amlang, C. J., Trujillo Diaz, D., & Louis, E. D. (2020). Essential tremor as a "waste basket" diagnosis: Diagnosing essential tremor remains a challenge. *Frontiers in Neurology*, *11*, 172. https://doi.org/10.3389/fneur.2020.00172.

Axelrad, J. E., Louis, E. D., Honig, L. S., Flores, I., Ross, G. W., Pahwa, R., et al. (2008). Reduced Purkinje cell number in essential tremor. *Archives of Neurology*, *65*(1). https://doi.org/10.1001/archneurol.2007.8.

Babij, R., Lee, M., Cortés, E., Vonsattel, J.-P. G., Faust, P. L., & Louis, E. D. (2013). Purkinje cell axonal anatomy: Quantifying morphometric changes in essential tremor versus control brains. *Brain*, *136*(10), 3051–3061. https://doi.org/10.1093/brain/awt238.

Bagepally, B. S., Bhatt, M. D., Chandran, V., Saini, J., Bharath, R. D., Vasudev, M., et al. (2012). Decrease in cerebral and cerebellar gray matter in essential tremor: A voxel-based morphometric analysis under 3T MRI. *Journal of Neuroimaging*, *22*(3), 275–278. https://doi.org/10.1111/j.1552-6569.2011.00598.x.

Bayramoglu, B., Emre, U., Erdal, Y., Demirhan, H., Yasak, I., & Yalin, O. O. (2021). Cervical vestibular-evoked myogenic potentials in patients with essential tremor. *Journal of Clinical Neuroscience*, *91*, 365–368. https://doi.org/10.1016/j.jocn.2021.07.015.

Bellows, S., & Jankovic, J. (2021). Parkinsonism and tremor syndromes. *Journal of the Neurological Sciences*, 120018. https://doi.org/10.1016/j.jns.2021.120018.

Bengtsson, F., Ekerot, C.-F., & Jörntell, H. (2011). In vivo analysis of inhibitory synaptic inputs and rebounds in deep cerebellar nuclear neurons. *PLoS One*, *6*(4), e18822. https://doi.org/10.1371/journal.pone.0018822.

Benito-León, J., Alvarez-Linera, J., Hernández-Tamames, J. A., Alonso-Navarro, H., Jiménez-Jiménez, F. J., & Louis, E. D. (2009). Brain structural changes in essential tremor: Voxel-based morphometry at 3-Tesla. *Journal of the Neurological Sciences*, *287*(1–2), 138–142. https://doi.org/10.1016/j.jns.2009.08.037.

Bhalsing, K. S., Upadhyay, N., Kumar, K. J., Saini, J., Yadav, R., Gupta, A. K., et al. (2014). Association between cortical volume loss and cognitive impairments in essential tremor. *European Journal of Neurology*, *21*(6), 874–883. https://doi.org/10.1111/ene.12399.

Bhatia, K. P., Bain, P., Bajaj, N., Elble, R. J., Hallett, M., Louis, E. D., et al. (2018). Consensus statement on the classification of tremors. From the task force on tremor of the International Parkinson and Movement Disorder Society. *Movement Disorders: Official Journal of the Movement Disorder Society*, *33*(1), 75–87. https://doi.org/10.1002/mds.27121.

Boecker, H., Weindl, A., Brooks, D. J., Ceballos-Baumann, A. O., Liedtke, C., Miederer, M., et al. (2010). GABAergic dysfunction in essential tremor: An 11 C-flumazenil PET study. *Journal of Nuclear Medicine*, *51*(7), 1030–1035. https://doi.org/10.2967/jnumed.109.074120.

Broersma, M., van der Stouwe, A. M. M., Buijink, A. W. G., de Jong, B. M., Groot, P. F. C., Speelman, J. D., et al. (2016). Bilateral cerebellar activation in unilaterally challenged essential tremor. *NeuroImage: Clinical*, *11*, 1–9. https://doi.org/10.1016/j.nicl.2015.12.011.

Brown, A. M., White, J. J., van der Heijden, M. E., Zhou, J., Lin, T., & Sillitoe, R. V. (2020). Purkinje cell misfiring generates high-amplitude action tremors that are corrected by cerebellar deep brain stimulation. *eLife*, *9*. https://doi.org/10.7554/eLife.51928.

Bucher, S. F., Seelos, K. C., Dodel, R. C., Reiser, M., & Oertel, W. H. (1997). Activation mapping in essential tremor with functional magnetic resonance imaging. *Annals of Neurology*, *41*(1), 32–40. https://doi.org/10.1002/ana.410410108.

Buijink, A. W. G., Broersma, M., van der Stouwe, A. M. M., Sharifi, S., Tijssen, M. A. J., Speelman, J. D., et al. (2016). Cerebellar atrophy in cortical myoclonic tremor and not in hereditary essential tremor—A voxel-based morphometry study. *The Cerebellum*, *15*(6), 696–704. https://doi.org/10.1007/s12311-015-0734-0.

Buijink, A. W. G., Prent, N., Puts, N. A., Schrantee, A., Potters, W. V., & van Rootselaar, A.-F. (2021). GABA, glutamate, and NAA levels in the deep cerebellar nuclei of essential tremor patients. *Frontiers in Neurology*, *12*. https://doi.org/10.3389/fneur.2021.664735.

Caillard, O., Moreno, H., Schwaller, B., Llano, I., Celio, M. R., & Marty, A. (2000). Role of the calcium-binding protein parvalbumin in short-term synaptic plasticity. *Proceedings of the National Academy of Sciences*, *97*(24), 13372–13377. https://doi.org/10.1073/pnas.230362997.

Cerasa, A., Messina, D., Nicoletti, G., Novellino, F., Lanza, P., Condino, F., et al. (2009). Cerebellar atrophy in essential tremor using an automated segmentation method. *American Journal of Neuroradiology*, *30*(6), 1240–1243. https://doi.org/10.3174/ajnr.A1544.

Choe, M., Cortés, E., Vonsattel, J.-P. G., Kuo, S.-H., Faust, P. L., & Louis, E. D. (2016). Purkinje cell loss in essential tremor: Random sampling quantification and nearest neighbor analysis. *Movement Disorders*, *31*(3), 393–401. https://doi.org/10.1002/mds.26490.

Choi, S. M., Kim, B. C., Chang, J., Choi, K. H., Nam, T. S., Kim, J. T., et al. (2015). Comparison of the brain volume in essential tremor and Parkinson's disease tremor using

an automated segmentation method. *European Neurology*, *73*(5–6), 303–309. https://doi.org/10.1159/000381708.
Coenen, V. A., Sajonz, B., Prokop, T., Reisert, M., Piroth, T., Urbach, H., et al. (2020). The dentato-rubro-thalamic tract as the potential common deep brain stimulation target for tremor of various origin: An observational case series. *Acta Neurochirurgica*, *162*(5), 1053–1066. https://doi.org/10.1007/s00701-020-04248-2.
Colebatch, J. G., Findley, L. J., Frackowiak, R. S., Marsden, C. D., & Brooks, D. J. (1990). Preliminary report: Activation of the cerebellum in essential tremor. *Lancet (London, England)*, *336*(8722), 1028–1030. https://doi.org/10.1016/0140-6736(90)92489-5.
Contarino, M. F., Groot, P. F. C., van der Meer, J. N., Bour, L. J., Speelman, J. D., Nederveen, A. J., et al. (2012). Is there a role for combined EMG-fMRI in exploring the pathophysiology of essential tremor and improving functional neurosurgery? *PLoS One*, 7(10), e46234. https://doi.org/10.1371/journal.pone.0046234.
Daniels, C., Peller, M., Wolff, S., Alfke, K., Witt, K., Gaser, C., et al. (2006). Voxel-based morphometry shows no decreases in cerebellar gray matter volume in essential tremor. *Neurology*, *67*(8), 1452–1456. https://doi.org/10.1212/01.wnl.0000240130.94408.99.
de Solages, C., Szapiro, G., Brunel, N., Hakim, V., Isope, P., Buisseret, P., et al. (2008). High-frequency organization and synchrony of activity in the Purkinje cell layer of the cerebellum. *Neuron*, *58*(5), 775–788. https://doi.org/10.1016/j.neuron.2008.05.008.
Dembek, T. A., Petry-Schmelzer, J. N., Reker, P., Wirths, J., Hamacher, S., Steffen, J., et al. (2020). PSA and VIM DBS efficiency in essential tremor depends on distance to the dentatorubrothalamic tract. *NeuroImage: Clinical*, *26*, 102235. https://doi.org/10.1016/j.nicl.2020.102235.
Deuschl, G. (2000). Essential tremor and cerebellar dysfunction clinical and kinematic analysis of intention tremor. *Brain*, *123*(8), 1568–1580. https://doi.org/10.1093/brain/123.8.1568.
Dyke, J. P., Cameron, E., Hernandez, N., Dydak, U., & Louis, E. D. (2017). Gray matter density loss in essential tremor: A lobule by lobule analysis of the cerebellum. *Cerebellum & Ataxias*, *4*(1), 10. https://doi.org/10.1186/s40673-017-0069-3.
Dykstra, S., Engbers, J. D. T., Bartoletti, T. M., & Turner, R. W. (2016). Determinants of rebound burst responses in rat cerebellar nuclear neurons to physiological stimuli. *The Journal of Physiology*, *594*(4), 985–1003. https://doi.org/10.1113/JP271894.
Elble, R. J. (2014). Mechanisms of deep brain stimulation for essential tremor. *Brain*, *137*(1), 4–6. https://doi.org/10.1093/brain/awt347.
Elkouzi, A., Kattah, J. C., & Elble, R. J. (2016). Hypertrophic olivary degeneration does not reduce essential tremor. *Movement Disorders Clinical Practice*, *3*(2), 209–211. https://doi.org/10.1002/mdc3.12275.
Farkas, Z., Szirmai, I., & Kamondi, A. (2006). Impaired rhythm generation in essential tremor. *Movement Disorders*, *21*(8), 1196–1199. https://doi.org/10.1002/mds.20934.
Fenoy, A. J., & Schiess, M. C. (2017). Deep brain stimulation of the dentato-rubro-thalamic tract: Outcomes of direct targeting for tremor. *Neuromodulation: Technology at the Neural Interface*, *20*(5), 429–436. https://doi.org/10.1111/ner.12585.
Gironell, A. (2014). The GABA hypothesis in essential tremor: Lights and shadows. *Tremor and Other Hyperkinetic Movements*, *4*, 254. https://doi.org/10.5334/tohm.229.
Gironell, A., & Marin-Lahoz, J. (2016). Ethosuximide for essential tremor: An open-label trial. *Tremor and Other Hyperkinetic Movements*, *6*, 378. https://doi.org/10.5334/tohm.306.
Grandes, P., Ortega, F., & Streit, P. (1994). Glutamate-immunoreactive climbing fibres in the cerebellar cortex of the rat. *Histochemistry*, *101*(6), 427–437. https://doi.org/10.1007/BF00269493.
Grüsser-Cornehls, U., Grüsser, C., & Bäurle, J. (1999). Vermectomy enhances parvalbumin expression and improves motor performance in weaver mutant mice: An animal model for cerebellar ataxia. *Neuroscience*, *91*(1), 315–326. https://doi.org/10.1016/S0306-4522(98)00618-6.

Hallett, M., & Dubinsky, R. M. (1993). Glucose metabolism in the brain of patients with essential tremor. *Journal of the Neurological Sciences, 114*(1), 45–48. https://doi.org/10.1016/0022-510X(93)90047-3.

Han, K.-S., Guo, C., Chen, C. H., Witter, L., Osorno, T., & Regehr, W. G. (2018). Ephaptic coupling promotes synchronous firing of cerebellar Purkinje cells. *Neuron, 100*(3), 564–578.e3. https://doi.org/10.1016/j.neuron.2018.09.018.

Hanajima, R., Tsutsumi, R., Shirota, Y., Shimizu, T., Tanaka, N., & Ugawa, Y. (2016). Cerebellar dysfunction in essential tremor. *Movement Disorders, 31*(8), 1230–1234. https://doi.org/10.1002/mds.26629.

Handforth, A. (2012). Harmaline tremor: Underlying mechanisms in a potential animal model of essential tremor. *Tremor and Other Hyperkinetic Movements, 2*, 02. https://doi.org/10.5334/tohm.108.

Handforth, A. (2016). Linking essential tremor to the cerebellum—Animal model evidence. *The Cerebellum, 15*(3), 285–298. https://doi.org/10.1007/s12311-015-0750-0.

Handforth, A., Homanics, G. E., Covey, D. F., Krishnan, K., Lee, J. Y., Sakimura, K., et al. (2010). T-type calcium channel antagonists suppress tremor in two mouse models of essential tremor. *Neuropharmacology, 59*(6), 380–387. https://doi.org/10.1016/j.neuropharm.2010.05.012.

Handforth, A., & Lang, E. J. (2021). Increased Purkinje cell complex spike and deep cerebellar nucleus synchrony as a potential basis for syndromic essential tremor. A review and synthesis of the literature. *The Cerebellum, 20*(2), 266–281. https://doi.org/10.1007/s12311-020-01197-5.

Handforth, A., Martin, F. C., Kang, G. A., & Vanek, Z. (2009). Zonisamide for essential tremor: An evaluator-blinded study. *Movement Disorders, 24*(3), 437–440. https://doi.org/10.1002/mds.22418.

Hashizume, M., Miyazaki, T., Sakimura, K., Watanabe, M., Kitamura, K., & Kano, M. (2013). Disruption of cerebellar microzonal organization in GluD2 (GluRδ2) knockout mouse. *Frontiers in Neural Circuits, 7*. https://doi.org/10.3389/fncir.2013.00130.

Helmchen, C., Hagenow, A., Miesner, J., Sprenger, A., Rambold, H., Wenzelburger, R., et al. (2003). Eye movement abnormalities in essential tremor may indicate cerebellar dysfunction. *Brain, 126*(6), 1319–1332. https://doi.org/10.1093/brain/awg132.

Helmich, R. C., Toni, I., Deuschl, G., & Bloem, B. R. (2013). The pathophysiology of essential tremor and Parkinson's tremor. *Current Neurology and Neuroscience Reports, 13*(9), 378. https://doi.org/10.1007/s11910-013-0378-8.

Hoang, H., Lang, E. J., Hirata, Y., Tokuda, I. T., Aihara, K., Toyama, K., et al. (2020). Electrical coupling controls dimensionality and chaotic firing of inferior olive neurons. *PLoS Computational Biology, 16*(7), e1008075. https://doi.org/10.1371/journal.pcbi.1008075.

Hoebeek, F. E., Witter, L., Ruigrok, T. J. H., & De Zeeuw, C. I. (2010). Differential olivo-cerebellar cortical control of rebound activity in the cerebellar nuclei. *Proceedings of the National Academy of Sciences, 107*(18), 8410–8415. https://doi.org/10.1073/pnas.0907118107.

Hoogland, T. M., De Gruijl, J. R., Witter, L., Canto, C. B., & De Zeeuw, C. I. (2015). Role of synchronous activation of cerebellar Purkinje cell ensembles in multi-joint movement control. *Current Biology, 25*(9), 1157–1165. https://doi.org/10.1016/j.cub.2015.03.009.

Ibrahim, M. F., Beevis, J. C., & Empson, R. M. (2020). Essential tremor—A cerebellar driven disorder? *Neuroscience*. https://doi.org/10.1016/j.neuroscience.2020.11.002.

Jenkins, I. H., Bain, P. G., Colebatch, J. G., Thompson, P. D., Findley, L. J., Frackowiak, R. S. J., et al. (1993). A positron emission tomography study of essential tremor: Evidence for overactivity of cerebellar connections. *Annals of Neurology, 34*(1), 82–90. https://doi.org/10.1002/ana.410340115.

Kano, M., Watanabe, T., Uesaka, N., & Watanabe, M. (2018). Multiple phases of climbing fiber synapse elimination in the developing cerebellum. *The Cerebellum*, *17*(6), 722–734. https://doi.org/10.1007/s12311-018-0964-z.

Kralic, J. E., Criswell, H. E., Osterman, J. L., O'Buckley, T. K., Wilkie, M. E., Matthews, D. B., et al. (2005). Genetic essential tremor in γ-aminobutyric acid A receptor α1 subunit knockout mice. *Journal of Clinical Investigation*, *115*(3), 774–779. https://doi.org/10.1172/JCI23625.

Kronenbuerger, M., Gerwig, M., Brol, B., Block, F., & Timmann, D. (2007). Eyeblink conditioning is impaired in subjects with essential tremor. *Brain*, *130*(6), 1538–1551. https://doi.org/10.1093/brain/awm081.

Kronenbuerger, M., Tronnier, V. M., Gerwig, M., Fromm, C., Coenen, V. A., Reinacher, P., et al. (2008). Thalamic deep brain stimulation improves eyeblink conditioning deficits in essential tremor. *Experimental Neurology*, *211*(2), 387–396. https://doi.org/10.1016/j.expneurol.2008.02.002.

Kuo, S., Lin, C., Wang, J., Liou, J., Pan, M., Louis, R. J., et al. (2016). Deep brain stimulation and climbing fiber synaptic pathology in essential tremor. *Annals of Neurology*, *80*(3), 461–465. https://doi.org/10.1002/ana.24728.

Kuo, S.-H., Lin, C.-Y., Wang, J., Sims, P. A., Pan, M.-K., Liou, J., et al. (2017). Climbing fiber-Purkinje cell synaptic pathology in tremor and cerebellar degenerative diseases. *Acta Neuropathologica*, *133*(1), 121–138. https://doi.org/10.1007/s00401-016-1626-1.

Lalouette, A., Guénet, J.-L., & Vriz, S. (1998). Hotfoot mouse mutations affect the δ2 glutamate receptor gene and are allelic to Lurcher. *Genomics*, *50*(1), 9–13. https://doi.org/10.1006/geno.1998.5314.

Lamarre, Y., & Mercier, L.-A. (1971). Neurophysiological studies of Harmaline-induced tremor in the cat. *Canadian Journal of Physiology and Pharmacology*, *49*(12), 1049–1058. https://doi.org/10.1139/y71-149.

Lang, E. J., Sugihara, I., & Llinas, R. (1996). GABAergic modulation of complex spike activity by the cerebellar nucleoolivary pathway in rat. *Journal of Neurophysiology*, *76*(1), 255–275. https://doi.org/10.1152/jn.1996.76.1.255.

Lee, D., Gan, S. R., Faust, P. L., Louis, E. D., & Kuo, S. H. (2018). Climbing fiber-Purkinje cell synaptic pathology across essential tremor subtypes. *Parkinsonism & Related Disorders*, *51*, 24–29. https://doi.org/10.1016/j.parkreldis.2018.02.032.

Lenka, A., & Louis, E. D. (2020). Do we belittle essential tremor by calling it a syndrome rather than a disease? Yes. *Frontiers in Neurology*, *11*, 522687. https://doi.org/10.3389/fneur.2020.522687.

Leznik, E., Makarenko, V., & Llinás, R. (2002). Electrotonically mediated oscillatory patterns in neuronal ensembles: An in vitro voltage-dependent dye-imaging study in the inferior olive. *The Journal of Neuroscience*, *22*(7), 2804–2815. https://doi.org/10.1523/JNEUROSCI.22-07-02804.2002.

Lin, C.-Y., Louis, E. D., Faust, P. L., Koeppen, A. H., Vonsattel, J.-P. G., & Kuo, S.-H. (2014). Abnormal climbing fibre-Purkinje cell synaptic connections in the essential tremor cerebellum. *Brain*, *137*(12), 3149–3159. https://doi.org/10.1093/brain/awu281.

Long, M. A., Deans, M. R., Paul, D. L., & Connors, B. W. (2002). Rhythmicity without synchrony in the electrically uncoupled inferior olive. *The Journal of Neuroscience*, *22*(24), 10898–10905. https://doi.org/10.1523/JNEUROSCI.22-24-10898.2002.

Louis, E. D. (2021). The essential tremors: Evolving concepts of a family of diseases. *Frontiers in Neurology*, *12*. https://doi.org/10.3389/fneur.2021.650601.

Louis, E. D., Babij, R., Cortés, E., Vonsattel, J.-P. G., & Faust, P. L. (2013). The inferior olivary nucleus: A postmortem study of essential tremor cases versus controls. *Movement Disorders*, *28*(6), 779–786. https://doi.org/10.1002/mds.25400.

Louis, E. D., Diaz, D. T., Kuo, S.-H., Gan, S.-R., Cortes, E. P., Vonsattel, J. P. G., et al. (2018). Inferior olivary nucleus degeneration does not lessen tremor in essential tremor. *Cerebellum & Ataxias*, *5*(1), 1. https://doi.org/10.1186/s40673-018-0080-3.

Louis, E. D., & Faust, P. L. (2020a). Essential tremor pathology: Neurodegeneration and reorganization of neuronal connections. *Nature Reviews Neurology*, *16*(2), 69–83. https://doi.org/10.1038/s41582-019-0302-1.

Louis, E. D., & Faust, P. L. (2020b). Essential tremor: The most common form of cerebellar degeneration? *Cerebellum & Ataxias*, 7(1), 12. https://doi.org/10.1186/s40673-020-00121-1.

Louis, E. D., Faust, P. L., Vonsattel, J.-P. G., Honig, L. S., Rajput, A., Robinson, C. A., et al. (2007). Neuropathological changes in essential tremor: 33 cases compared with 21 controls. *Brain*, *130*(12), 3297–3307. https://doi.org/10.1093/brain/awm266.

Louis, E. D., Frucht, S. J., & Rios, E. (2009). Intention tremor in essential tremor: Prevalence and association with disease duration. *Movement Disorders*, *24*(4), 626–627. https://doi.org/10.1002/mds.22370.

Louis, E. D., Hernandez, N., Chen, K. P., Naranjo, K. V., Park, J., Clark, L. N., et al. (2017). Familial aggregation of the cerebellar signs in familial essential tremor. *Tremor and Other Hyperkinetic Movements*, 7. https://doi.org/10.7916/D8KK9C8Q.

Louis, E. D., Hernandez, N., Dyke, J. P., Ma, R. E., & Dydak, U. (2018). In vivo dentate nucleus gamma-aminobutyric acid concentration in essential tremor vs. controls. *The Cerebellum*, *17*(2), 165–172. https://doi.org/10.1007/s12311-017-0891-4.

Louis, E. D., Huang, C. C., Dyke, J. P., Long, Z., & Dydak, U. (2014). Neuroimaging studies of essential tremor: How well do these studies support/refute the neurodegenerative hypothesis? *Tremor and Other Hyperkinetic Movements*, *4*, 235. https://doi.org/10.5334/tohm.224.

Louis, E. D., & Lenka, A. (2017). The Olivary hypothesis of essential tremor: Time to lay this model to rest? *Tremor and Other Hyperkinetic Movements*, 7, 473. https://doi.org/10.5334/tohm.380.

Louis, R. J., Lin, C.-Y., Faust, P. L., Koeppen, A. H., & Kuo, S.-H. (2015). Climbing fiber synaptic changes correlate with clinical features in essential tremor. *Neurology*, *84*(22), 2284–2286. https://doi.org/10.1212/WNL.0000000000001636.

Louis, E. D., Rios, E., & Rao, A. K. (2010). Tandem gait performance in essential tremor: Clinical correlates and association with midline tremors. *Movement Disorders*, *25*(11), 1633–1638. https://doi.org/10.1002/mds.23144.

Marsden, J. F., Ashby, P., Limousin-Dowsey, P., Rothwell, J. C., & Brown, P. (2000). Coherence between cerebellar thalamus, cortex and muscle in man: Cerebellar thalamus interactions. *Brain: A Journal of Neurology*, *123*(Pt 7(7)), 1459–1470. https://doi.org/10.1093/brain/123.7.1459.

Martin, F. C., & Handforth, A. (2006). Carbenoxolone and mefloquine suppress tremor in the harmaline mouse model of essential tremor. *Movement Disorders*, *21*(10), 1641–1649. https://doi.org/10.1002/mds.20940.

McAuley, J. H. (2000). Physiological and pathological tremors and rhythmic central motor control. *Brain*, *123*(8), 1545–1567. https://doi.org/10.1093/brain/123.8.1545.

Miyazaki, T., Yamasaki, M., Takeuchi, T., Sakimura, K., Mishina, M., & Watanabe, M. (2010). Ablation of glutamate receptor GluR 2 in adult Purkinje cells causes multiple innervation of climbing fibers by inducing aberrant invasion to parallel fiber innervation territory. *Journal of Neuroscience*, *30*(45), 15196–15209. https://doi.org/10.1523/JNEUROSCI.0934-10.2010.

Najac, M., & Raman, I. M. (2015). Integration of Purkinje cell inhibition by cerebellar nucleo-olivary neurons. *The Journal of Neuroscience*, *35*(2), 544–549. https://doi.org/10.1523/JNEUROSCI.3583-14.2015.

Pan, M.-K., Li, Y.-S., Wong, S.-B., Ni, C.-L., Wang, Y.-M., Liu, W.-C., et al. (2020). Cerebellar oscillations driven by synaptic pruning deficits of cerebellar climbing fibers contribute to tremor pathophysiology. *Science Translational Medicine*, *12*(526), eaay1769. https://doi.org/10.1126/scitranslmed.aay1769.

Papapetropoulos, S., Lee, M. S., Versavel, S., Newbold, E., Jinnah, H. A., Pahwa, R., et al. (2021). A phase 2 proof-of-concept, randomized, placebo-controlled trial of CX-8998 in essential tremor. *Movement Disorders*, *36*(8), 1944–1949. https://doi.org/10.1002/mds.28584.

Park, Y.-G., Park, H.-Y., Lee, C. J., Choi, S., Jo, S., Choi, H., et al. (2010). CaV3.1 is a tremor rhythm pacemaker in the inferior olive. *Proceedings of the National Academy of Sciences*, *107*(23), 10731–10736. https://doi.org/10.1073/pnas.1002995107.

Pedrosa, D. J., Nelles, C., Brown, P., Volz, L. J., Pelzer, E. A., Tittgemeyer, M., et al. (2017). The differentiated networks related to essential tremor onset and its amplitude modulation after alcohol intake. *Experimental Neurology*, *297*, 50–61. https://doi.org/10.1016/j.expneurol.2017.07.013.

Pedrosa, D. J., Quatuor, E.-L., Reck, C., Pauls, K. A. M., Huber, C. A., Visser-Vandewalle, V., et al. (2014). Thalamomuscular coherence in essential tremor: Hen or egg in the emergence of tremor? *Journal of Neuroscience*, *34*(43), 14475–14483. https://doi.org/10.1523/JNEUROSCI.0087-14.2014.

Person, A. L., & Raman, I. M. (2012). Purkinje neuron synchrony elicits time-locked spiking in the cerebellar nuclei. *Nature*, *481*(7382), 502–505. https://doi.org/10.1038/nature10732.

Pietracupa, S., Bologna, M., Bharti, K., Pasqua, G., Tommasin, S., Elifani, F., et al. (2019). White matter rather than gray matter damage characterizes essential tremor. *European Radiology*, *29*(12), 6634–6642. https://doi.org/10.1007/s00330-019-06267-9.

Pietracupa, S., Bologna, M., Tommasin, S., Berardelli, A., & Pantano, P. (2021). The contribution of neuroimaging to the understanding of essential tremor pathophysiology: A systematic review. *The Cerebellum*.. https://doi.org/10.1007/s12311-021-01335-7.

Quattrone, A., Cerasa, A., Messina, D., Nicoletti, G., Hagberg, G. E., Lemieux, L., et al. (2008). Essential head tremor is associated with cerebellar vermis atrophy: A volumetric and voxel-based morphometry MR imaging study. *American Journal of Neuroradiology*, *29*(9), 1692–1697. https://doi.org/10.3174/ajnr.A1190.

Raethjen, J., & Deuschl, G. (2012). The oscillating central network of essential tremor. *Clinical Neurophysiology*, *123*(1), 61–64. https://doi.org/10.1016/j.clinph.2011.09.024.

Raethjen, J., Govindan, R. B., Kopper, F., Muthuraman, M., & Deuschl, G. (2007). Cortical involvement in the generation of essential tremor. *Journal of Neurophysiology*, *97*(5), 3219–3228. https://doi.org/10.1152/jn.00477.2006.

Rajput, A. H., Robinson, C. A., Rajput, M. L., Robinson, S. L., & Rajput, A. (2012). Essential tremor is not dependent upon cerebellar Purkinje cell loss. *Parkinsonism & Related Disorders*, *18*(5), 626–628. https://doi.org/10.1016/j.parkreldis.2012.01.013.

Rao, A. K., & Louis, E. D. (2019). Ataxic gait in essential tremor: A disease-associated feature? *Tremor and Other Hyperkinetic Movements (New York, N.Y.)*, *9*. https://doi.org/10.7916/d8-28jq-8t52.

Rossi, F., Wiklund, L., van der Want, J. J. L., & Strata, P. (1991). Reinnervation of cerebellar Purkinje cells by climbing fibres surviving a subtotal lesion of the inferior olive in the adult rat. I. Development of new collateral branches and terminal plexuses. *The Journal of Comparative Neurology*, *308*(4), 513–535. https://doi.org/10.1002/cne.903080403.

Schnitzler, A., Münks, C., Butz, M., Timmermann, L., & Gross, J. (2009). Synchronized brain network associated with essential tremor as revealed by magnetoencephalography. *Movement Disorders*, *24*(11), 1629–1635. https://doi.org/10.1002/mds.22633.

Schreglmann, S. R., Wang, D., Peach, R. L., Li, J., Zhang, X., Latorre, A., et al. (2021). Non-invasive suppression of essential tremor via phase-locked disruption of its temporal coherence. *Nature Communications*, *12*(1), 363. https://doi.org/10.1038/s41467-020-20581-7.

Sharabi, S., Daniels, D., Last, D., Guez, D., Zivli, Z., Castel, D., et al. (2019). Non-thermal focused ultrasound induced reversible reduction of essential tremor in a rat model. *Brain Stimulation, 12*(1), 1–8. https://doi.org/10.1016/j.brs.2018.08.014.

Shiihara, T., Kato, M., Konno, A., Takahashi, Y., & Hayasaka, K. (2007). Acute cerebellar ataxia and consecutive cerebellitis produced by glutamate receptor δ2 autoantibody. *Brain and Development, 29*(4), 254–256. https://doi.org/10.1016/j.braindev.2006.09.004.

Shill, H. A., Adler, C. H., Beach, T. G., Lue, L.-F., Caviness, J. N., Sabbagh, M. N., et al. (2012). Brain biochemistry in autopsied patients with essential tremor. *Movement Disorders, 27*(1), 113–117. https://doi.org/10.1002/mds.24004.

Shin, H., Lee, D.-K., Lee, J.-M., Huh, Y.-E., Youn, J., Louis, E. D., et al. (2016). Atrophy of the cerebellar vermis in essential tremor: Segmental volumetric MRI analysis. *The Cerebellum, 15*(2), 174–181. https://doi.org/10.1007/s12311-015-0682-8.

Slemmer, J. E., De Zeeuw, C. I., & Weber, J. T. (2005). Don't get too excited: Mechanisms of glutamate-mediated Purkinje cell death. *Progress in Brain Research, 148*, 367–390. https://doi.org/10.1016/S0079-6123(04)48029-7.

Stolze, H. (2001). The gait disorder of advanced essential tremor. *Brain, 124*(11), 2278–2286. https://doi.org/10.1093/brain/124.11.2278.

Streng, M. L., Popa, L. S., & Ebner, T. J. (2018). Complex spike wars: A new Hope. *The Cerebellum, 17*(6), 735–746. https://doi.org/10.1007/s12311-018-0960-3.

Symanski, C., Shill, H. A., Dugger, B., Hentz, J. G., Adler, C. H., Jacobson, S. A., et al. (2014). Essential tremor is not associated with cerebellar Purkinje cell loss. *Movement Disorders: Official Journal of the Movement Disorder Society, 29*(4), 496–500. https://doi.org/10.1002/mds.25845.

Tanaka, H., Ishikawa, T., & Kakei, S. (2019). Neural evidence of the cerebellum as a state predictor. *The Cerebellum, 18*(3), 349–371. https://doi.org/10.1007/s12311-018-0996-4.

Tang, T., Xiao, J., Suh, C. Y., Burroughs, A., Cerminara, N. L., Jia, L., et al. (2017). Heterogeneity of Purkinje cell simple spike–complex spike interactions: Zebrin- and non-zebrin-related variations. *The Journal of Physiology, 595*(15), 5341–5357. https://doi.org/10.1113/JP274252.

Tapper, S., Göransson, N., Lundberg, P., Tisell, A., & Zsigmond, P. (2020). A pilot study of essential tremor: Cerebellar GABA+/Glx ratio is correlated with tremor severity. *Cerebellum & Ataxias*, 7(1), 8. https://doi.org/10.1186/s40673-020-00116-y.

Trillenberg, P., Führer, J., Sprenger, A., Hagenow, A., Kömpf, D., Wenzelburger, R., et al. (2006). Eye-hand coordination in essential tremor. *Movement Disorders, 21*(3), 373–379. https://doi.org/10.1002/mds.20729.

Uyaroglu, F. G., Ucar, R., Acarer, A., & Celebisoy, N. (2021). What might cervical vestibular–evoked myogenic potential abnormalities mean in essential tremor? *Neurological Sciences*. https://doi.org/10.1007/s10072-021-05248-5.

Wadiche, J. I., & Jahr, C. E. (2001). Multivesicular release at climbing fiber-Purkinje cell synapses. *Neuron, 32*(2), 301–313. https://doi.org/10.1016/S0896-6273(01)00488-3.

Watanabe, M. (2008). Molecular mechanisms governing competitive synaptic wiring in cerebellar Purkinje cells. *The Tohoku Journal of Experimental Medicine, 214*(3), 175–190. https://doi.org/10.1620/tjem.214.175.

Wills, A. J., Jenkins, I. H., Thompson, P. D., Findley, L. J., & Brooks, D. J. (1994). Red nuclear and cerebellar but no olivary activation associated with essential tremor: A positron emission tomoraphic study. *Annals of Neurology, 36*(4), 636–642. https://doi.org/10.1002/ana.410360413.

Wills, A. J., Jenkins, L. H., Thompson, P. D., Findley, L. J., & Brooks, D. J. (1995). A positron emission tomography study of cerebral activation associated with essential and writing tremor. *Archives of Neurology, 52*(3), 299–305. https://doi.org/10.1001/archneur.1995.00540270095025.

Wong, S.-B., Wang, Y.-M., Lin, C.-C., Geng, S. K., Vanegas-Arroyave, N., Pullman, S. L., et al. (2021). Cerebellar oscillations in familial and sporadic essential tremor. *The Cerebellum*. https://doi.org/10.1007/s12311-021-01309-9.

Wu, Y., Louis, E. D., Gionco, J., Pan, M., Faust, P. L., & Kuo, S. (2021). Increased climbing fiber lateral crossings on Purkinje cell dendrites in the cerebellar hemisphere in essential tremor. *Movement Disorders, mds.*, 28502. https://doi.org/10.1002/mds.28502.

Yoshida, T., Katoh, A., Ohtsuki, G., Mishina, M., & Hirano, T. (2004). Oscillating Purkinje neuron activity causing involuntary eye movement in a mutant mouse deficient in the glutamate receptor delta2 subunit. *The Journal of Neuroscience: The Official Journal of the Society for Neuroscience*, *24*(10), 2440–2448. https://doi.org/10.1523/JNEUROSCI.0783-03.2004.

Zesiewicz, T. A., Elble, R. J., Louis, E. D., Gronseth, G. S., Ondo, W. G., Dewey, R. B., et al. (2011). Evidence-based guideline update: Treatment of essential tremor: Report of the Quality Standards Subcommittee of the American Academy of Neurology. *Neurology*, 77(19), 1752–1755. https://doi.org/10.1212/WNL.0b013e318236f0fd.

Zesiewicz, T. A., Ward, C. L., Hauser, R. A., Sanchez-Ramos, J., Staffetti, J. F., & Sullivan, K. L. (2007). A double-blind placebo-controlled trial of zonisamide (zonegran) in the treatment of essential tremor. *Movement Disorders*, *22*(2), 279–282. https://doi.org/10.1002/mds.21282.

SECTION VI

The future of pharmacotherapies for essential tremor: Enhancing GABA neurotransmission or reducing neuronal hyperexcitability?

Sheng-Han Kuo[a,b] and Elan D. Louis[c]

[a]Department of Neurology, College of Physicians and Surgeons, Columbia University, New York, NY, United States
[b]Initiative for Columbia Ataxia and Tremor, Columbia University, New York, NY, United States
[c]Department of Neurology, University of Texas Southwestern Medical Center, Dallas, TX, United States

As part of the current series, which highlights current controversies in the essential tremor (ET) field, we commissioned authors to submit papers with opposing viewpoints on potential strategies to use in the development of future therapies in ET, one using available data to put forth the argument that enhancing gamma aminobutyric acid (GABA)ergic neurotransmission holds promise for future ET therapies (Ondo, 2022) and the other, using available data to put forth the argument that reducing neuronal hyper-excitability is the key for future novel ET therapies (Shukla, 2022). We are grateful to both sets of authors for thoughtfully making their cases and here we provide additional commentary.

The theory that ET is a disorder of insufficient GABA comes from several clinical observations, including the one that about two-thirds of ET patients report some degree of tremor responsiveness to ethanol (Lou & Jankovic, 1991), presumably acting via enhancing the GABAergic system. Dysfunction of GABAergic system in ET is further supported by neuropathological findings. The ET cerebellum has reduced GABA receptor levels in the dentate nucleus (Paris-Robidas, Brochu, Sintes, et al., 2012), indicating insufficient post-synaptic GABA responses, and the ET cerebellum also evidences abundant Purkinje cell axonal degenerative changes (Babij, Lee, Cortés, Vonsattel, et al., 2013; Louis, Faust, Vonsattel, et al., 2007), suggesting a decrease in pre-synaptic GABAergic inputs at the Purkinje cell-deep cerebellar nuclear synapses. These pathological findings are summarized by Dr. Faust in this series (Faust, 2022) and further support the proposition that enhancing and/or replacing GABA neurotransmission in the cerebellum could modulate or normalize circuitry dysfunction and improve tremor in ET patients.

ET has also traditionally been thought to be a disorder of increased neuronal oscillatory activity in the cerebello-thalamo-cortical loop (Neely, Kurani, & Shukla, 2015; Schreglmann, Wang, & Peach, 2021), which implies some degree of neuronal hyperexcitability in this system. Elegant physiological studies have demonstrated that harmaline, a chemical that enhances gap junction coupling in the inferior olive neurons can induce tremor that resembles ET by setting up neuronal hyper-excitability in the olivocerebellum (Linas & Volkind, 1973; Llinas & Yarom, 1986). Furthermore, increased excitatory synaptic inputs to Purkinje cells by climbing fibers have been identified in a postmortem study of the ET cerebellum (Lin, Louis, Faust, et al., 2014), and such a pathological feature is sufficient to drive cerebellar oscillatory activity and tremor in a mouse model of tremor, *hotfoot17J* (Pan, Li, Wong, et al., 2020). Therefore, neuronal hyperexcitability in the olivocerebellum can create tremor that shares features with ET in animal models, and this concept is further summarized by Lang and Handforth in this series (Lang and Handforth, 2022).

These animal studies provide evidence that targeting neuronal hyper-excitability could be an appealing strategy for ET therapeutic development.

Several pharmaceutical agents for ET are recently being developed in clinical trials. These trials have centered around these two mechanisms: (1) enhancing GABA neurotransmission by activating extra-synaptic GABA receptors (Paskavitz, Bullock, Nguyen, et al., 2019), (2) reducing neuronal hyper-excitability by modulating either T-type calcium channels or calcium-activated potassium channels (SK channels) (Papapetropoulos, Lee, Versavel, et al., 2021), two types of channels that modulate Purkinje cell firing (Womack & Khodakhah, 2004; Yazdi, Janahmadi, & Behzadi, 2007). In the current series, authors present data and scientific rationales to support each strategy and also prior successes or failures of clinical trials using these strategies. Collectively, the information will provide critical context for future ET therapeutic development and also for the consideration of patient selection as well as choices of physiological markers in trial design.

Since ET is likely to be a heterogeneous disorder or set of disorders, perhaps it is important to ask what proportion of ET population will respond to each therapeutic strategy, and whether we should develop tools to detect these physiological endophenotypes as a biomarker to improve trial success. For example, ethanol responsiveness could serve as a marker for ET patients with GABAergic dysfunction whereas cerebellar oscillatory activity, detected by cerebellar encephalogram (Pan et al., 2020), could conceivably be used to select ET patients with neuronal hyper-excitability. These notions should be further tested. Another important consideration is that these two therapeutic strategies may not be mutually exclusive and can be overlapping mechanisms. For example, progressive loss of $GABA_A$ receptors in the dentate nucleus can cause a dysfunctional cerebello-thalamo-cortical loop, leading to tremor-related oscillatory activity and neuronal burst firing (Zhang & Santaniello, 2019). In addition, tremor from harmaline-induced neuronal hyper-excitability can be suppressed by ethanol (Martin, Thu Le, & Handforth, 2005) and other GABAergic drugs (Paterson, Malekiani, & Foreman, 2009). Finally, it is also crucial to recognize detailed pathophysiology studies will help us further understand the circuit dysfunction in tremor. Both Purkinje cells and deep cerebellar neurons have a very high baseline firing frequency, and the firing patterns of the deep cerebellar neurons are regulated by transient and also convergent GABAergic transmission from Purkinje cells with exquisite precision of timing and rhythmicity (Tang, Blenkinsop, & Lang, 2019). Therefore, the term "GABA deficiency" does not capture the whole picture of possible GABAergic dysfunction in such synapses. On the other hand, Purkinje cell can also have different firing modes and burst patterns (Engbers, Fernandez, & Turner, 2013) and thus "hyperexcitability" may be a general term that does not specifically point to neuronal firing related to tremor generation. More in-depth understanding of tremor pathophysiology will help to develop specific therapies to treat ET.

References

Babij, R., Lee, M., Cortés, E., Vonsattel, J. P., et al. (2013). Purkinje cell axonal anatomy: Quantifying morphometric changes in essential tremor versus control brains. *Brain, 136*, 3051–3061.

Engbers, J. D. T., Fernandez, F. R., & Turner, R. W. (2013). Bistability in Purkinje neurons: Ups and downs in cerebellar research. *Neural Networks, 47*, 18–31.

Faust, P. L. (2022). Is essential tremor a degenerative disorder or an electric disorder? Degenerative disorder. *International Review of Neurobiology, 163*, 65–101.

Lang, E. J., & Handforth, A. (2022). Is the inferior olive central to essential tremor? Yes. *International Review of Neurobiology, 163*, 133–165.

Lin, C. Y., Louis, E. D., Faust, P. L., et al. (2014). Abnormal climbing fibre-Purkinje cell synaptic connections in the essential tremor cerebellum. *Brain, 137*, 3149–3159.

Linas, R., & Volkind, R. A. (1973). The olivo-cerebellar system: Functional properties as revealed by harmaline-induced tremor. *Experimental Brain Research, 18*, 69–87.

Llinas, R., & Yarom, Y. (1986). Oscillatory properties of guinea-pig inferior olivary neurones and their pharmacological modulation: An in vitro study. *The Journal of Physiology, 376*, 163–182.

Lou, J. S., & Jankovic, J. (1991). Essential tremor: Clinical correlates in 350 patients. *Neurology, 41*, 234–238.

Louis, E. D., Faust, P. L., Vonsattel, J. P., et al. (2007). Neuropathological changes in essential tremor: 33 cases compared with 21 controls. *Brain, 130*, 3297–3307.

Martin, F. C., Thu Le, A., & Handforth, A. (2005). Harmaline-induced tremor as a potential preclinical screening method for essential tremor medications. *Movement Disorders, 20*, 298–305.

Neely, K. A., Kurani, A. S., & Shukla, P. (2015). Functional brain activity relates to 0-3 and 3-8 Hz force oscillations in essential tremor. *Cerebral Cortex, 25*, 4191–4202.

Ondo, W. (2022). Enhancing GABA inhibition is the next generation of medications for essential tremor. *International Review of Neurobiology, 163*, 317–334.

Pan, M. K., Li, Y. S., Wong, S. B., et al. (2020). Cerebellar oscillations driven by synaptic pruning deficits of cerebellar climbing fibers contribute to tremor pathophysiology. *Science Translational Medicine, 12*, eaay1769.

Papapetropoulos, S., Lee, M. S., Versavel, S., et al. (2021). *Movement Disorders, 36*, 1944–1949.

Paris-Robidas, S., Brochu, E., Sintes, M., et al. (2012). Defective dentate nucleus GABA receptors in essential tremor. *Brain, 135*, 105–116.

Paskavitz, J., Bullock, A., Nguyen, D., et al. (2019). GABA-A receptor positive allosteric modulators: Phase 2 proof of concept studies of brexanolone injection and SAGE-217 in essential tremor. *MDS abstract.* https://www.mdsabstracts.org/abstract/gaba-a-receptor-positive-allosteric-modulators-phase-2-proof-of-concept-studies-of-brexanolone-injection-and-sage-217-in-essential-tremor/.

Paterson, N. E., Malekiani, S. A., & Foreman, M. M. (2009). Pharmacological characterization of harmaline-induced tremor activity in mice. *European Journal of Pharmacology, 616*, 73–80.

Schreglmann, S. R., Wang, D., & Peach, R. L. (2021). Non-invasive suppression of essential tremor via phase-locked disruption of its temporal coherence. *Nature Communications*, *12*, 363.
Shukla, A. W. (2022). Reduction of neuronal hyperexcitability with modulation of T-type calcium channel or SK channel in essential tremor. *International Review of Neurobiology*, *163*, 335–356.
Tang, T., Blenkinsop, T. A., & Lang, E. J. (2019). Complex spike synchrony dependent modulation of rat deep cerebellar nuclear activity. *eLife*, *8*, e40101.
Womack, M. D., & Khodakhah, K. (2004). Dendritic control of spontaneous bursting in cerebellar Purkinje cells. *The Journal of Neuroscience*, *24*, 3511–3521.
Yazdi, H. H., Janahmadi, M., & Behzadi, G. (2007). The role of small-conductance Ca2 +−activated K + channels in the modulation of 4-aminopyridine-induced burst firing in rat cerebellar Purkinje cells. *Brain Research*, *1156*, 59–66.
Zhang, X., & Santaniello, S. (2019). Role of cerebellar GABAergic dysfunctions in the origins of essential tremor. *Proceedings of the National Academy of Sciences of the United States of America*, *116*, 13592–13601.

CHAPTER ELEVEN

Enhancing GABA inhibition is the next generation of medications for essential tremor

William Ondo*
Houston Methodist Hospital, Weill Cornel Medical School, Houston, TX, United States
*Corresponding author: e-mail address: wondo@houstonmethodist.org

Contents

Abstract

γ-Aminobutyric acid (GABA) is the most prevalent inhibitory CNS neurotransmitter. Activating GABA-A receptors hyperpolarizes cells via Cl^- influx, which inhibits action potentials. Although the exact pathophysiologies of tremor are incompletely understood, proposed neuroanatomy extensively implicates GABA pathways. Pathological studies and imaging studies also show GABA abnormalities in patients with ET. Most importantly, medications that activate GABA-A receptors, such as primidone, often improve tremor. Ongoing clinical trials and physiology research should further refine

International Review of Neurobiology, Volume 163
ISSN 0074-7742
https://doi.org/10.1016/bs.irn.2022.02.007

potential future GABAergic targets and treatments, which are currently the most promising targets for pharmacological intervention.

1. Introduction

The pathophysiology of essential tremor (ET) is only partially understood. There is some understanding of culpable macro-circuitry, mostly based on functional PET/SPECT studies, functional MRI, tractography, transcranial stimulation, and other electrophysiological techniques (Buijink et al., 2015). However, there is less understanding at the cellular level. Post-mortem pathology of ET is inconsistent; variably being normal, demonstrating Lewy body pathology, or more commonly cerebellar Purkinje cell loss and degeneration, and altered inferior olivary climbing fiber synapses to the Purkinje cell layer (Erickson-Davis et al., 2010; Louis, Lin, Faust, Koeppen, & Kuo, 2015; Rajput, Adler, Shill, & Rajput, 2012). That said, a large body of evidence directly or indirectly supports involvement of γ-aminobutyric acid (GABA) systems, and argues that targeted pharmacological intervention at GABA receptors would most likely lessen tremor.

2. Essential tremor macro and microanatomy

Two interconnected neuropathways have been most implicated in ET; the Cerebellar (deep nuclei)-Thalamic-Cortical Circuit, and Mollaret's Triangle (inferior olive-cerebellar cortex-red nucleus). These two circuits intersect at the GABAergic Purkinje cells of the cerebellar cortex.

The inferior olivary nucleus (ION) in the superior medulla is the most rostral anatomy of interest and has been long implicated in ET due to its robust auto-oscillatory firing (Fig. 1).

Extensive Gap junctions within the nucleus result in variable 5–10 Hz sub-threshold oscillations. Climbing fibers, mostly from the ION send glutaminergic projections to Purkinje cells in the contralateral cerebellar cortex, but also directly to deep cerebellar nuclei. The ION climbing fibers form a robust 1:1 relationship with Purkinje cells, and single climbing fiber discharges create Purkinje excitatory postsynaptic potentials. Purkinje cells receive numerous other excitatory and inhibitory inputs but none of these other sources have a 1:1 synaptic structure. They also receive GABAergic interneuron connections within the cerebellar cortex from Basket cells

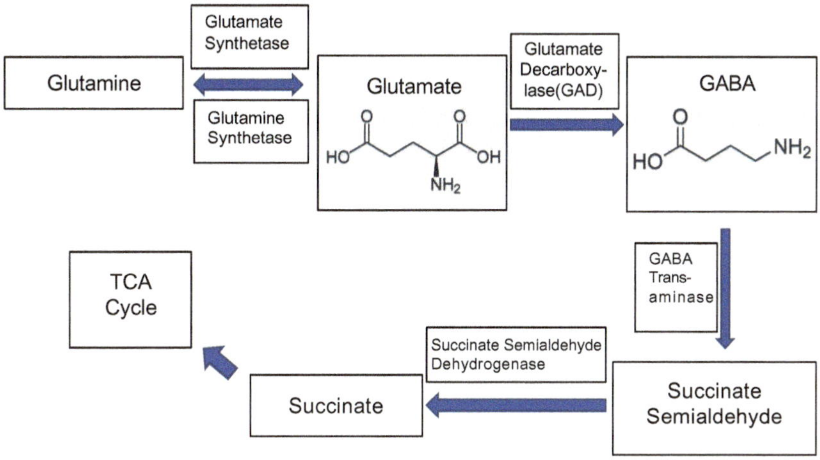

Fig. 1 Major metabolic pathway for GABA. GABA can also be derived from other pathways and precursors including putrescine, spermine, spermidine, and ornithine. Glutamate is also derived from α-ketoglutarate. Glutamate decarboxylase has two active forms, GAD65 and GAD67, and uses pyridoxine (Vitamin B6) as a co-factor.

and Stellate cells. Conceptually the cerebellar cortex seems to integrate information from spinal, vestibular and other sensory inputs before the Purkinje cells send extensive GABAergic inhibitory synapses to the deep cerebellar nuclei including the dentate (largest and most lateral), but also the emboliform, globose, and fastigial nuclei. The deep cerebellar nuclei complete a reciprocal inhibitory circuit via descending inhibitory GABAergic projections back to the ION. GABA has been shown to increase ION synchrony and action potential formation, possibly through enhancement of gap junction signaling (Lang, 2002). Therefore, cerebellar injury will subsequently impact ION neurons and vice versa.

The deep nuclei are the main outflow tracts of the cerebellum. The dentate nucleus projects through the superior peduncle contralaterally to the thalamus, specifically the ventral lateral, and ventral intermediate nucleus (VIM), and to a lesser extent to the ventral anterior nucleus of the thalamus. The VIM is the part of the ventral posterior nucleus that is adjacent to the ventral lateral nucleus. Often anatomic studies do not isolate the VIM, complicating the literature. The VIM and ventral lateral nuclei send diffuse projections mostly to the primary motor cortex, whereas the ventral anterior nucleus projects to the pre-motor cortex. In ET, the VIM demonstrates marked rhythmic oscillation that correlates with clinical tremor.

3. γ-aminobutyric acid (GABA) and GABA receptors

GABA is chemically an amino acid, but not an alpha-amino acid (amino group is not attached to the alpha carbon) and thus GABA is not incorporated into proteins. It is mostly synthesized in astrocytes by glutamate decarboxylase (GAD) and pyridoxal phosphate (B6) cofactor, from glutamate, which is the most abundant excitatory neurotransmitter (Fig. 2). It is primarily metabolized by GABA transaminase, eventually into succinic acid. GABA is found in plants and supplements but has poor blood brain barrier penetration.

GABA type A receptors are transmembrane, ionotropic ligand gaited ion channels found extensively throughout the central nervous system. When stimulated by endogenous GABA, a central pore opens resulting in chloride (Cl^-), and to a lesser extent bicarbonate (HCO_3-), influx that hyperpolarizes (stabilizes) cell resting potential, making it much more difficult to create an action potential, and therefore inhibits neuronal cell firing.

Endogenous GABA is predominately released in vesicles, and therefore phasically inhibits neuron output. However, some GABA trans-membrane leak can result in mild tonic inhibition. The GABA binding site on the GABA-A receptor is near the narrowest portion of the ion pore. Interestingly, most medications that target GABA receptors have different binding sites, however muscimol and gaboxadol do target the endogenous

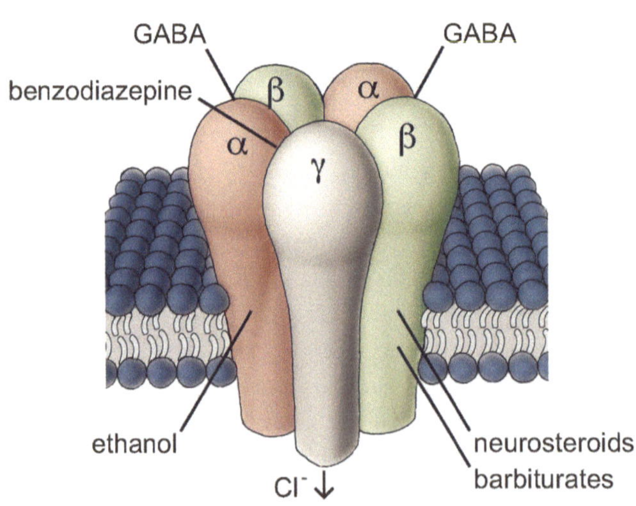

Fig. 2 Stylized GABA-A receptor showing engagement site of different drug classes.

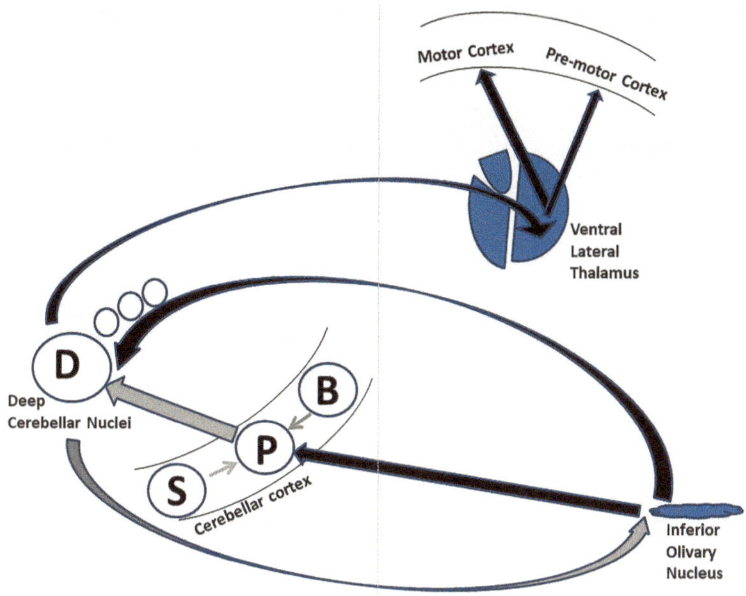

Fig. 3 Anatomy most associated with essential tremor. Inhibitory GABAergic pathways include Stellate cells (S) to Purkinje cells (P) and Basket cells (B) to Purkinje cells (P) within the cerebellar cortex; Purkinje cells to deep cerebellar nuclei including the Dentate (D); and deep cerebellar nuclei to the Inferior Olivary nucleus. Light grey tracts are GABAergic (inhibitory) and black tracts are excitatory.

GABA binding site. These drugs are not currently clinically used but were reported to cause euphoria, sedation, and amnesia. Any impact on tremor was never evaluated.

The GABA-A receptor is a pentamer, variably employing 5 of 19 different proteins (alpha 1-6, beta 1-3, gamma 1-3, delta, epsilon, theta, pi, rho 1-3) (Fig. 3). By definition, a GABA receptor must include at least one alpha and one beta protein. The most common arrangement is two alphas, two betas, and one gamma. Alpha-1 is by far the most common sub-unit and is especially present in synaptic receptors. Extra-synaptic GABA-A receptors, which may modulate baseline activity, actually have higher affinity for endogenous GABA, and more delta subunits. Although GABA-A receptors are distributed throughout the nervous system and extra brain organs e.g. liver, placenta, bone growth plates, certain specific protein configurations may be more localized. For example, alpha-6 is particularly associated with the cerebellar nuclei and alpha-1 is especially associated with Purkinje cells. Although some generalizations regarding specific protein constituents are known, detailed knowledge differentiating the roles of the many thousands of possible protein combinations in GABA-A receptors in lacking.

4. Anatomic and pathological abnormalities in ET are associated with GABA

Gross anatomy studies in ET are largely unremarkable. Standard voxel-based MRI imaging of the cerebellum have been inconsistent, but one recent meta-analysis found grey matter atrophy in hemispheric and vermic lobule 4/5 (Mavroudis et al., 2021), but no overall atrophy. Some studies do associate cerebellar atrophy specifically with head tremor (Holtbernd & Shah, 2021). Other areas of the brain show no structural imaging abnormalities.

Microscopic pathological studies in ET mostly report Purkinje cell loss and other markers of cellular stress in the cerebellar cortex (Louis et al., 2007; Louis & Faust, 2020). More recently, abnormalities in climbing fiber Purkinje synapses have been identified (Kuo et al., 2016). ET patients had decreased climbing fiber-Purkinje cell synaptic density, more climbing fibers extending to the outer portion of the cerebellar cortex molecular layer, and more climbing fiber-Purkinje cell synapses on the thin Purkinje cell spiny branchlet, which are usually occupied by parallel fibers (Lin et al., 2014). Additionally, ET patients demonstrate greater lateral crossings in these climbing fiber connections (Wu et al., 2021). Genetically modified animal model with these abnormal climbing-Purkinje synapsis cause cerebellar oscillations and tremor (Pan et al., 2020).

Structural pathologic changes have not been found in the Dentate nuclei (Hartstone et al., 2021). However, one post-mortem study on 10 ET patients found at 35% reduction in GABA-A receptors and a 22–31% reduction of GABA-B receptors on the Dentate nucleus compared to controls (Paris-Robidas et al., 2012).

5. The role of inferior olivary nucleus/cerebellar connections

In ET patients, no pathological changes, nor consistent macrophysiologic changes have been identified in the ION (Elkouzi, Kattah, & Elble, 2016; Louis et al., 2018; Wills, Jenkins, Thompson, Findley, & Brooks, 1994). Nevertheless, the robust auto-oscillatory signaling draws attention to this anatomy. ION firing is at least partially controlled by calcium T-type (CAV3) receptors (Park et al., 2010), which is the main rational for trials of T-Type Calcium antagonists in human tremor.

Harmaline, an alkaloid reversible monoamine oxidase-A inhibitor, accentuates ION signaling and causes tremor in rodents (Handforth, 2012). Many medications including GABA-A agonists, and T-type calcium channel blockers reduce this clinical tremor, but some medications that clearly do not improve human ET or even cause tremor, such as lithium and valproate, also improve harmaline induced tremor, arguing against it truly replicating ET pathophysiology (Paterson, Malekiani, Foreman, Olivier, & Hanania, 2009). Whether or not the ION is the ET generator, part of the ET pathway, or not involved at all, is still debated.

ION/cerebellar interactive anatomy has its own closed loop triangular circuit: ION to Purkinje, Purkinje to deep cerebellar nuclei, deep cerebellar nuclei to ION. Purkinje cell loss would result in less GABAergic inhibition of deep cerebelli nuclei. The subsequent increased GABAergic inhibitory signaling from the deep cerebelli nuclei will inhibit ION cells, resulting in less excitatory glutamatergic signaling back to the Purkinje cells, resulting in even less GABAergic inhibition of deep cerebelli nuclei, which continues to perpetuate this pathological closed loop firing pattern. Intra-ION injection of the GABA-A receptor antagonist picrotoxin increases complex spike synchrony, and induces tremor in animal models, although to our knowledge, similar experiments with GABA-A agonists have not been done to see if they reduce tremor (Handforth & Lang, 2021). In relation to this physiology, GABA-A agonists could disrupt this pathologic firing pattern at the Purkinje cell to deep cerebelli nuclei synapse or the deep cerebelli nuclei to ION synapse. Furthermore, inhibition of output from cerebellar cortical interneurons, which GABAergically inhibit Purkinje cells, also alter Purkinje cell firing (Brown et al., 2019), add a third possible synapse whereby GABA medications could mitigate tremor.

6. Functional imaging studies

Functional imaging studies of the cerebellum consistently show increased bilateral metabolic activity (Lan et al., 2021). In contrast, functional imaging studies have not demonstrated increased ION metabolic activity (Wills et al., 1994). This is consistent with the presented physiology that decreased inhibition of the DCN (increased activity) results in increased inhibition (reduced activity) in the ION, rather than increased activation of the ION.

[^{11}C] Flumazenil is a radiotracer that binds to GABA-A receptors. One study found increased binding, consistent with decreased endogenous GABA, in the cerebellum, ventrolateral thalamus, and pre-frontal cortex compared to controls (Boecker et al., 2010). A second [^{11}C] Flumazenil study correlated tremor severity with binding abnormalities (Gironell et al., 2012). Magnetic resonance spectroscopy study that have looked at GABA+: glutamate/glutamine ratios in the cerebellum have shown no difference compared to controls (Buijink et al., 2021; Tapper, Goransson, Lundberg, Tisell, & Zsigmond, 2020) but did correlate with tremor severity (Tapper et al., 2020).

Some cerebral spinal fluid (CSF) studies in ET patients suggest decreased GABA levels and increased in glutamate (Gironell, 2014; Mally, Baranyi, & Vizi, 1996).

7. GABAergic genetics and genetic manipulation in ET

Despite the obvious genetic contribution in ET, human genetic studies have identified relatively few genetic markers and no highly penetrant gene abnormalities (Jimenez-Jimenez et al., 2021). To date no genetic human abnormality directly involving GABAergic proteins is identified (Siokas et al., 2020). However, genetic knock-out models, especially in GABA-A receptors do create a tremor phenotype. Alpha-1 protein (the most abundant component of the GABA-A receptor) knock-out mice demonstrate a high frequency tremor, which improves with some tremor medications (Kralic et al., 2005). Selective GABA-A alpha-1 protein knock-out in cerebellar Purkinje cells also causes tremor, but not ataxia (Nietz et al., 2020). Tremor from both general and cerebellar selective knock-outs improve with ethanol, suggesting GABA-A receptors without alpha-1 are still impacted by ethanol.

Completely inhibiting Purkinje cell GABAergic output by genetically knocking out vesicular GABA transporter (VGAT) causes ataxia but not tremor in mice (Brown et al., 2020). However, this knock-out will block harmaline induced tremor generated in the ION (Martin, Thu Le, & Handforth, 2005). Brown et al. also reported a complex optogenetic study where rhythm induced firing of Purkinje cells, especially around 10 Hz and 3 Hz, does result in rhythmic deep cerebellar nuclei firing and clinical tremor. Moreover, this tremor can be stopped by high frequency deep brain stimulation of deep cerebellar nuclei (Brown et al., 2020).

Computer modelling of gradual loss of Purkinje output and specifically GABA-A alpha -1 receptor subunits also predicts oscillation throughout the cerebellar-thalamic-cortical networks (Zhang & Santaniello, 2019). This is predicated on alterations in opening times of the GABA-A receptor channels in the dentate nucleus, which could be altered by GABAergic medications.

Taken in whole this strongly implicates the Purkinje cell to deep cerebelli nuclei GABAergic synapse, as either the originator or at least gait keep for tremor.

8. Most medications that improve tremor target GABA-A receptors

Potentially the most compelling evidence supporting the future development of medication that target GABA systems for tremor are the efficacy of current medications that target GABA systems. Alcohols, primidone and other barbiturates, and benzodiazepines all reduce tremor, presumably via GABA-A receptor activation.

8.1 Alcohols

Ethanol is arguably the most potent tremor suppressor. Other alcohols such as 1-octanol may also improve tremor, at least in subjects who respond to ethanol (Frucht, Houghton, Bordelon, Greene, & Louis, 2005; Haubenberger et al., 2013; Nahab et al., 2011). However, 1-octanol is rapidly metabolized to octanoic acid, a carboxylic acid, which is now thought to be the active therapeutic component. Methylpentynol, a six-carbon chain alcohol proved clinically ineffective when compared to placebo (Teravainen, Huttunen, & Lewitt, 1986). Therefore, it is not clear whether alcohols improve tremor as a class or whether it is specific to ethanol.

Ethanol improves tremor at relatively low levels, usually within 20 min for 3–4 h, sometimes followed by a rebound tremor augmentation. Improvement mostly originates in the central nervous system based on weight loading studies and reduced cerebellar activity on PET following ingestion (Boecker et al., 1996).

Ethanol potentiates the inhibition of cortical neurons by GABA (Nestoros, 1980) and at low doses, consistent with levels that reduce tremor, ethanol has greatest affinity for GABA-A receptor rich in delta sub-units. Anatomically, these are most abundant in the pre-frontal cortex, hippocampus, and cerebellum. That said, ethanol has widespread effects throughout the entire brain and in general is a cellular decoupling agent. Ethanol also specifically inhibits neurons by specific G-protein-activated inwardly rectifying potassium (GIRK) channels, alteration in NMDA receptors and various effects on glycine receptors (Kumar et al., 2009). Whether ethanol improves tremor via GABA-A receptor, other mechanisms, or multiple mechanisms is not definitively known.

8.2 Benzodiazepines

Benzodiazepines bind to GABA-A receptors at the alpha-gamma interface. In contrast to ethanol, benzodiazepines have their greatest affinity to GABA-A pentamers with two alpha, two beta, and one gamma sub-unit, and less to those with delta sub-units, although specific affinities of different benzodiazepines varies somewhat (Sankar, 2012). Affinities to alpha sub-units in general is associated with anti-convulsive affects; alpha-1 sub-units have been specifically associated with sedation/amnestic effects, and alpha-2 with anxiolytic properties (Sankar, 2012). GABA-A receptors containing alpha-4 and alpha-6 units are generally insensitive to benzodiazepines (Kumar et al., 2009). No particular drug or specific affinity is clearly elucidated for tremor suppression, as there is little comparative data among benzodiazepines. TPA023, a GABA-A alpha-2, 3 subtype-selective partial agonist did show relatively little effect on tremor, at least compared to ethanol (de Haas et al., 2012). The ideal GABA-A specific protein affinities that would maximize tremor suppression is not known.

8.3 Barbiturates

Barbiturates such as phenobarbital, and primidone probably improve tremor more than benzodiazepines, although formal comparative data is lacking. Barbiturates have a distinct GABA-A binding site that increase the duration of receptor opening, as opposed to increasing the frequency of openings. These appear relatively non-specific for different sub-units and will bind to any pentamer with an alpha or beta sub-unit. Primidone probably acts via this mechanism although there is remarkably little published data on primidone's MOA, because the anti-epileptic effects were historically ascribed to its metabolite, phenobarbital. At least one study found primidone superior to phenobarbital against tremor, and primidone improved tremor within an hour of initial ingestion, prior to the formation of any phenobarbital metabolite (Koller & Royse, 1986). Therefore, it is possible primidone has another unknown mechanism of action or stimulates GABA-A receptors somewhat differently than phenobarbital, but this is unknown.

8.4 Drugs that increase GABA availability or stimulate the endogenous GABA site

Despite the apparent impact of drugs that potentiate GABA-A receptors have on tremor, there is less evidence supporting a role for manipulating endogenous GABA. Endogenous GABA in the CNS is converted from glutamate by L-glutamic acid decarboxylase (GAD) using pyridoxal phosphate (the active form of vitamin B6) as a cofactor. Drugs that increase GABA availability (tiagabine, vigabatrin, valproate) have not been associated with tremor improvement, and may even exacerbate tremor in some cases (Gironell, Martinez-Corral, Pagonabarraga, & Kulisevsky, 2008; Zesiewicz, Sullivan, Ward, & Hauser, 2007). Tiagabine, which increases GABA levels via blocking the synaptic GABA transporter (GAT-1), has been specifically reported to worsen tremor. Vigabatrin, which prevents GABA degradation via irreversible inhibition of the mitochondrial enzyme GABA transaminase, has not demonstrated tremor benefit, and is reported to cause tremor as an adverse event (Walker & Kalviainen, 2011). Valproate, which increases GABA turnover and synthesis, along with several other MOA, often causes tremor.

In contrast, topiramate, which does often improve tremor, may increase GABA-A receptor density, although several other potential tremorlytic mechanisms also exist (Ondo et al., 2006). Felbamate (tremor response unknown) also increases receptor density but has not been evaluated for tremor and is seldom used. The pure GABA-A agonist progabide, which binds at the endogenous GABA site, failed to improve tremor (Koller, Rubino, & Gupta, 1987; Mondrup, Dupont, & Pedersen, 1983) although the GABA-A agonist muscimol improved tremor when injected directly into the VIM nucleus in 6 subjects undergoing deep brain stimulation (DBS) (Pahapill et al., 1999).

8.5 Positive allosteric GABA-A modulators

Sage Therapeutics developed a series of positive allosteric steroid modulators of GABA-A receptors, several of which have been evaluated for ET or are currently being evaluated in ET. These drugs potentiate both synaptic and extra-synaptic receptors, in contrast to benzodiazepines and barbiturates, which largely stimulate only synaptic receptors. The unique active site transverses alpha and beta sub-units at a point within the cellular membrane, and distinct from that of benzodiazepines, barbiturates, or endogenous GABA.

Physiologically, this class reduces occipital *beta* activity, a possibly biomarker for tremor (K. Takahashi, Personal Communication 2020) Brexanolone (Sage 547), an intravenous preparation and Zuranolone (Sage 217), an oral preparation, both underwent small exploratory ET trials (ClinicalTrials.gov Identifier: NCT02277106). Brexanolone is current approved for use in post-partum depression but future tremor studies are not planned for either drug (Gunduz-Bruce, Takahashi, & Huang, 2021). Sage 324, an oral allosteric modulator with a longer $T_{1/2}$ showed improved tremor scores in a 69 subject parallel trial. Somnolence and dizziness were the most common adverse events (ClinicalTrials.gov Identifier: NCT04305275). Previously, a small six-subject open label trial showed improved accelerometry scores, which correlated with reduction in *beta* frequency occipital EEG activity (K. Takahashi, Personal Communication 2020).

8.6 Carbonic anhydrase inhibitors impact GABA-A receptors

Potent specific carbonic anhydrase inhibitors including acetazolamide and methazolamide, and milder less specific carbonic anhydrase inhibitors such as topiramate and zonisamide, may improve tremor, although data supporting the more specific carbonic anhydrase inhibitors is modest (Busenbark et al., 1993; Busenbark, Pahwa, Hubble, & Koller, 1992; Busenbark, Ramig, Dromey, & Koller, 1996; Gunal, Afsar, Bekiroglu, & Aktan, 2000). These drugs locally acidify brain areas and increase bicarbonate ion concentrations. PH buffering of extra- and intracellular spaces is mainly carried out by the CO(2)/HCO(3)(−) buffer. Acidification facilitates GABA-A receptor receptors, and reduces NMDA receptor activity. HCO_3 can also travel though the "chloride" GABA-A pore. Therefore, acidification and increased HCO3 production could also potentiate cellular hyperpolarization via GABA-A receptors, further supporting a role for GABA-A intervention.

Overall, multiple drugs that activate GABA-A receptors seem to improve clinical tremor and physiologic assessments of tremor. Interestingly, to date increasing GABA availability via several different mechanisms, or even targeting the endogenous GABA receptor site have shown less promise than targeting GABA receptor at novel sites. This implies that GABA the neurotransmitter may not be culpable for ET since increasing GABA doesn't seem to improve tremor. Rather GABA systems that normally inhibit CNS oscillation activity, diminish in their capacity to suppress tremor. Whether this fails because of continued worsening of upstream endogenous oscillatory activity or physiologic tolerance in the GABAergic system over time, or some combination, is not known. Either way, stimulating GABA receptors via a different mechanism would more likely impact the physiology as opposed to augmenting the endogenous GABAergic system in a fatigued system. The configuration of GABA-A protein constituents most associated with tremor suppression, and the best specific binding site on that GABA-A receptor to target to improve tremor both remain unknown.

9. GABA-B receptors

GABA-B receptors are actually quite different from GABA-A receptors and have been less implicated in tremor. They are mostly pre-synaptic, are metabotropic G protein-coupled receptors, and inhibits action potentials by opening potassium channels. Baclofen, a GABA-B agonist, is not thought to improve ET, although a single case reported improved cerebellar outflow tremor with intrathecal baclofen (Weiss, North, Ohara, & Lenz, 2003). Sodium oxybate (γ-hydroxybutyrate, GHB) is a very powerful cellular decoupler that may improve tremor, at least in those who are also responsive to ethanol (Frucht, Bordelon, Houghton, & Reardan, 2005). GHB modulates GABA-B receptors, but also inhibits glutamate transmission via stimulation of a relatively novel GHB receptor (Li, Kuhn, Wilson, & Lewis, 2007) and inhibits dopamine release. No other GABA-B agonist is known to improve tremor, therefore the mechanism by which sodium oxybate may improve tremor, and whether GABA-B mechanisms are involved, is not known.

10. Future directions

Trials of several existing drugs could further elucidate the role of GABA in ET treatment. Clobazam is a novel long acting 1,5-benzodiazepine with somewhat unique GABA-A affinities (much greater alpha2 to alpha-1 affinity), which is thus less sedating, has never been assessed for tremor. Stiripentol, an AED, approved in many countries for Dravet's syndrome, which increases the duration of opening of GABA-A receptor, similar to barbiturates has never been assessed. Even carefully comparing tremor response in existing benzodiazepines with known specific GABA-A protein affinities could improve understanding of optimal targeting and facilitate development of new drugs.

Additional physiology and pathology discoveries may further elucidate the role of GABAergic systems in tremor, especially whether one particular protein configuration of GABA-A receptor could most improve tremor. It is unlikely that more potent non-selective GABA agonists will advance tremor management, as side effects would be prohibitive, but drugs with more selectivity to cerebellar GABA receptors, or some yet unknown properties may prove more successful. Overall, it is likely that an "optimized" GABA-A receptor agonist will prove to be the best oral drug for ET.

References

Boecker, H., Weindl, A., Brooks, D. J., Ceballos-Baumann, A. O., Liedtke, C., Miederer, M., et al. (2010). GABAergic dysfunction in essential tremor: An 11C-flumazenil PET study. *Journal of Nuclear Medicine, 51*(7), 1030–1035. https://doi.org/10.2967/jnumed.109.074120.

Boecker, H., Wills, A. J., Ceballos-Baumann, A., Samuel, M., Thompson, P. D., Findley, L. J., et al. (1996). The effect of ethanol on alcohol-responsive essential tremor: A positron emission tomography study. *Annals of Neurology, 39*(5), 650–658. https://doi.org/10.1002/ana.410390515.

Brown, A. M., Arancillo, M., Lin, T., Catt, D. R., Zhou, J., Lackey, E. P., et al. (2019). Molecular layer interneurons shape the spike activity of cerebellar Purkinje cells. *Scientific Reports, 9*(1), 1742. https://doi.org/10.1038/s41598-018-38264-1.

Brown, A. M., White, J. J., van der Heijden, M. E., Zhou, J., Lin, T., & Sillitoe, R. V. (2020). Purkinje cell misfiring generates high-amplitude action tremors that are corrected by cerebellar deep brain stimulation. *eLife, 9*. https://doi.org/10.7554/eLife.51928.

Buijink, A. W. G., Prent, N., Puts, N. A., Schrantee, A., Potters, W. V., & van Rootselaar, A. F. (2021). GABA, glutamate, and NAA levels in the deep cerebellar nuclei of essential tremor patients. *Frontiers in Neurology, 12*, 664735. https://doi.org/10.3389/fneur.2021.664735.

Buijink, A. W., van der Stouwe, A. M., Broersma, M., Sharifi, S., Groot, P. F., Speelman, J. D., et al. (2015). Motor network disruption in essential tremor: A functional and effective connectivity study. *Brain, 138*(Pt. 10), 2934–2947. https://doi.org/10.1093/brain/awv225.

Busenbark, K., Pahwa, R., Hubble, J., Hopfensperger, K., Koller, W., & Pogrebra, K. (1993). Double-blind controlled study of methazolamide in the treatment of essential tremor. *Neurology, 43*(5), 1045–1047. Retrieved from http://www.ncbi.nlm.nih.gov/pubmed/8492925.

Busenbark, K., Pahwa, R., Hubble, J., & Koller, W. (1992). The effect of acetazolamide on essential tremor: An open-label trial. *Neurology, 42*(7), 1394–1395. Retrieved from http://www.ncbi.nlm.nih.gov/pubmed/1620352.

Busenbark, K., Ramig, L., Dromey, C., & Koller, W. C. (1996). Methazolamide for essential voice tremor. *Neurology, 47*(5), 1331–1332. Retrieved from http://www.ncbi.nlm.nih.gov/pubmed/8909454.

de Haas, S. L., Zoethout, R. W., Van Dyck, K., De Smet, M., Rosen, L. B., Murphy, M. G., et al. (2012). The effects of TPA023, a GABAAalpha2,3 subtype-selective partial agonist, on essential tremor in comparison to alcohol. *Journal of Psychopharmacology, 26*(2), 282–291. https://doi.org/10.1177/0269881111415731.

Elkouzi, A., Kattah, J. C., & Elble, R. J. (2016). Hypertrophic olivary degeneration does not reduce essential tremor. *Movement Disorders Clinical Practice*, *3*(2), 209–211. https://doi.org/10.1002/mdc3.12275.

Erickson-Davis, C. R., Faust, P. L., Vonsattel, J. P., Gupta, S., Honig, L. S., & Louis, E. D. (2010). "Hairy baskets" associated with degenerative Purkinje cell changes in essential tremor. *Journal of Neuropathology and Experimental Neurology*, *69*(3), 262–271. https://doi.org/10.1097/NEN.0b013e3181d1ad04.

Frucht, S. J., Bordelon, Y., Houghton, W. H., & Reardan, D. (2005). A pilot tolerability and efficacy trial of sodium oxybate in ethanol-responsive movement disorders. *Movement Disorders*, *20*(10), 1330–1337. Retrieved from http://www.ncbi.nlm.nih.gov/entrez/query.fcgi?cmd=Retrieve&db=PubMed&dopt=Citation&list_uids=15986420.

Frucht, S. J., Houghton, W. C., Bordelon, Y., Greene, P. E., & Louis, E. D. (2005). A single-blind, open-label trial of sodium oxybate for myoclonus and essential tremor. *Neurology*. Retrieved from http://www.ncbi.nlm.nih.gov/entrez/query.fcgi?cmd=Retrieve&db=PubMed&dopt=Citation&list_uids=16267211.

Gironell, A. (2014). The GABA hypothesis in essential tremor: Lights and shadows. *Tremor and Other Hyperkinetic Movements (New York, N.Y.)*, *4*, 254. https://doi.org/10.7916/D8SF2T9C.

Gironell, A., Figueiras, F. P., Pagonabarraga, J., Herance, J. R., Pascual-Sedano, B., Trampal, C., et al. (2012). Gaba and serotonin molecular neuroimaging in essential tremor: A clinical correlation study. *Parkinsonism & Related Disorders*, *18*(7), 876–880. https://doi.org/10.1016/j.parkreldis.2012.04.024.

Gironell, A., Martinez-Corral, M., Pagonabarraga, X., & Kulisevsky, J. (2008). Tiagabine for essential tremor: An open-label trial. *Movement Disorders*, *23*(13), 1955–1956. https://doi.org/10.1002/mds.22094.

Gunal, D. I., Afsar, N., Bekiroglu, N., & Aktan, S. (2000). New alternative agents in essential tremor therapy: Double-blind placebo-controlled study of alprazolam and acetazolamide. *Neurological Sciences*, *21*(5), 315–317. Retrieved from http://www.ncbi.nlm.nih.gov/entrez/query.fcgi?cmd=Retrieve&db=PubMed&dopt=Citation&list_uids=11286044.

Gunduz-Bruce, H., Takahashi, K., & Huang, M. Y. (2021). Development of neuroactive steroids for the treatment of postpartum depression. *Journal of Neuroendocrinology*, e13019. https://doi.org/10.1111/jne.13019.

Handforth, A. (2012). Harmaline tremor: Underlying mechanisms in a potential animal model of essential tremor. *Tremor and Other Hyperkinetic Movements (New York, N.Y.)*, *2*. Retrieved from http://www.ncbi.nlm.nih.gov/pubmed/23440018.

Handforth, A., & Lang, E. J. (2021). Increased Purkinje cell complex spike and deep cerebellar nucleus synchrony as a potential basis for syndromic essential tremor. A review and synthesis of the literature. *Cerebellum*, *20*(2), 266–281. https://doi.org/10.1007/s12311-020-01197-5.

Hartstone, W. G., Brown, M. H., Kelly, G. C., Tate, W. J., Kuo, S. H., Dwork, A. J., et al. (2021). Dentate nucleus neuronal density: A postmortem study of essential tremor versus control brains. *Movement Disorders*, *36*(4), 995–999. https://doi.org/10.1002/mds.28402.

Haubenberger, D., McCrossin, G., Lungu, C., Considine, E., Toro, C., Nahab, F. B., et al. (2013). Octanoic acid in alcohol-responsive essential tremor: A randomized controlled study. *Neurology*. https://doi.org/10.1212/WNL.0b013e3182840c4f.

Holtbernd, F., & Shah, N. J. (2021). Imaging the pathophysiology of essential tremor—A systematic review. *Frontiers in Neurology*, *12*, 680254. https://doi.org/10.3389/fneur.2021.680254.

Jimenez-Jimenez, F. J., Alonso-Navarro, H., Garcia-Martin, E., Alvarez, I., Pastor, P., & Agundez, J. A. G. (2021). Genomic markers for essential tremor. *Pharmaceuticals (Basel)*, *14*(6). https://doi.org/10.3390/ph14060516.

Koller, W. C., & Royse, V. L. (1986). Efficacy of primidone in essential tremor. *Neurology, 36*(1), 121–124. Retrieved from http://www.ncbi.nlm.nih.gov/pubmed/3941767.

Koller, W. C., Rubino, F., & Gupta, S. (1987). Pharmacologic probe with progabide of GABA mechanisms in essential tremor. *Archives of Neurology, 44*(9), 905–906. Retrieved from http://www.ncbi.nlm.nih.gov/pubmed/3619708.

Kralic, J. E., Criswell, H. E., Osterman, J. L., O'Buckley, T. K., Wilkie, M. E., Matthews, D. B., et al. (2005). Genetic essential tremor in gamma-aminobutyric acidA receptor alpha1 subunit knockout mice. *The Journal of Clinical Investigation, 115*(3), 774–779. https://doi.org/10.1172/JCI23625.

Kumar, S., Porcu, P., Werner, D. F., Matthews, D. B., Diaz-Granados, J. L., Helfand, R. S., et al. (2009). The role of GABA(A) receptors in the acute and chronic effects of ethanol: A decade of progress. *Psychopharmacology, 205*(4), 529–564. https://doi.org/10.1007/s00213-009-1562-z.

Kuo, S. H., Lin, C. Y., Wang, J., Liou, J. Y., Pan, M. K., Louis, R. J., et al. (2016). Deep brain stimulation and climbing fiber synaptic pathology in essential tremor. *Annals of Neurology, 80*(3), 461–465. https://doi.org/10.1002/ana.24728.

Lan, H., Suo, X., Li, W., Li, N., Li, J., Peng, J., et al. (2021). Abnormalities of intrinsic brain activity in essential tremor: A meta-analysis of resting-state functional imaging. *Human Brain Mapping, 42*(10), 3156–3167. https://doi.org/10.1002/hbm.25425.

Lang, E. J. (2002). GABAergic and glutamatergic modulation of spontaneous and motor-cortex-evoked complex spike activity. *Journal of Neurophysiology, 87*(4), 1993–2008. https://doi.org/10.1152/jn.00477.2001.

Li, Q., Kuhn, C. M., Wilson, W. A., & Lewis, D. V. (2007). Effects of gamma hydroxybutyric acid on inhibition and excitation in rat neocortex. *Neuroscience, 150*(1), 82–92. https://doi.org/10.1016/j.neuroscience.2007.08.023.

Lin, C. Y., Louis, E. D., Faust, P. L., Koeppen, A. H., Vonsattel, J. P., & Kuo, S. H. (2014). Abnormal climbing fibre-Purkinje cell synaptic connections in the essential tremor cerebellum. *Brain, 137*(Pt. 12), 3149–3159. https://doi.org/10.1093/brain/awu281.

Louis, E. D., Diaz, D. T., Kuo, S. H., Gan, S. R., Cortes, E. P., Vonsattel, J. P. G., et al. (2018). Inferior olivary nucleus degeneration does not lessen tremor in essential tremor. *Cerebellum Ataxias, 5*, 1. https://doi.org/10.1186/s40673-018-0080-3.

Louis, E. D., & Faust, P. L. (2020). Essential tremor pathology: Neurodegeneration and reorganization of neuronal connections. *Nature Reviews. Neurology, 16*(2), 69–83. https://doi.org/10.1038/s41582-019-0302-1.

Louis, E. D., Faust, P. L., Vonsattel, J. P., Honig, L. S., Rajput, A., Robinson, C. A., et al. (2007). Neuropathological changes in essential tremor: 33 cases compared with 21 controls. *Brain, 130*(Pt 12), 3297–3307. https://doi.org/10.1093/brain/awm266.

Louis, R. J., Lin, C. Y., Faust, P. L., Koeppen, A. H., & Kuo, S. H. (2015). Climbing fiber synaptic changes correlate with clinical features in essential tremor. *Neurology, 84*(22), 2284–2286. https://doi.org/10.1212/WNL.0000000000001636.

Mally, J., Baranyi, M., & Vizi, E. S. (1996). Change in the concentrations of amino acids in CSF and serum of patients with essential tremor. *Journal of Neural Transmission, 103*(5), 555–560. Retrieved from http://www.ncbi.nlm.nih.gov/pubmed/8811501.

Martin, F. C., Thu Le, A., & Handforth, A. (2005). Harmaline-induced tremor as a potential preclinical screening method for essential tremor medications. *Movement Disorders, 20*(3), 298–305. Retrieved from http://www.ncbi.nlm.nih.gov/entrez/query.fcgi?cmd=Retrieve&db=PubMed&dopt=Citation&list_uids=15580562.

Mavroudis, I., Petrides, F., Karantali, E., Chatzikonstantinou, S., McKenna, J., Ciobica, A., et al. (2021). A voxel-wise meta-analysis on the cerebellum in essential tremor. *Medicina (Kaunas, Lithuania), 57*(3). https://doi.org/10.3390/medicina57030264.

Mondrup, K., Dupont, E., & Pedersen, E. (1983). The effect of the GABA-agonist, progabide, on benign essential tremor. A controlled clinical trial. *Acta Neurologica Scandinavica*, *68*(4), 248–252. Retrieved from http://www.ncbi.nlm.nih.gov/pubmed/6362317.

Nahab, F. B., Wittevrongel, L., Ippolito, D., Toro, C., Grimes, G. J., Starling, J., et al. (2011). An open-label, single-dose, crossover study of the pharmacokinetics and metabolism of two oral formulations of 1-octanol in patients with essential tremor. *Neurotherapeutics*, *8*(4), 753–762. https://doi.org/10.1007/s13311-011-0045-1.

Nestoros, J. N. (1980). Ethanol specifically potentiates GABA-mediated neurotransmission in feline cerebral cortex. *Science*, *209*(4457), 708–710. Retrieved from http://www.ncbi.nlm.nih.gov/pubmed/7394531.

Nietz, A., Krook-Magnuson, C., Gutierrez, H., Klein, J., Sauve, C., Hoff, I., et al. (2020). Selective loss of the GABAAalpha1 subunit from Purkinje cells is sufficient to induce a tremor phenotype. *Journal of Neurophysiology*, *124*(4), 1183–1197. https://doi.org/10.1152/jn.00100.2020.

Ondo, W. G., Jankovic, J., Connor, G. S., Pahwa, R., Elble, R., Stacy, M. A., et al. (2006). Topiramate in essential tremor: a double-blind, placebo-controlled trial. *Neurology*, *66*(5), 672–677. https://doi.org/10.1212/01.wnl.0000200779.03748.0f.

Pahapill, P. A., Levy, R., Dostrovsky, J. O., Davis, K. D., Rezai, A. R., Tasker, R. R., et al. (1999). Tremor arrest with thalamic microinjections of muscimol in patients with essential tremor. *Annals of Neurology*, *46*(2), 249–252. Retrieved from http://www.ncbi.nlm.nih.gov/pubmed/10443891.

Pan, M. K., Li, Y. S., Wong, S. B., Ni, C. L., Wang, Y. M., Liu, W. C., et al. (2020). Cerebellar oscillations driven by synaptic pruning deficits of cerebellar climbing fibers contribute to tremor pathophysiology. *Science Translational Medicine*, *12*(526). https://doi.org/10.1126/scitranslmed.aay1769.

Paris-Robidas, S., Brochu, E., Sintes, M., Emond, V., Bousquet, M., Vandal, M., et al. (2012). Defective dentate nucleus GABA receptors in essential tremor. *Brain*, *135*(Pt. 1), 105–116. https://doi.org/10.1093/brain/awr301.

Park, Y. G., Park, H. Y., Lee, C. J., Choi, S., Jo, S., Choi, H., et al. (2010). Ca(V)3.1 is a tremor rhythm pacemaker in the inferior olive. *Proceedings of the National Academy of Sciences of the United States of America*, *107*(23), 10731–10736. https://doi.org/10.1073/pnas.1002995107.

Paterson, N. E., Malekiani, S. A., Foreman, M. M., Olivier, B., & Hanania, T. (2009). Pharmacological characterization of harmaline-induced tremor activity in mice. *European Journal of Pharmacology*, *616*(1-3), 73–80. https://doi.org/10.1016/j.ejphar.2009.05.031.

Rajput, A. H., Adler, C. H., Shill, H. A., & Rajput, A. (2012). Essential tremor is not a neurodegenerative disease. *Neurodegenerative Disease Management*, *2*(3), 259–268. https://doi.org/10.2217/nmt.12.23.

Sankar, R. (2012). GABA(A) receptor physiology and its relationship to the mechanism of action of the 1,5-benzodiazepine clobazam. *CNS Drugs*, *26*(3), 229–244. https://doi.org/10.2165/11599020-000000000-00000.

Siokas, V., Aloizou, A. M., Tsouris, Z., Liampas, I., Aslanidou, P., Dastamani, M., et al. (2020). Genetic risk factors for essential tremor: A review. *Tremor Other Hyperkinetic Movements (New York, N.Y.)*, *10*, 4. https://doi.org/10.5334/tohm.67.

Tapper, S., Goransson, N., Lundberg, P., Tisell, A., & Zsigmond, P. (2020). A pilot study of essential tremor: Cerebellar GABA+/Glx ratio is correlated with tremor severity. *Cerebellum Ataxias*, 7, 8. https://doi.org/10.1186/s40673-020-00116-y.

Teravainen, H., Huttunen, J., & Lewitt, P. (1986). Ineffective treatment of essential tremor with an alcohol, methylpentynol. *Journal of Neurology, Neurosurgery, and Psychiatry*, *49*(2), 198–199. Retrieved from http://www.ncbi.nlm.nih.gov/pubmed/3512778.

Walker, S. D., & Kalviainen, R. (2011). Non-vision adverse events with vigabatrin therapy. *Acta Neurologica Scandinavica. Supplementum*, *192*, 72–82. https://doi.org/10.1111/j.1600-0404.2011.01602.x.

Weiss, N., North, R. B., Ohara, S., & Lenz, F. A. (2003). Attenuation of cerebellar tremor with implantation of an intrathecal baclofen pump: The role of gamma-aminobutyric acidergic pathways. Case report. *Journal of Neurosurgery*, *99*(4), 768–771. https://doi.org/10.3171/jns.2003.99.4.0768.

Wills, A. J., Jenkins, I. H., Thompson, P. D., Findley, L. J., & Brooks, D. J. (1994). Red nuclear and cerebellar but no olivary activation associated with essential tremor: a positron emission tomographic study. *Annals of Neurology*, *36*(4), 636–642.

Wu, Y. C., Louis, E. D., Gionco, J., Pan, M. K., Faust, P. L., & Kuo, S. H. (2021). Increased climbing fiber lateral crossings on Purkinje cell dendrites in the cerebellar hemisphere in essential tremor. *Movement Disorders*, *36*(6), 1440–1445. https://doi.org/10.1002/mds.28502.

Zesiewicz, T. A., Sullivan, K. L., Ward, C. L., & Hauser, R. A. (2007). Tiagabine and exacerbation of essential tremor. *Movement Disorders*, *22*(14), 2132–2133. https://doi.org/10.1002/mds.21559.

Zhang, X., & Santaniello, S. (2019). Role of cerebellar GABAergic dysfunctions in the origins of essential tremor. *Proceedings of the National Academy of Sciences of the United States of America*. https://doi.org/10.1073/pnas.1817689116.

CHAPTER TWELVE

Reduction of neuronal hyperexcitability with modulation of T-type calcium channel or SK channel in essential tremor

Aparna Wagle Shukla*
Department of Neurology, Fixel Institute for Neurological Diseases, University of Florida, Gainesville, FL, United States
*Corresponding author: e-mail address: aparna.shukla@neurology.ufl.edu

Contents

Abstract

Essential tremor is one of the most prevalent movement disorders. Propranolol and primidone are the first-line pharmacological therapies. They provide symptomatic control in less than 50% of patients. Topiramate, alprazolam, clonazepam, gabapentin, and botulinum toxin injections are the next line of treatments. These medications lead to modest improvements and are therefore commonly used as add-on agents. Surgical therapies, including deep brain stimulation (DBS) surgery and focused ultrasound beam targeted to the thalamus, are considered for treating tremor refractory to medications and lead to greater than 75% improvements in tremor symptoms. However, DBS is a costly and an invasive procedure; some patients report tolerance to benefits.

International Review of Neurobiology, Volume 163
ISSN 0074-7742
https://doi.org/10.1016/bs.irn.2022.02.008

Focused ultrasound therapy leading to brain lesions is associated with a possibility for permanent clinical deficits. Therefore, research efforts to develop the next generation of oral medications with greater benefits and lesser adverse effects are warranted. There is considerable evidence that the increased functions of calcium channels (P/Q-type and T-type channels) and reduced functions of calcium-activated potassium channels (SK channels) located in the neuronal membranes lead to tremor oscillations. Consequently, many new pharmacological studies have targeted these channels to leverage better clinical outcomes. The current review will discuss the pathophysiology, the specific importance of these channels, and the early clinical experience of using compounds targeting these channels to treat essential tremor.

1. Introduction

Essential tremor is one of the most prevalent movement disorders in the world, affecting about 4% of the population after the age of 40 years (Dogu, Sevim, Camdeviren, et al., 2003; Louis, 2008; Louis & Ferreira, 2010; Louis & McCreary, 2021). The oscillatory movements in the hands presenting during everyday motor tasks can be functionally disabling and socially embarrassing (Haubenberger & Hallett, 2018). The currently available treatments for essential tremor are mainly symptomatic, with propranolol and primidone as the first-line drugs. Clinical evidence indicates that only 50% of patients benefit from one or both of these medications. Propranolol, a beta-blocker approved by the FDA, blocks the beta-2 receptors found in the peripheral muscle spindles that send feedback signals to the brain; however, aside from peripheral actions, the medication also directly affects the central brain circuitries (Abila, Wilson, Marshall, & Richens, 1985; Routledge & Shand, 1979). Primidone, on the other hand, has pure central actions. The main compound regulates the sodium and calcium channels in the neuronal membrane, whereas phenobarbital, a metabolite acts on the inhibitory GABA-A receptors (Guan & Peroutka, 1990). The next line of treatment includes topiramate, alprazolam, clonazepam, gabapentin and botulinum toxin injections (Hedera, Cibulčík, & Davis, 2013). These medications commonly used as add-on agents reveal a lower degree of improvements in the range of 30–40% (Hopfner & Deuschl, 2020). Surgical therapies, including deep brain stimulation (DBS) surgery (Peng-Chen, Morishita, Vaillancourt, et al., 2013; Wagle Shukla & Okun, 2016) and focused ultrasound beam targeted to the thalamus, are considered when the tremor persists despite multiple medication trials or when there are intolerable adverse effects (Wang, Ren, Chiu, et al., 2020).

While the surgical therapies are no doubt powerful; they have limitations. For example, DBS surgery is invasive and costly, and in the long term, some patients have diminished clinical response due to tolerance to stimulation effects. Then, focused ultrasound therapy leads to the creation of brain lesions, potentially resulting in permanent clinical deficits. Thus, the next generation of oral medications with greater benefits and lesser adverse effects is highly warranted.

In the last decade, there has been a significant advancement in the neuronal and network-level understanding of the pathophysiology of essential tremor. In addition to the P/Q-type calcium channels, T-type calcium channels and calcium-activated potassium channels (SK channels) present in the neuronal membranes are increasingly recognized to generate tremor oscillations within the brain network. There are ongoing efforts to develop pharmacological compounds targeting these channels to gain better clinical outcomes. The current review will discuss the pathophysiology, the role of the P/Q-type calcium channel, and the evolving experience with novel compounds targeting specific subtypes of T-type calcium channels and SK channels for treating essential tremor.

2. Pathophysiology of essential tremor

2.1 Network model theory

Many studies have shown that an oscillatory dysfunction involving the olivo-cerebello-thalamo-cortical network is the pathophysiological hallmark for essential tremor (Buijink, van der Stouwe, Broersma, et al., 2015; Helmich, Toni, Deuschl, & Bloem, 2013; Hopfner, Haubenberger, Galpern, et al., 2016; Muthuraman et al., 2015; Raethjen & Muthuraman, 2015). The electrical properties of neurons in the olivary and cerebellar nuclei and the thalamus that relay to the motor cortex have unique oscillatory properties. Membrane hyperpolarization causes a rebound effect that leads to an oscillatory firing of the neurons at a given frequency (Jahnsen, 1986; Llinas, 1988). Cellular bursts in the cerebellar receiving zone of the thalamus (ventral intermediate nucleus or Vim) have been found to correlate strongly with the peripheral tremor signal (Hua & Lenz, 2005). Magnetoencephalography and electroencephalography studies have shown the coherence between the contralateral motor cortex, thalamus and ipsilateral cerebellum and the peripheral tremor signal to be significant (Muthuraman, Raethjen, Koirala, et al., 2018). The inferior olivary nucleus, cerebellum, Vim and motor cortex together form the network leading to generation of tremor oscillations.

2.2 Are there single or multiple oscillators?

Membrane hyperpolarization of pathogenic neurons leads to independent oscillatory behavior at a given frequency (Jahnsen, 1986; Llinas, 1988). While this theory is well accepted, it is still not clear whether a single oscillator or multiple sources of oscillations are involved in the pathophysiology. Microelectrode recordings from the thalamus during the DBS surgery have shown multiple spatially separated clusters of tremor cells (Pedrosa, Reck, Florin, et al., 2012). A recent study found that essential tremor demonstrates a low tolerance for frequency change which means that the oscillating system does not have the capability of maintaining several resonant frequencies. The multiple pathogenic oscillators within the tremor network generate a more finely tuned convergent drive to the peripheral muscles (Brittain et al., 2015; di Biase, Brittain, Shah, et al., 2017). Although the coherence between the motor cortex neurons and peripheral tremor activity could be intermittent (Pedrosa, Quatuor, Reck, et al., 2014; Raethjen, Govindan, Kopper, Muthuraman, & Deuschl, 2007), the multiple oscillators are observed to couple during the performance of the motor tasks.

2.3 Origins of tremor oscillations: An inferior olivary nucleus or the cerebellum?

There is consensual agreement that the olivo-cerebello-thalamo-cortical network is dysfunctional in essential tremor; however, the exact origins of the oscillatory signals remain to be confirmed (Helmich et al., 2013). Previously, the proximal elements of the network, such as the inferior olivary nucleus and the cerebellar cortex, were supposedly the main potential pacemakers. Depolarization of the inferior olivary neurons that depends on the high-threshold and low-threshold calcium and potassium currents led to generation of tremor oscillations (Llinás & Yarom, 1981). In early animal models of tremor, exposure to harmaline, a plant alkaloid resulted in hyperpolarization of the inferior olivary neurons (Llinás & Yarom, 1986; Park, Park, Lee, et al., 2010). The hyperpolarization led to the generation of 4– to 10– Hz burst oscillations in the olivocerebellar pathways or the climbing fibers (Lee, Kim, Lee, et al., 2018; Raethjen & Deuschl, 2012). These oscillations were then transmitted to the cerebellar cortex, eventually manifesting behaviorally as an action tremor (Bernard, Buisseret-Delmas, Compoint, & Laplante, 1984; Kuo, Louis, Faust, et al., 2019).

Recently concerns were raised about whether the inferior olivary neurons played any pacemaker role as most evidence to date is from animal models with no concordant evidence in humans (Louis & Lenka, 2017). For example, imaging studies in patients with essential tremor found clear abnormalities in the cerebellum, thalamus and the motor cortex but structural or functional abnormalities affecting the inferior olivary nucleus could not be confirmed (Fang, Lv, Luo, et al., 2013). Archer et al. found that visual feedback increased tremor amplitude and the tremor-related activity in the cerebello-thalamo-cortical network (Archer, Coombes, Chu, et al., 2018). These abnormalities extended to the supplementary motor area, visual and parietal areas, but the inferior olive did not reveal any change. Furthermore, a post-mortem study found that the pathological examination of the inferior olivary nucleus in essential tremor patients was not significantly different from the healthy controls (Louis, Diaz, Kuo, et al., 2018; Rajput, Robinson, & Rajput, 2004). Therefore, the role of the olivary nucleus in the context of essential tremor pathogenesis has become increasingly questionable.

On the other hand, several lines of evidence indicate that the cerebellum is the primary pathophysiological substrate for essential tremor. Quantitative gait and limb movement assessment has revealed coordination abnormalities similar to ataxia in patients with essential tremor (Deuschl, Wenzelburger, Loffler, Raethjen, & Stolze, 2000; Stolze, Petersen, Raethjen, Wenzelburger, & Deuschl, 2001). While magnetic resonance spectroscopy has revealed decreased levels of *N*-acetyl aspartate in the cerebellum, a finding that indicates loss or dysfunction of neurons (Pagan, Butman, Dambrosia, & Hallett, 2003), PET studies have shown increased cerebellar metabolism in essential tremor patients (Jenkins, Bain, Colebatch, et al., 1993; Louis et al., 2002). Pathological studies have shown axonal swellings in the Purkinje cells, swellings in and regression of the Purkinje cell dendritic arbor and Purkinje cell death (Brown et al., 2020; Kuo, Wang, Tate, et al., 2017; Louis & Faust, 2020; Louis, Kerridge, Chatterjee, et al., 2019; Louis, Lee, Babij, et al., 2014). Some investigators advocate that the cerebellum decouples or disconnects from the thalamus and the motor cortex (cerebellar decoupling hypothesis), leading to pathological oscillations (Madelein van der Stouwe, Nieuwhof, & Helmich, 2020). In a functional MRI study, a significantly decreased connectivity between the dentate nucleus and cortical and subcortical brain regions was seen, indicating a functional disconnection of the dentate nucleus from the rest of the network (Tikoo, Pietracupa, Tommasin, et al., 2020). In other structural MRI studies, white matter integrity of the cerebellar peduncles in patients with clinical tremor was reduced providing further support that the cerebellum is disconnected or decoupled from the rest of the network (Juttukonda, Franco, Englot, et al., 2019; Pietracupa, Bologna, Bharti, et al., 2019). In contrast to the decoupling hypothesis, the oscillator hypothesis proposes the cerebellum as the primary driving oscillator. In an fMRI study, patients were examined in the scanner with DBS turned on and off and with arms examined at rest and postural elevation (Awad, Blomstedt, Westling, & Eriksson, 2020). DBS therapy led to normalization of cerebellar activity within the sensorimotor cerebellar lobules (IV/V and VIII), supporting a cerebellar oscillator or cerebellar hyperexcitability hypothesis (Broersma, van der Stouwe, Buijink, et al., 2016; Buijink et al., 2015).

2.4 Purkinje cell neurons in the cerebellar cortex: Hyperexcitability hypothesis

Purkinje cells are large neurons in the cerebellar cortex that characteristically have an elaborate fan-shaped dendritic arbor with a large number of spines. Dendritic spines receive glutamatergic excitatory fibers from two sources; the climbing and the parallel fibers (Llinás, 2013). The climbing fibers originate from the inferior olivary nucleus and they form excitatory synapses with dendritic spines that are located proximal to the Purkinje cell soma. Parallel fibers originate from the granule cells in the cerebellar cortex; they pass orthogonally through the dendritic arbor to form a granule-cell-Purkinje-cell synapse. In contrast to the parallel and climbing fibers, basket and stellate cells (found in the cerebellar molecular layer) provide inhibitory (GABAergic) input to the Purkinje cell (Louis, 2018).

At birth, multiple climbing fibers innervate the Purkinje cells. However, the maturation of microcircuitries during the early postnatal period leads to eliminating surplus climbing fibers, also known as "pruning" (Kano, Watanabe, Uesaka, & Watanabe, 2018). The remaining single climbing fiber ends up providing strong excitatory input to the Purkinje cell neuron whereas the parallel fibers that are as many as 100,000 in number provide weak excitatory input (Palay, 1974). In patients with essential tremor, pruning of climbing fiber - Purkinje cell synapses is reportedly deficient (Ito, 1984) whereas the density, counts and integrity of parallel fiber remain intact compared to healthy control brains (Kuo, Faust, Vonsattel, Ma, & Louis, 2011). In a mouse model study involving optogenetic manipulation, excessive activity at the climbing fiber—Purkinje cell synapse was found to correlate with excessive oscillatory activity (Pan, Li, Wong, et al., 2020). These mouse model findings were confirmed by another group of investigators who found patterned firings of Purkinje cell neurons were generated in response to optogenetics and these led to powerful oscillatory signals (Brown et al., 2020). Thus, growing and compelling evidence points to abnormal Purkinje cell firing as a critical neural substrate for essential tremor (Handforth, 2016). Whether these neurons generate oscillatory signals or they propagate existing tremor signals needs further examination as some studies have found abnormalities in cells directly upstream (Erickson-Davis et al., 2010) as well as in the cerebellar nuclei directly downstream (Paris-Robidas, Brochu, Sintes, et al., 2012).

Purkinje cells generate *simple spikes* and *complex spikes* (highly stereotyped burst of decrementing spikes) to communicate with the downstream output neurons of the cerebellar nuclei (Llinás, 2013; Yamamoto, Fukuda, & Llinás, 2001). Simple spikes are high frequency 50 Hz spikes driven by intrinsic ion channels expressed by the Purkinje cells whereas complex spikes are driven by large voltage-gated calcium channels in the dendrites of neurons that respond to excitatory input from the climbing fiber (Otis, 2016) (Fig. 1). Another striking feature of Purkinje cell is the tendency to fire in bursts. Bursts consist of 3–20 spikes with a stereotypical waveform. The rate and pattern of firing of Purkinje cells are controlled both by synaptic input and by intrinsic ion channels that allow the neurons to fire spontaneously (Streng, Popa, & Ebner, 2017). Purkinje cells express several classes of

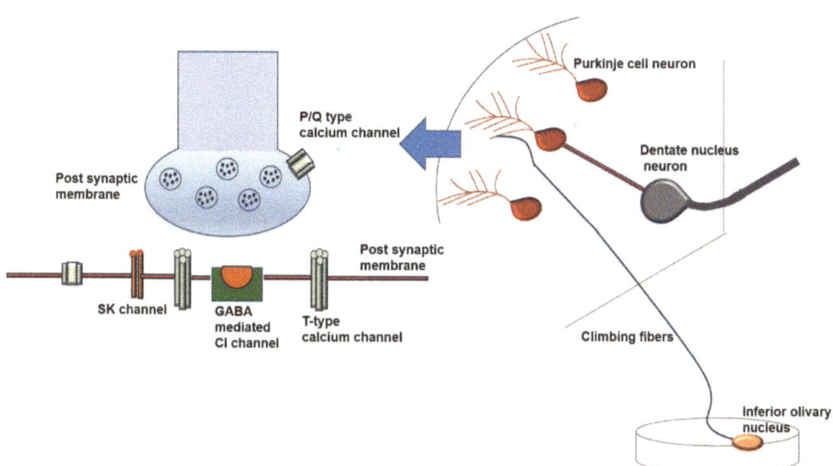

Fig. 1 Climbing fibers synapse with Purkinje cell dendrites. In the presynaptic terminal, activation of P/Q-type calcium channels releases excitatory neurotransmitters, leading to the firing of the Purkinje cell. In the postsynaptic membrane, activation of T-type calcium channels and inhibition of calcium-activated potassium channels (SK channel) or GABA receptor activity leads to increase in Purkinje cell firing.

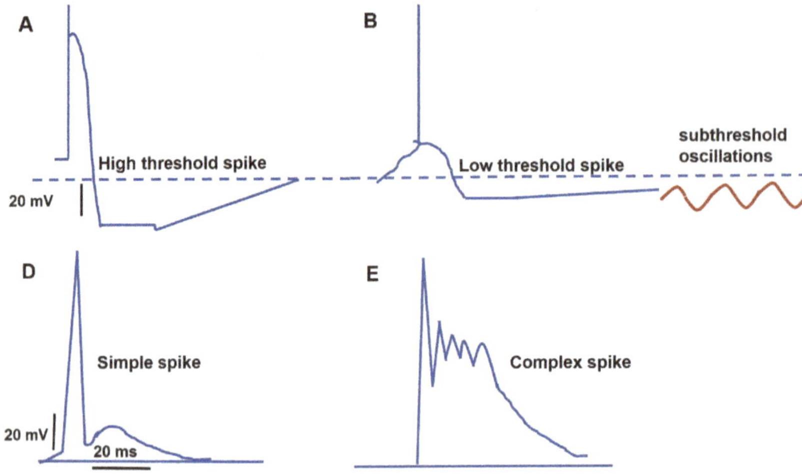

Fig. 2 Purkinje cell firing and spikes. High threshold depolarization is depolarization when the rest membrane potential is more negative compared to the threshold (A); low threshold depolarization is depolarization when the rest membrane potential is less negative compared to the threshold (B); subthreshold oscillations are fluctuations occurring below the threshold potential (C); simple spike is a single spike (D); complex spike is a stereotyped burst of decrementing spikes (E).

voltage-gated calcium channels such as the P/Q type calcium, T-type calcium, and calcium-activated potassium channels (SK channels) that contribute to burst firing (Womack, Chevez, & Khodakhah, 2004). (Fig. 2).

3. Reducing hyperexcitability of Purkinje cell and other hyperexcitable neurons

3.1 Targeting P/Q type calcium channels in essential tremor

The P/Q-type calcium channel (also referred to as $Ca_v2.1$) is a presynaptic high-voltage-gated calcium channel that couples neuronal excitation to the secretion of neurotransmitters (Ishikawa, Kaneko, Shin, & Takahashi, 2005). The ion-conducting pore is formed by four domains of the α_{1A} subunit, whereas accessory subunits (β, $\alpha 2\delta$) modulate channel kinetics and the level of expression on the neuronal membrane. Although P-type currents are mainly identified in Purkinje cells (Choi, Yu, Kim, et al., 2010) and Q-type currents are identified in cerebellar granule neurons (Randall & Tsien, 1995), they are referred together as P/Q-type channels since the distinction between these channels is not always clear.

Primidone, topiramate, gabapentin, pregabalin, and levetiracetam interact with P/Q-type calcium channels. While the exact mechanism of action in essential tremor is not completely elucidated, primidone alters the functions of transmembrane P/Q-type calcium channels. Phenobarbital is a metabolite of primidone that increases the duration of GABA receptor opening but plays a lesser role in the therapeutic control of tremor (Sasso, Perucca, & Calzetti, 1988). Clinical studies, including double-blind placebo-controlled trials, have shown consistent improvements in the patient-reported, physician-based, and electrophysiologic assessments (Findley, Cleeves, & Calzetti, 1985; Koller & Royse, 1986; O'Suilleabhain & Dewey, 2002). However, the overall efficacy is 50–60% improvements and multiple adverse effects, including the first dose acute toxic reaction, sedation, daytime sleepiness, tiredness, nausea, ataxia, dizziness, and confusion, frequently leading to discontinuation of the therapy (Ferreira, Mestre, Lyons, et al., 2019).

Unlike clonazepam and alprazolam which have pure GABAergic actions, topiramate has multiple actions, including augmenting activity at the $GABA_A$ receptors, antagonizing the glutamate receptors, and inhibiting the carbonic anhydrase enzyme (Hedera et al., 2013). Topiramate also blocks voltage-gated sodium channels and P/Q-type calcium channels. It remains unknown which of these mechanisms play the key role in tremor control. Many randomized placebo-controlled studies involving more than 300 patients found topiramate as an effective monotherapy or as an add-on treatment (Connor, Edwards, & Tarsy, 2008; Frima & Grünewald, 2006; Ondo, Jankovic, Connor, et al., 2006). There is an overall 30–40% reduction of tremor with paresthesia, concentration/attention difficulty, appetite suppression/weight loss, and nausea, which are among the most common adverse events (Ferreira et al., 2019).

Then, gabapentin, an inhibitory neurotransmitter in the brain, was developed as a structural analog of gamma-aminobutyric acid (GABA). However, gabapentin does not bind to $GABA_A$ or $GABA_B$ receptors (Taylor, Gee, Su, et al., 1998). Instead, it interacts with the $\alpha 2\delta$ subunit of the presynaptic voltage-gated P/Q-type calcium channels leading to reduced neurotransmitters in the synaptic space (McLean, 1994). Despite high lipophilicity and easy crossing of the blood-brain barrier, clinical studies have revealed mixed efficacies (Ondo, Hunter, Vuong, Schwartz, & Jankovic, 2000). While one study found a nearly 40–50% reduction of tremor amplitude, other trials did not confirm these findings (Gironell et al., 1999; Pahwa, Lyons, Hubble, et al., 1998). Moreover, dosing at 1200–3600 mg was found to lead to side effects such as sleepiness, dizziness and nausea in a third of patients. Pregabalin is another GABA analog that also has no affinity to GABAergic receptors (Ben-Menachem, 2004), but binds to $\alpha2\delta$ subunit of voltage-gated calcium channels. Although pregabalin can reduce neuronal hyperexcitability theoretically, clinical studies did not reveal significant benefits. Similarly, the use of levetiracetam (an anti-epileptic drug) did not lead to a consistent reduction of tremor (Elble, Lyons, & Pahwa, 2007). A recent systematic review concluded that topiramate was efficacious; however, the limiting factors were side effects such as appetite suppression, weight loss, cognitive impairment, and paresthesia. While gabapentin lacked sufficient evidence, pregabalin and levetiracetam did not reveal significant benefits for the treatment of essential tremor (Ferreira et al., 2019).

3.2 Targeting T-type calcium channels

T-type voltage-gated calcium channels were identified around three decades ago. Three distinct isoforms of T-type calcium channels, Ca_v 3.1, 3.2, and 3.3 (Lacinová, 2011) are expressed in cerebellar motor circuits (Apps & Garwicz, 2005; Talley et al., 1999; Wolpert, Miall, & Kawato, 1998). Of particular importance is the Ca_v3.1 subtype which is highly expressed in the tremor network (Miwa & Kondo, 2011). Dendrites of Purkinje cell show a high density of voltage-gated calcium channels (Usowicz, Sugimori, Cherksey, & Llinás, 1992) Khodakhah et al. have shown that the dendritic calcium spike is the key factor leading to a spontaneous burst firing of Purkinje cells. T-type calcium channels with a low threshold for activation (Llinás & Jahnsen, 1982) reveal a tendency to generate a spontaneous rhythmic subthreshold oscillations (STO) of the neuronal membrane. These subthreshold oscillations can potentially convert to rhythmic 9–12 Hz burst-firing (Llinás & Yarom, 1981). A low threshold for activation also leads to the generation of low-threshold spikes (LTS) of calcium and sodium (Crunelli, Lightowler, & Pollard, 1989; Perez-Reyes, 2003). Hyperpolarization of the neuronal membrane increases the possibility of generating LTS through activation of the T-type calcium channels. LTS regulates neural oscillations, resonance (Crunelli et al., 1989; Llinás & Yarom, 1986), and synchronous firing of multiple neurons. The membrane threshold is easily pushed by the LTS to generate the next spike and ultimately the development of bursts of action potentials (Handforth, 2012). Each burst developed with an increase in firing rate is terminated by a more rapid increase in firing rate and a decrease in spike height. Electrical coupling between multiple neurons leads to the formation of ensembles and clusters (Alviña, Ellis-Davies, & Khodakhah, 2009; Walter, Alviña, Womack, Chevez, & Khodakhah, 2006). Patch-clamp recordings have found that harmaline exposure under certain physiological conditions leads to net potentiation of T-type calcium channel (Park et al., 2010). Thus, drug compounds directed against the $Ca_v3.1$ subtype are highly desirable to control the Purkinje cell firing.

Despite conflicting data from early mouse model experiments, several studies pursued the development of compounds targeting T-type calcium channels (Handforth, Homanics, Covey, et al., 2010; Miwa & Kondo, 2011; Womack, Hoang, & Khodakhah, 2009). Ethosuximide and zonisamide, antiepileptic drugs were initially considered as potential therapeutic agents as they had antagonistic properties against T-type calcium channels (Handforth, Homanics, Covey, et al., 2010). Ethosuximide was primarily examined in the MPTP primate model (Gomez-Mancilla, Latulippe, Boucher, & Bédard, 1992) and the rat model for parkinsonian tremor (Miwa, Koh, Kajimoto, & Kondo, 2011). Unfortunately, clinical studies revealed mostly null results (Gironell & Marin-Lahoz, 2016; Pourcher, Gomez-Mancilla, & Bédard, 1992). While the initial data with zonisamide was promising a subsequent double-blind and placebo-controlled study did not confirm the clinical benefits (Bermejo, 2007; Handforth, Martin, Kang, & Vanek, 2009; Morita, Miwa, & Kondo, 2005; Ondo, 2007; Zesiewicz et al., 2007). The double-blind and placebo-controlled study was small; it included patients with mild tremor severity and did not achieve adequate dosing with careful escalation was not achieved; thus, precluding a definitive conclusion. Future studies with larger cohorts are warranted to investigate the role of zonisamide further. Flunarizine is another T-type calcium channel blocker prescribed for migraine and vertigo and was tried in multiple small studies in patients with essential tremor (Biary, al Deeb, & Bahou, 1995; Jiménez-Jiménez & García-Ruiz, 1993). Overall, the drug was found to be ineffective, with studies reporting drug-induced movement disorders (orofacial dyskinesia and parkinsonism) in nearly 30% of patients (Jhang, Huang, Nfor, et al., 2019; Micheli, Pardal, Gatto, et al., 1987; Negrotti & Calzetti, 1997).

Many patients with essential tremor report that ingestion of small amounts of alcohol leads to improvements of tremor (Haubenberger, Nahab, Voller, & Hallett, 2014). Objective assessments have revealed that nearly 50% of patients demonstrate an improvement when subjected to an alcohol challenge test (Knudsen, Lorenz, & Deuschl, 2011). Alcohol effects are mainly mediated through the GABA receptors located in extra synaptic neuronal membrane and some through the T-type calcium channels located in postsynaptic neuronal membrane. Acute ethanol administration has been shown to disrupt the native calcium currents from activation of T-type calcium channels (Ca_V 3.2 isoform; Mu, Carden, Kurukulasuriya, Alexander, & Godwin, 2003; Shan, Hammarback, & Godwin, 2013). Chronic exposure to intermittent ethanol followed by withdrawal can also disrupt calcium channel expression and function (Wu, Tang, Chen, & Cao, 2020).

Although ethanol leads to tremor improvement, the dose required for beneficial effects has been found to be close to the dose needed for inducing toxicity effects (a narrow therapeutic range) (Haubenberger et al., 2014). But 1-octanol, the long-chain version of alcohol, was found to reduce tremor without causing intoxication or other clinically relevant adverse effects (Shill, Bushara, Mari, Reich, & Hallett, 2004). In a placebo-controlled trial involving 12 essential tremor patients, a single dose (1 mg/kg) of 1-octanol was found to lead to a distinct taste of orange peel. The side effect of dysgeusia was expected as the 1-octanol compound is a natural flavoring substance present in orange peel (Bushara, Goldstein, Grimes, Burstein, & Hallett, 2004). Overall, 1-octanol is well-tolerated, but the benefits last only 15 to 90 min and relatively large volumes are necessary to formulate capsules for oral administration; thus, further studies do not seem feasible (Haubenberger et al., 2014).

Octanoic acid, a metabolite of 1-octanol and approved as a food additive has better pharmacokinetics (Haubenberger et al., 2014). Since the bioavailability is nearly 100%, lower oral dose is needed compared to the 1-octanol compound (Schwabe, Bennett, & Bowman, 1964; Traul, Driedger, Ingle, & Nakhasi, 2000). After revealing benefits in the harmaline model for essential tremor, octanoic acid was examined in a pilot cohort of essential tremor. An accelerometer-based assessment observed that the drug was tolerated at a low dose (4 mg/kg) and reduced tremors. However, there was no improvement in patient-reported activities of daily living and the digital spiral analysis (Haubenberger, McCrossin, Lungu, et al., 2013).

Recently, CX-8998 (a compound) was developed that has a specific potency against the Ca_v3 subtype of T-type calcium channels (Papapetropoulos, Lee, Versavel, et al., 2021). The compound at single and multiple doses in healthy volunteers and patient populations had revealed adequate safety. Following the promising efficacy in early clinical studies, a phase 2 proof of concept study was launched in essential tremor comparing an active group ($n=39$) with a placebo ($n=44$) group. Patients with moderate to severe tremor severity were randomized 1:1 to CX-8998 and placebo groups. The CX-8998 compound was titrated to 10 mg twice a day for 4 weeks. The primary endpoint was the change from baseline in the blinded video ratings of the Essential Tremor Rating Assessment Scale performance subscale scored by two groups of raters. The first group comprised experts from the tremor research group (central raters) and the other group comprised the investigators from the individual sites. Secondary outcomes included blinded live ratings of The Essential Tremor Rating Assessment Scale performance subscale, assessment of the activities of daily living and tremor severity with Kinesia ONE accelerometry. Overall the study revealed mixed findings. The central ratings of the videos were not significant, but the investigator ratings were statistically significant ($P=0.017$). The patients reported improvements in the activities of daily living ($P=0.049$), but the objective sensor-based assessment was not significant ($P=0.421$). Dizziness (21%), headache (8%), euphoric mood (6%), and insomnia (6%) were common adverse effects reported by the participants (Papapetropoulos, Lee, Versavel, et al., 2021).

3.3 Targeting potassium channels

Calcium-activated potassium (SK) channels are widely distributed in the central nervous system. While P/Q-type and T-type voltage-gated calcium channels contribute to burst firing of the Purkinje cells, SK channels contribute to interspike and interburst intervals and termination of burst firing (Womack et al., 2004, 2009). In one study, a partial block of P/Q-type calcium channels eliminated calcium spikes and caused a switch from regular bursting to tonic firing or irregular bursting (Womack & Khodakhah, 2004). An increase in intracellular calcium activates these channels which leads to potassium efflux, thereby repolarizing or hyperpolarizing the synaptic membrane. Protein kinases and phosphatases regulate the synaptic levels of SKs and their sensitivity to calcium levels (Womack et al., 2009). These channels play a critical role in synaptic transmission and thus transmit information along the neuronal circuits expressing them. These channels are key for learning, memory, and rhythmic activities in synaptic plasticity. In the context of essential tremor, an increased function of SK channels would lead to an overall reduced and regularized firing of neurons within the tremor network (Sun, Liu, Baudry, & Bi, 2020). SKs are critically involved in coordinating diverse calcium signaling pathways and controlling calcium signal amplitude and duration (Sun et al., 2020). In a recent open-label study, CAD-1883, a compound with positive allosteric modulation of the SK channel, was tested to evaluate the safety, tolerability, and efficacy. The oral compound was administered twice daily to adult participants with essential tremor to increase the sensitivity of SK channels to calcium (ClinicalTrials.gov Identifier: NCT03688685). The trial results will pave a novel pathway for targeting the Purkinje cell hyperexcitability.

Finally, hyperpolarization-activated cyclic nucleotide-gated (HCN) channels are cation channels in the voltage-gated potassium channel superfamily (Lee & MacKinnon, 2017). In a rat model study, a missense mutation in the *Hcn1* gene that led to activation of these channels resulted in membrane hyperpolarization and tremor oscillations. The tremor phenotype of these rats was similar to essential tremor. Propranolol was found to lead to alleviation of tremors though the modulation of HCN channels. It is postulated that the central mechanism of action for this drug involves the HCN channel but further studies are needed for definitive evidence (Ohno, Shimizu, Tatara, et al., 2015).

3.4 Targeting for optimal therapy

In summary, the last decade has seen a significant increase in the pathophysiological understanding of essential tremor. Growing evidence supports that the inferior olivary nucleus may not play as important a pathogenic role as previously thought. Instead, a defective climbing fiber afferent input leading to Purkinje cell burst firing is likely the key component of the pathophysiological network. Purkinje cells express several voltage-gated calcium channels and potassium channels that contribute to burst firing, such as the P/Q-type, T-type, and SK channels. As summarized in Fig. 3, treatment for essential tremor involves either reducing the activation of calcium channels or increasing the functions of SK channels. The older drugs, including the primidone and topiramate, have partial actions against the P/Q-type channels in the Purkinje cell dendrites. Primidone and topiramate also potentiate GABA receptor activation in the dentate neurons which is another important contributor to pathophysiology. The current drugs available for clinical use frequently lead to less than 50% control of symptoms with many dose-limiting side effects. There is mounting preclinical evidence to show that an increased T-type calcium channel activation and a reduced SK channel activation are critical for the hyperexcitability of the Purkinje cell neurons. Thus, many new drug developments in the pipeline are

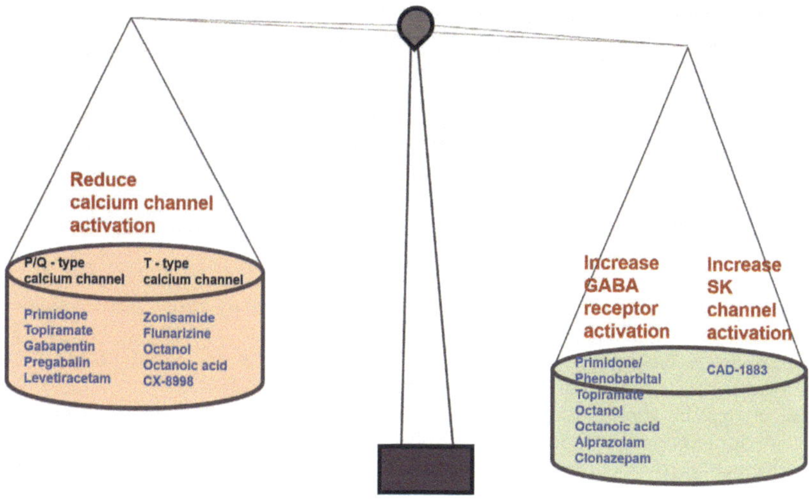

Fig. 3 Drugs reducing the calcium channel functions or increasing the SK channel or the GABA receptor functions are shown in the figure. A balance of activation will lead to effective tremor suppression.

focusing on the T-type calcium channel and SK channel activity. These next generations of drugs will likely lead to more effective tremor control with greater tolerance and a lower risk of undesirable side effects.

References

Abila, B., Wilson, J. F., Marshall, R. W., & Richens, A. (1985). The tremorolytic action of beta-adrenoceptor blockers in essential, physiological and isoprenaline-induced tremor is mediated by beta-adrenoceptors located in a deep peripheral compartment. *British Journal of Clinical Pharmacology*, *20*, 369–376.

Alviña, K., Ellis-Davies, G., & Khodakhah, K. (2009). T-type calcium channels mediate rebound firing in intact deep cerebellar neurons. *Neuroscience*, *158*, 635–641.

Apps, R., & Garwicz, M. (2005). Anatomical and physiological foundations of cerebellar information processing. *Nature Reviews Neuroscience*, *6*, 297–311.

Archer, D. B., Coombes, S. A., Chu, W. T., et al. (2018). A widespread visually-sensitive functional network relates to symptoms in essential tremor. *Brain : A Journal of Neurology*, *141*, 472–485.

Awad, A., Blomstedt, P., Westling, G., & Eriksson, J. (2020). Deep brain stimulation in the caudal zona incerta modulates the sensorimotor cerebello-cerebral circuit in essential tremor. *NeuroImage*, *209*, 116511.

Ben-Menachem, E. (2004). Pregabalin pharmacology and its relevance to clinical practice. *Epilepsia*, *45*(Suppl 6), 13–18.

Bermejo, P. E. (2007). Zonisamide in patients with essential tremor and Parkinson's disease. *Movement Disorders : Official Journal of the Movement Disorder Society*, *22*, 2137–2138.

Bernard, J. F., Buisseret-Delmas, C., Compoint, C., & Laplante, S. (1984). Harmaline induced tremor. III. A combined simple units, horseradish peroxidase, and 2-deoxyglucose study of the olivocerebellar system in the rat. *Experimental Brain Research*, *57*, 128–137.

Biary, N., al Deeb, S. M., & Bahou, Y. (1995). Long-term therapy of essential tremor with flunarizine. *European Neurology*, *35*, 217–219.

Brittain, J.-S., Cagnan, H., Mehta, A. R., Saifee, T. A., Edwards, M. J., & Brown, P. (2015). Distinguishing the central drive to tremor in Parkinson's disease and essential tremor. *The Journal of Neuroscience : The Official Journal of the Society for Neuroscience*, *35*, 795–806.

Broersma, M., van der Stouwe, A. M. M., Buijink, A. W. G., et al. (2016). Bilateral cerebellar activation in unilaterally challenged essential tremor. *NeuroImage Clinical*, *11*, 1–9.

Brown, A. M., White, J. J., van der Heijden, M. E., Zhou, J., Lin, T., & Sillitoe, R. V. (2020). Purkinje cell misfiring generates high-amplitude action tremors that are corrected by cerebellar deep brain stimulation. *eLife*, *9*.

Buijink, A. W., van der Stouwe, A. M., Broersma, M., et al. (2015). Motor network disruption in essential tremor: A functional and effective connectivity study. *Brain : A Journal of Neurology*, *138*, 2934–2947.

Bushara, K. O., Goldstein, S. R., Grimes, G. J., Jr., Burstein, A. H., & Hallett, M. (2004). Pilot trial of 1-octanol in essential tremor. *Neurology*, *62*, 122–124.

Choi, S., Yu, E., Kim, D., et al. (2010). Subthreshold membrane potential oscillations in inferior olive neurons are dynamically regulated by P/Q- and T-type calcium channels: A study in mutant mice. *The Journal of Physiology*, *588*, 3031–3043.

Connor, G. S., Edwards, K., & Tarsy, D. (2008). Topiramate in essential tremor: Findings from double-blind, placebo-controlled, crossover trials. *Clinical Neuropharmacology*, *31*, 97–103.

Crunelli, V., Lightowler, S., & Pollard, C. E. (1989). A T-type Ca2+ current underlies low-threshold Ca2+ potentials in cells of the cat and rat lateral geniculate nucleus. *The Journal of Physiology*, *413*, 543–561.

Deuschl, G., Wenzelburger, R., Loffler, K., Raethjen, J., & Stolze, H. (2000). Essential tremor and cerebellar dysfunction clinical and kinematic analysis of intention tremor. *Brain : A Journal of Neurology*, *123*(Pt 8), 1568–1580.

di Biase, L., Brittain, J.-S., Shah, S. A., et al. (2017). Tremor stability index: A new tool for differential diagnosis in tremor syndromes. *Brain : A Journal of Neurology*, *140*, 1977–1986.

Dogu, O., Sevim, S., Camdeviren, H., et al. (2003). Prevalence of essential tremor: Door-to-door neurologic exams in Mersin Province, Turkey. *Neurology*, *61*, 1804–1806.

Elble, R. J., Lyons, K. E., & Pahwa, R. (2007). Levetiracetam is not effective for essential tremor. *Clinical Neuropharmacology*, *30*, 350–356.

Erickson-Davis, C. R., Faust, P. L., Vonsattel, J. P., Gupta, S., Honig, L. S., & Louis, E. D. (2010). "hairy baskets" associated with degenerative Purkinje cell changes in essential tremor. *Journal of Neuropathology and Experimental Neurology*, *69*, 262–271.

Fang, W., Lv, F., Luo, T., et al. (2013). Abnormal regional homogeneity in patients with essential tremor revealed by resting-state functional MRI. *PLoS One*, *8*, e69199.

Ferreira, J. J., Mestre, T. A., Lyons, K. E., et al. (2019). MDS evidence-based review of treatments for essential tremor. *Movement Disorders : Official Journal of the Movement Disorder Society*, *34*, 950–958.

Findley, L. J., Cleeves, L., & Calzetti, S. (1985). Primidone in essential tremor of the hands and head: A double blind controlled clinical study. *Journal of Neurology, Neurosurgery, and Psychiatry*, *48*, 911–915.

Frima, N., & Grünewald, R. A. (2006). A double-blind, placebo-controlled, crossover trial of topiramate in essential tremor. *Clinical Neuropharmacology*, *29*, 94–96.

Gironell, A., Kulisevsky, J., Barbanoj, M., López-Villegas, D., Hernández, G., & Pascual-Sedano, B. (1999). A randomized placebo-controlled comparative trial of gabapentin and propranolol in essential tremor. *Archives of Neurology*, *56*, 475–480.

Gironell, A., & Marin-Lahoz, J. (2016). Ethosuximide for essential tremor: An open-label trial. *Tremor and Other Hyperkinetic Movements (New York, NY)*, *6*, 378.

Gomez-Mancilla, B., Latulippe, J. F., Boucher, R., & Bédard, P. J. (1992). Effect of ethosuximide on rest tremor in the MPTP monkey model. *Movement Disorders : Official Journal of the Movement Disorder Society*, *7*, 137–141.

Guan, X. M., & Peroutka, S. J. (1990). Basic mechanisms of action of drugs used in the treatment of essential tremor. *Clinical Neuropharmacology*, *13*, 210–223.

Handforth, A. (2012). Harmaline tremor: Underlying mechanisms in a potential animal model of essential tremor. *Tremor and Other Hyperkinetic Movements (New York, NY)*, *2*. 02-92-769-1.

Handforth, A. (2016). Linking essential tremor to the cerebellum-animal model evidence. *Cerebellum (London, England)*, *15*, 285–298.

Handforth, A., Homanics, G. E., Covey, D. F., et al. (2010). T-type calcium channel antagonists suppress tremor in two mouse models of essential tremor. *Neuropharmacology*, *59*, 380–387.

Handforth, A., Martin, F. C., Kang, G. A., & Vanek, Z. (2009). Zonisamide for essential tremor: an evaluator-blinded study. *Movement Disorders : Official Journal of the Movement Disorder Society*, *24*, 437–440.

Haubenberger, D., & Hallett, M. (2018). Essential Tremor. *The New England Journal of Medicine*, *378*, 1802–1810.

Haubenberger, D., McCrossin, G., Lungu, C., et al. (2013). Octanoic acid in alcohol-responsive essential tremor: A randomized controlled study. *Neurology*, *80*, 933–940.

Haubenberger, D., Nahab, F. B., Voller, B., & Hallett, M. (2014). Treatment of essential tremor with long-chain alcohols: Still experimental or ready for prime time? *Tremor and Other Hyperkinetic Movements (New York, NY)*, *4*. tre-04-211-4673-2.

Hedera, P., Cibulčík, F., & Davis, T. L. (2013). Pharmacotherapy of essential tremor. *Journal of Central Nervous System Disease*, *5*, 43–55.

Helmich, R. C., Toni, I., Deuschl, G., & Bloem, B. R. (2013). The pathophysiology of essential tremor and Parkinson's tremor. *Current Neurology and Neuroscience Reports*, *13*, 378.

Hopfner, F., & Deuschl, G. (2020). Managing Essential Tremor. *Neurotherapeutics : The Journal of the American Society for Experimental NeuroTherapeutics*, *17*, 1603–1621.

Hopfner, F., Haubenberger, D., Galpern, W. R., et al. (2016). Knowledge gaps and research recommendations for essential tremor. *Parkinsonism & Related Disorders*, *33*, 27–35.

Hua, S. E., & Lenz, F. A. (2005). Posture-related oscillations in human cerebellar thalamus in essential tremor are enabled by voluntary motor circuits. *Journal of Neurophysiology*, *93*, 117–127.

Ishikawa, T., Kaneko, M., Shin, H. S., & Takahashi, T. (2005). Presynaptic N-type and P/Q-type Ca2+ channels mediating synaptic transmission at the calyx of held of mice. *The Journal of Physiology*, *568*, 199–209.

Ito, M. (1984). *The cerebellum and neural control*. New York: Raven Press.

Jahnsen, H. (1986). Electrophysiological characteristics of neurones in the guinea-pig deep cerebellar nuclei in vitro. *The Journal of Physiology*, *372*, 129–147.

Jenkins, I. H., Bain, P. G., Colebatch, J. G., et al. (1993). A positron emission tomography study of essential tremor: Evidence for overactivity of cerebellar connections. *Annals of Neurology*, *34*, 82–90.

Jhang, K. M., Huang, J. Y., Nfor, O. N., et al. (2019). Flunarizine related movement disorders: A nationwide population-based study. *Scientific Reports*, *9*, 1705.

Jiménez-Jiménez, F. J., & García-Ruiz, P. J. (1993). Flunarizine and essential tremor. *Neurology*, *43*, 239.

Juttukonda, M. R., Franco, G., Englot, D. J., et al. (2019). White matter differences between essential tremor and Parkinson disease. *Neurology*, *92*, e30–e39.

Kano, M., Watanabe, T., Uesaka, N., & Watanabe, M. (2018). Multiple phases of climbing Fiber synapse elimination in the developing cerebellum. *Cerebellum (London, England)*, *17*, 722–734.

Knudsen, K., Lorenz, D., & Deuschl, G. (2011). A clinical test for the alcohol sensitivity of essential tremor. *Movement Disorders : Official Journal of the Movement Disorder Society*, *26*, 2291–2295.

Koller, W. C., & Royse, V. L. (1986). Efficacy of primidone in essential tremor. *Neurology*, *36*, 121–124.

Kuo, S. H., Faust, P. L., Vonsattel, J. P., Ma, K., & Louis, E. D. (2011). Parallel fiber counts and parallel fiber integrated density are similar in essential tremor cases and controls. *Acta Neuropathologica*, *121*, 287–289.

Kuo, S. H., Louis, E. D., Faust, P. L., et al. (2019). Current opinions and consensus for studying tremor in animal models. *Cerebellum (London, England)*, *18*, 1036–1063.

Kuo, S. H., Wang, J., Tate, W. J., et al. (2017). Cerebellar pathology in early onset and late onset essential tremor. *Cerebellum (London, England)*, *16*, 473–482.

Lacinová, L. (2011). T-type calcium channel blockers - new and notable. *General Physiology and Biophysics*, *30*, 403–409.

Lee, J., Kim, I., Lee, J., et al. (2018). Development of Harmaline-induced tremor in a swine model. *Tremor and Other Hyperkinetic Movements (New York, NY)*, *8*, 532.

Lee, C.-H., & MacKinnon, R. (2017). Structures of the human HCN1 hyperpolarization-activated channel. *Cell*, *168*, 111–120.e111.

Llinas, R. R. (1988). The intrinsic electrophysiological properties of mammalian neurons: Insights into central nervous system function. *Science (New York, N.Y.)*, *242*, 1654–1664.

Llinás, R. R. (2013). The olivo-cerebellar system: A key to understanding the functional significance of intrinsic oscillatory brain properties. *Frontiers in Neural Circuits*, *7*, 96.

Llinás, R., & Jahnsen, H. (1982). Electrophysiology of mammalian thalamic neurones in vitro. *Nature*, *297*, 406–408.

Llinás, R., & Yarom, Y. (1981). Electrophysiology of mammalian inferior olivary neurones in vitro. Different types of voltage-dependent ionic conductances. *The Journal of Physiology*, *315*, 549–567.

Llinás, R., & Yarom, Y. (1986). Oscillatory properties of guinea-pig inferior olivary neurones and their pharmacological modulation: An in vitro study. *The Journal of Physiology*, *376*, 163–182.

Louis, E. D. (2008). Environmental epidemiology of essential tremor. *Neuroepidemiology*, *31*, 139–149.

Louis, E. D. (2018). Essential tremor and the cerebellum. *Handbook of Clinical Neurology*, *155*, 245–258.

Louis, E. D., Diaz, D. T., Kuo, S. H., et al. (2018). Inferior Olivary nucleus degeneration does not lessen tremor in essential tremor. *Cerebellum & Ataxias*, *5*, 1.

Louis, E. D., & Faust, P. L. (2020). Essential tremor pathology: Neurodegeneration and reorganization of neuronal connections. *Nature Reviews Neurology*, *16*, 69–83.

Louis, E. D., & Ferreira, J. J. (2010). How common is the most common adult movement disorder? Update on the worldwide prevalence of essential tremor. *Movement Disorders : Official Journal of the Movement Disorder Society*, *25*, 534–541.

Louis, E. D., Kerridge, C. A., Chatterjee, D., et al. (2019). Contextualizing the pathology in the essential tremor cerebellar cortex: A patholog-omics approach. *Acta Neuropathologica*, *138*, 859–876.

Louis, E. D., Lee, M., Babij, R., et al. (2014). Reduced Purkinje cell dendritic arborization and loss of dendritic spines in essential tremor. *Brain : A Journal of Neurology*, *137*, 3142–3148.

Louis, E. D., & Lenka, A. (2017). The Olivary hypothesis of essential tremor: Time to lay this model to rest? *Tremor and Other Hyperkinetic Movements (New York, NY)*, *7*, 473.

Louis, E. D., & McCreary, M. (2021). How common is essential tremor? Update on the worldwide prevalence of essential tremor. *Tremor and Other Hyperkinetic Movements (New York, NY), 11*, 28.

Louis, E. D., Shungu, D. C., Chan, S., Mao, X., Jurewicz, E. C., & Watner, D. (2002). Metabolic abnormality in the cerebellum in patients with essential tremor: A proton magnetic resonance spectroscopic imaging study. *Neuroscience Letters, 333*, 17–20.

Madelein van der Stouwe, A. M., Nieuwhof, F., & Helmich, R. C. (2020). Tremor pathophysiology: Lessons from neuroimaging. *Current Opinion in Neurology, 33*, 474–481.

McLean, M. J. (1994). Clinical pharmacokinetics of gabapentin. *Neurology, 44* (S17–22; discussion S31-12).

Micheli, F., Pardal, M. F., Gatto, M., et al. (1987). Flunarizine- and cinnarizine-induced extrapyramidal reactions. *Neurology, 37*, 881–884.

Miwa, H., Koh, J., Kajimoto, Y., & Kondo, T. (2011). Effects of T-type calcium channel blockers on a parkinsonian tremor model in rats. *Pharmacology, Biochemistry, and Behavior, 97*, 656–659.

Miwa, H., & Kondo, T. (2011). T-type calcium channel as a new therapeutic target for tremor. *Cerebellum (London, England), 10*, 563–569.

Morita, S., Miwa, H., & Kondo, T. (2005). Effect of zonisamide on essential tremor: A pilot crossover study in comparison with arotinolol. *Parkinsonism & Related Disorders, 11*, 101–103.

Mu, J., Carden, W. B., Kurukulasuriya, N. C., Alexander, G. M., & Godwin, D. W. (2003). Ethanol influences on native T-type calcium current in thalamic sleep circuitry. *The Journal of Pharmacology and Experimental Therapeutics, 307*, 197–204.

Muthuraman, M., Deuschl, G., Anwar, A. R., Mideksa, K. G., von Helmolt, F., & Schneider, S. A. (2015). Essential and aging-related tremor: Differences of central control. *Movement Disorders : Official Journal of the Movement Disorder Society, 30*, 1673–1680.

Muthuraman, M., Raethjen, J., Koirala, N., et al. (2018). Cerebello-cortical network fingerprints differ between essential, Parkinson's and mimicked tremors. *Brain : A Journal of Neurology, 141*, 1770–1781.

Negrotti, A., & Calzetti, S. (1997). A long-term follow-up study of cinnarizine- and flunarizine-induced parkinsonism. *Movement Disorders : Official Journal of the Movement Disorder Society, 12*, 107–110.

Ohno, Y., Shimizu, S., Tatara, A., et al. (2015). Hcn1 is a tremorgenic genetic component in a rat model of essential tremor. *PLoS One, 10*, e0123529.

Ondo, W. G. (2007). Zonisamide for essential tremor. *Clinical Neuropharmacology, 30*, 345–349.

Ondo, W., Hunter, C., Vuong, K. D., Schwartz, K., & Jankovic, J. (2000). Gabapentin for essential tremor: A multiple-dose, double-blind, placebo-controlled trial. *Movement Disorders : Official journal of the Movement Disorder Society, 15*, 678–682.

Ondo, W. G., Jankovic, J., Connor, G. S., et al. (2006). Topiramate in essential tremor: A double-blind, placebo-controlled trial. *Neurology, 66*, 672–677.

O'Suilleabhain, P., & Dewey, R. B., Jr. (2002). Randomized trial comparing primidone initiation schedules for treating essential tremor. *Movement Disorders : Official Journal of the Movement Disorder Society, 17*, 382–386.

Otis, T. S. (2016). Simple spikes and complex spikes. In *Essentials of cerebellum and cerebellar disorders. A primer for graduate students* (pp. 299–303).

Pagan, F. L., Butman, J. A., Dambrosia, J. M., & Hallett, M. (2003). Evaluation of essential tremor with multi-voxel magnetic resonance spectroscopy. *Neurology, 60*, 1344–1347.

Pahwa, R., Lyons, K., Hubble, J. P., et al. (1998). Double-blind controlled trial of gabapentin in essential tremor. *Movement Disorders : Official Journal of the Movement Disorder Society, 13*, 465–467.

Palay, S. L. C.-P. V. (1974). *Cerebellar cortex: Cytology and organization*. New York: Springer-Verlag.

Pan, M. K., Li, Y. S., Wong, S. B., et al. (2020). Cerebellar oscillations driven by synaptic pruning deficits of cerebellar climbing fibers contribute to tremor pathophysiology. *Science Translational Medicine, 12*.

Papapetropoulos, S., Lee, M. S., Versavel, S., et al. (2021). A phase 2 proof-of-concept, randomized, placebo-controlled trial of CX-8998 in essential tremor. *Movement Disorders : Official Journal of the Movement Disorder Society, 36*, 1944–1949.

Paris-Robidas, S., Brochu, E., Sintes, M., et al. (2012). Defective dentate nucleus GABA receptors in essential tremor. *Brain : A Journal of Neurology, 135*, 105–116.

Park, Y. G., Park, H. Y., Lee, C. J., et al. (2010). Ca(V)3.1 is a tremor rhythm pacemaker in the inferior olive. *Proceedings of the National Academy of Sciences of the United States of America, 107*, 10731–10736.

Pedrosa, D. J., Quatuor, E. L., Reck, C., et al. (2014). Thalamomuscular coherence in essential tremor: Hen or egg in the emergence of tremor? *The Journal of Neuroscience : The Official Journal of the Society for Neuroscience, 34*, 14475–14483.

Pedrosa, D. J., Reck, C., Florin, E., et al. (2012). Essential tremor and tremor in Parkinson's disease are associated with distinct 'tremor clusters' in the ventral thalamus. *Experimental Neurology, 237*, 435–443.

Peng-Chen, Z., Morishita, T., Vaillancourt, D., et al. (2013). Unilateral thalamic deep brain stimulation in essential tremor demonstrates long-term ipsilateral effects. *Parkinsonism & Related Disorders, 19*, 1113–1117.

Perez-Reyes, E. (2003). Molecular physiology of low-voltage-activated t-type calcium channels. *Physiological Reviews, 83*, 117–161.

Pietracupa, S., Bologna, M., Bharti, K., et al. (2019). White matter rather than gray matter damage characterizes essential tremor. *European Radiology, 29*, 6634–6642.

Pourcher, E., Gomez-Mancilla, B., & Bédard, P. J. (1992). Ethosuximide and tremor in Parkinson's disease: A pilot study. *Movement Disorders : Official Journal of the Movement Disorder Society*, 7, 132–136.

Raethjen, J., & Deuschl, G. (2012). The oscillating central network of essential tremor. *Clinical Neurophysiology : Official Journal of the International Federation of Clinical Neurophysiology, 123*, 61–64.

Raethjen, J., Govindan, R. B., Kopper, F., Muthuraman, M., & Deuschl, G. (2007). Cortical involvement in the generation of essential tremor. *Journal of Neurophysiology, 97*, 3219–3228.

Raethjen, J., & Muthuraman, M. (2015). Cause or compensation? Complex changes in cerebello-thalamo-cortical networks in pathological action tremor. *Brain : A Journal of Neurology, 138*, 2808–2810.

Rajput, A., Robinson, C. A., & Rajput, A. H. (2004). Essential tremor course and disability: A clinicopathologic study of 20 cases. *Neurology, 62*, 932–936.

Randall, A., & Tsien, R. W. (1995). Pharmacological dissection of multiple types of Ca2+ channel currents in rat cerebellar granule neurons. *The Journal of Neuroscience : The Official Journal of the Society for Neuroscience, 15*, 2995–3012.

Routledge, P. A., & Shand, D. G. (1979). Clinical pharmacokinetics of propranolol. *Clinical Pharmacokinetics, 4*, 73–90.

Sasso, E., Perucca, E., & Calzetti, S. (1988). Double-blind comparison of primidone and phenobarbital in essential tremor. *Neurology, 38*, 808–810.

Schwabe, A. D., Bennett, L. R., & Bowman, L. P. (1964). Octanoic acid absorption and oxidation in humans. *Journal of Applied Physiology, 19*, 335–337.

Shan, H. Q., Hammarback, J. A., & Godwin, D. W. (2013). Ethanol inhibition of a T-type Ca^2+ channel through activity of protein kinase C. *Alcoholism, Clinical and Experimental Research, 37*, 1333–1342.

Shill, H. A., Bushara, K. O., Mari, Z., Reich, M., & Hallett, M. (2004). Open-label dose-escalation study of oral 1-octanol in patients with essential tremor. *Neurology, 62*, 2320–2322.

Stolze, H., Petersen, G., Raethjen, J., Wenzelburger, R., & Deuschl, G. (2001). The gait disorder of advanced essential tremor. *Brain : A Journal of Neurology*, *124*, 2278–2286.

Streng, M. L., Popa, L. S., & Ebner, T. J. (2017). Climbing fibers control Purkinje cell representations of behavior. *The Journal of Neuroscience : The Official Journal of the Society for Neuroscience*, *37*, 1997–2009.

Sun, J., Liu, Y., Baudry, M., & Bi, X. (2020). SK2 channel regulation of neuronal excitability, synaptic transmission, and brain rhythmic activity in health and diseases. *Biochimica et Biophysica Acta Molecular Cell Research*, *1867*, 118834.

Talley, E. M., Cribbs, L. L., Lee, J. H., Daud, A., Perez-Reyes, E., & Bayliss, D. A. (1999). Differential distribution of three members of a gene family encoding low voltage-activated (T-type) calcium channels. *The Journal of Neuroscience : The Official Journal of the Society for Neuroscience*, *19*, 1895–1911.

Taylor, C. P., Gee, N. S., Su, T. Z., et al. (1998). A summary of mechanistic hypotheses of gabapentin pharmacology. *Epilepsy Research*, *29*, 233–249.

Tikoo, S., Pietracupa, S., Tommasin, S., et al. (2020). Functional disconnection of the dentate nucleus in essential tremor. *Journal of Neurology*, *267*, 1358–1367.

Traul, K. A., Driedger, A., Ingle, D. L., & Nakhasi, D. (2000). Review of the toxicologic properties of medium-chain triglycerides. *Food and Chemical Toxicology : An International Journal Published for the British Industrial Biological Research Association*, *38*, 79–98.

Usowicz, M. M., Sugimori, M., Cherksey, B., & Llinás, R. (1992). P-type calcium channels in the somata and dendrites of adult cerebellar Purkinje cells. *Neuron*, *9*, 1185–1199.

Wagle Shukla, A., & Okun, M. S. (2016). State of the art for deep brain stimulation therapy in movement disorders: A clinical and technological perspective. *IEEE Reviews in Biomedical Engineering*, *9*, 219–233.

Walter, J. T., Alviña, K., Womack, M. D., Chevez, C., & Khodakhah, K. (2006). Decreases in the precision of Purkinje cell pacemaking cause cerebellar dysfunction and ataxia. *Nature Neuroscience*, *9*, 389–397.

Wang, K. L., Ren, Q., Chiu, S., et al. (2020). Deep brain stimulation and other surgical modalities for the management of essential tremor. *Expert Review of Medical Devices*, *17*, 817–833.

Wolpert, D. M., Miall, R. C., & Kawato, M. (1998). Internal models in the cerebellum. *Trends in Cognitive Sciences*, *2*, 338–347.

Womack, M. D., Chevez, C., & Khodakhah, K. (2004). Calcium-activated potassium channels are selectively coupled to P/Q-type calcium channels in cerebellar Purkinje neurons. *The Journal of Neuroscience : The Official Journal of the Society for Neuroscience*, *24*, 8818–8822.

Womack, M. D., Hoang, C., & Khodakhah, K. (2009). Large conductance calcium-activated potassium channels affect both spontaneous firing and intracellular calcium concentration in cerebellar Purkinje neurons. *Neuroscience*, *162*, 989–1000.

Womack, M. D., & Khodakhah, K. (2004). Dendritic control of spontaneous bursting in cerebellar Purkinje cells. *The Journal of Neuroscience : The Official Journal of the Society for Neuroscience*, *24*, 3511–3521.

Wu, J., Tang, H., Chen, S., & Cao, L. (2020). Mechanisms and pharmacotherapy for ethanol-responsive movement disorders. *Frontiers in Neurology*, *11*, 892.

Yamamoto, T., Fukuda, M., & Llinás, R. (2001). Bilaterally synchronous complex spike Purkinje cell activity in the mammalian cerebellum. *The European Journal of Neuroscience*, *13*, 327–339.

Zesiewicz, T. A., Ward, C. L., Hauser, R. A., Sanchez-Ramos, J., Staffetti, J. F., & Sullivan, K. L. (2007). A double-blind placebo-controlled trial of zonisamide (zonegran) in the treatment of essential tremor. *Movement Disorders : Official Journal of the Movement Disorder Society*, *22*, 279–282.

FSC
www.fsc.org
MIX
Papier | Fördert
gute Waldnutzung
FSC® C083411

Kolibri 360 GmbH
Ferdinand-Jühlke-Straße 7
99095 Erfurt, Deutschland
produktsicherheit@kolibri360.de